TRENDS AND CHANGES IN HYDROCLIMATIC VARIABLES

Links to Climate Variability and Change

Edited by

RAMESH TEEGAVARAPU

ELSEVIER

Elsevier
Radarweg 29, PO Box 211, 1000 AE Amsterdam, Netherlands
The Boulevard, Langford Lane, Kidlington, Oxford OX5 1GB, United Kingdom
50 Hampshire Street, 5th Floor, Cambridge, MA 02139, United States

Library of Congress Cataloging-in-Publication Data
A catalog record for this book is available from the Library of Congress

British Library Cataloguing-in-Publication Data
A catalogue record for this book is available from the British Library

ISBN: 978-0-12-810985-4

Publisher: Candice Janco
Acquisition Editor: Louisa Hutchins
Editorial Project Manager: Hilary Carr
Production Project Manager: Divya KrishnaKumar
Cover Designer: Greg Harris

Typeset by TNQ Technologies

TRENDS AND CHANGES IN HYDROCLIMATIC VARIABLES

CONTENTS

LIST OF CONTRIBUTORS

Abhishek Gaur
Construction Research Centre, National Research Council Canada, Ottawa, ON, Canada

Rao S. Govindaraju
School of Civil Engineering, Purdue University, West Lafayette, IN, United States

Ganeshchandra Mallya
School of Civil Engineering, Purdue University, West Lafayette, IN, United States

P.P. Mujumdar
Department of Civil Engineering, Indian Institute of Science, Bangalore, India

Chandra R. Rupa
Department of Civil Engineering, Indian Institute of Science, Bangalore, India

Alejandra R. Schmidt
Department of Civil, Environmental and Geomatics Engineering, Florida Atlantic University, Boca Raton, FL, United States

Ashish Sharma
School of Civil and Environmental Engineering, University of New South Wales, Sydney, NSW, Australia

Slobodan P. Simonovic
Facility for Intelligent Decision Support, Western University, London, ON, Canada

Ramesh S.V. Teegavarapu
Department of Civil, Environmental and Geomatics Engineering, Florida Atlantic University, Boca Raton, FL, United States

Shivam Tripathi
Department of Civil Engineering, Indian Institute of Technology, Kanpur, India

Conrad Wasko
School of Civil and Environmental Engineering, University of New South Wales, Sydney, NSW, Australia

CHAPTER 1

Methods for Analysis of Trends and Changes in Hydroclimatological Time-Series

Ramesh S.V. Teegavarapu
Department of Civil, Environmental and Geomatics Engineering, Florida Atlantic University, Boca Raton, FL, United States

Contents

Trends and Changes in Hydroclimatic Variables
ISBN 978-0-12-810985-4
https://doi.org/10.1016/B978-0-12-810985-4.00001-3

1. CHANGES AND TRENDS IN HYDROCLIMATOLOGICAL TIME-SERIES

Variations in hydroclimatological time-series data can be noticed considering linear and nonlinear (monotonic) trends, and also step changes in the values over time or across two temporal windows. Three major components of any time-series are of interest for analysis and they are (1) trend; (2) seasonality; (3) noise. In the evaluation of any time-series, the time becomes an explanatory variable. Before any analysis of time-series can be taken up, the quality assessment of time-series to ensure serially continuous, error-, and gap-free data should be undertaken. Analysis of any hydrological time-series will involve evaluation of: (1) general characteristics of the variables; (2) trends and variations in variables at different temporal scales; (3) interannual and intraannual variations; (4) changing extremes over time; (4) changes in temporal occurrences of these extremes; (5) changes and trends in temporally aggregated variable values; (6) spatial variation of trends; (7) variations in the summary statistics of observations in two or more temporal windows; (8) indices that are specific to a particular hydroclimatic variable. While the methods described in this chapter are useful for evaluation of changes in any hydroclimatalogical variable over time, illustrative examples describing the use of these methods provided are limited mostly to precipitation datasets.

Gilbert (1987) classifies different types of time-series as: (1) random variations with no trend; (2) random variations with cycles and with no trend; (3) trend with random variations; (4) random variations with cyclical changes and trend; (5) trend with no random variations; (6) random variations with impulse (episodic variation); (7) step change with random variations and (7) random variations with trend. While discussion of these types of time-series was primarily in the context of environmental quality constituents by Gilbert (1987), these types of trends are also common in hydroclimatic variables.

Cavadias (1992) indicated that statistical tests for evaluating changes in hydroclimatic or any time-series depends on types of variations and they include: (1) abrupt change in the mean; (2) gradual change in the mean; (3) shifting levels (more than one change in the mean); (4) continuous trend in the mean; (5) cyclical patterns and (6) changes in variability. Literature is replete with parametric and nonparametric tests for assessment of trends, changes in the characteristics of time-series with step change (when the time of change is known and unknown). To understand the data and type of test the can be used, initial evaluation of the data needs to be conducted using several tools which is referred to as Exploratory Data Analysis (EDA).

2. EXPLORATORY DATA ANALYSIS (EDA)

Exploratory data analysis (EDA) is the first step in the basic understanding of the available data or hydrometeorological time-series. It is a critical component in statistical analysis that will be carried out in future. EDA involves mostly visual or graphical description of data using time-series plots, bivariate plots, histograms, autocorrelation functions, spatial plots of data (trends and magnitudes), box plots, and probability plots and smoothing curves based on scatter plots or time-series plots and kernel density estimates for nonparametric characterization of data. Grubb and Robson (2000) provide an excellent treatise on exploratory and visual data analysis. They illustrate the utility of EDA for understanding temporal patterns, seasonal variations, regional and spatial variations, data problems, correlations among variables, independent and auto-correlated variables and details of the seasonal structure. EPA (2006a) and EPA (2006b), EPA (2000) provides extensive guidance on data analysis with relevance to environmental data. Readers are referred to books by McCuen (2003), Rodda and Little (2015) and Machiwal and Jha (2012), Von Storch and Zwiers (1999) for basic approaches for analyzing changes in hydrological data. NIST (2018) provides general guidelines related to EDA. Many data analysis and data mining books (e.g., Tan et al., 2006; Han and Kamber, 2006; Myatt and Johnson, 2009) provide exhaustive details of EDA methods and techniques. Young et al. (2006) discuss several methods for visualizing data with dynamic interactive graphics.

World Climate Application Program (WCAP) WMO (1998) recommended a number of statistics to be computed from the available datasets that are beneficial for climate variability studies and they include: (1) mean; (2) standard error of mean; (3) standard deviation; (4) coefficient of variation; (5) coefficient of skew; (6) coefficient of kurtosis; (7) ranks for each month; (8) coefficient of autocorrelation; (9) standard error of coefficient of autocorrelation; (10) cumulative periodogram; (11) variance spectrum; (12) confidence intervals for variance spectrum; (13) rescaled range; (14) Hurst's coefficient; (15) number of runs; (16) trend in the mean; (17) trend in variance; (18) equality of subperiod means; (19) equality of subperiod variances; (20) jump in the mean, and (21) Gaussian filter. It was also recommended that computations of these statistics be carried out for subsamples (or non-overlapping subperiods) of 5, 10, 20, and 30 years of length derived from the original time-series.

3. NUMERICAL SUMMARY STATISTICS

Several summary statistics are used to describe different characteristics of time series and the main ones are: (1) central tendency; (2) dispersion; and (3) shape. These statistics are conceptually simple and are evaluated to gain a basic understanding of any dataset.

3.1 Measures of Central Tendency

Evaluation of central tendency can be carried out using mean, trimmed mean, mode, median, and other variants of arithmetic mean such as geometric mean and harmonic mean. These are referred to as measures of location. Details of these measures can be found in any standard statistics text book (e.g., Wilks, 2011). Some of these standard measures may not provide accurate information for several positively skewed hydroclimatic datasets.

3.2 Measures of Dispersion

Evaluation of variation (or spread) in the data can be carried out using several measures such as interquartile range (IQR), range, and mean absolute deviation. A brief description of these measures is provided in this section.

3.2.1 Interquartile Range

IQR refers to the difference between 75th and 25th percentiles (Q_3 and Q_1 respectively) of a variable. The interquartile range is an alternative to the standard deviation and is less affected by extremes than the standard deviation.

$$\mathrm{IQR} = Q_3 - Q_1 \tag{1.1}$$

3.2.2 Range

The range (R) provides information about the difference between the maximum and the minimum of a sample dataset.

$$R = \max(\theta_i) - \min(\theta_i) \tag{1.2}$$

3.2.3 Mean Absolute Deviation

The mean absolute deviation (MAD) is the mean ($\overline{\theta}$) of deviations of observations from mean of sample dataset given by Eq. (1.3).

$$\mathrm{MAD} = \frac{1}{N}\sum_{i=1}^{N}\left|\theta_i - \overline{\theta}\right| \tag{1.3}$$

The mean absolute deviation can also be calculated using the mean of deviation observations from a median value of the sample dataset (θ_M). This statistic is also referred to as MAD_M as given in Eq. (1.4).

$$\mathrm{MAD}_M = \frac{1}{N}\sum_{i=1}^{N}\left|\theta_i - \theta_M\right| \tag{1.4}$$

4. MEASURES OF SHAPE

Measures of shape are evaluated using skewness coefficient (g) and kurtosis (k) parameters of the dataset. These measures are estimated using the Eqs. (1.5) and (1.6) respectively. Sample size needs to be considered when interpreting skewness and kurtosis values.

$$g = \frac{\sum_{i=1}^{N} \left(\theta_i - \overline{\theta}\right)^3}{(N-1)S^3} \tag{1.5}$$

$$k = \frac{\sum_{i=1}^{N} \left(\theta_i - \overline{\theta}\right)^4}{(N-1)S^4} \tag{1.6}$$

Another measure of skewness of a dataset is the Yule coefficient (C_Y) (Dodge, 2008), which uses values of first quartile (Q_1) and third quartile (Q_3) and median (M). The Yule coefficient is given by Eq. (1.7).

$$C_Y = \frac{Q_3 - Q_1 - 2M}{Q_3 - Q_1} \tag{1.7}$$

When the C_Y is positive, then the distribution is positively skewed and when it is negative the distribution is negatively skewed. This coefficient is also referred to as quartile skew coefficient (Machiwal and Jha, 2012).

5. SERIAL AUTOCORRELATION

The autocorrelation coefficient is also referred to as serial correlation coefficient. The first-order autocorrelation coefficient can be referred to as correlation coefficient of the first $N-1$ observations, $\theta_1 \ldots \theta_{N-1}$, and the next $N-1$ observations, $\theta_2 \ldots \theta_N$. These two series are used for calculations of average values and they are referred to as $\overline{\theta}_{(1)}$ and $\overline{\theta}_{(2)}$, respectively. The autocorrelation values can be obtained for different lag (t) values as given by Eq. (1.8).

$$\rho_t = \frac{\sum_{i=1}^{N-t} \left(\theta_i - \overline{\theta}_{(1)}\right)\left(\theta_{i+t} - \overline{\theta}_{(2)}\right)}{\sqrt{\sum_{i=1}^{N-t} \left(\theta_i - \overline{\theta}_{(1)}\right)^2} \sqrt{\sum_{i=2}^{N} \left(\theta_i - \overline{\theta}_{(2)}\right)^2}} \tag{1.8}$$

For sufficiently large N, the autocorrelation at a specific lag can be defined by the Eq. (1.9).

$$\rho_t = \frac{\sum_{i=1}^{N-t} \left(\theta_i - \overline{\theta}\right)\left(\theta_{i+t} - \overline{\theta}\right)}{\sum_{i=1}^{N-t} \left(\theta_i - \overline{\theta}\right)^2} \tag{1.9}$$

The variable $\overline{\theta}$ is the mean (average) of the entire available time-series for data. Autocorrelation diagrams referred to as autocorrelograms can be developed for different datasets. An autocorrelogram is a plot of lagged time interval values and autocorrelation values at these intervals.

6. QUANTILE-QUANTILE PLOTS

A quantile-quantile plot (Wilks, 2011) which is also referred to as q-q plot is a visual way for comparing the marginal cumulative probability distributions of two datasets. A q-q plot will be linear if both the data samples come from the same distribution.

7. EVALUATING CHANGES IN HYDROLOGICAL TIME-SERIES: VISUAL DATA ASSESSMENT METHODS

Graphical techniques employed in EDA are often quite simple, consisting of various tasks such as plotting the raw data using histograms, bihistograms, probability plots, lag plots, and Youden plots when analyzing data from multiple laboratories, simple statistical graphical summaries described by mean, standard deviation, and box plots. Young et al. (2006) discuss different methods for visualizing univariate, bivariate, and multivariate datasets. They provide details about dotplots, boxplots, and diamond plots; histogram and frequency polygons; cumulative distribution plots; lag and sequence plots; scatter, distribution comparison and parallel-coordinate plots; orthogonal-axes, parallel-axes, paired-axes plots; orbitplots, biplots, wiggle-worm plots, and spinplots; and parallel-coordinates plot matrices. Young et al. (2006) also discuss the use of boxplots, scatter plots to visualize and mark imputed (i.e., filled) subsets of data in univariate and multivariate time-series. Appropriate selection of temporal window (i.e., interval or period of record) and use of more than one plot side-by-side might enhance the understanding of variations in the time-series. The following lists contain a number of plots or graphical summaries for univariate and multivariate time-series that are recommended for EDA.

7.1 Univariate Dataset

- Run-sequence plot (or time-series plot)
- Box and whisker plot or boxplot
- Histogram
- Autocorrelation plot
- Dot diagram
- Lag plot
- Stem and leaf plot
- Quantile plot and ranked data plot
- Bar graph
- Line graph
- Probability density function (PDF) plot (parametric)
- Cumulative distribution function (CDF) plot (nonparametric)
- Kernel density estimate (KDE) plot (nonparametric)

7.2 Bivariate Datasets

- Bihistogram (two histograms together—one below the other)
- Scatter plot
- Scatter histogram

A list of exploratory data analysis tools and their uses are provided in Table 1.1. The list is not comprehensive or exhaustive in describing all the available tools. The list also provides tools and utilities based on data type. The Table 1.1 is adopted with modifications from Teegavarapu (2013).

Wilks (2011), Machiwal and Jha (2012), and Naghettini (2017) provide an exhaustive descriptions of all the plots and graphical summaries reported in Table 1.1 and their uses for evaluation of atmospheric and hydrological data. The readers are also recommended to review material provided online from NIST (2018) for more details on each type of plot. EDA techniques can also be used for four aspects of preliminary data analysis. These four aspects (Chakrabarti et al., 2009) are: cleaning, integration, transformation, and reduction. Cleaning mainly refers to filling missing values, smoothing out noise with identification of outliers, and correcting inconsistencies in data. Data integration is a process of combining data from multiple sources to obtain coherent datasets. Transformation (or re-expression) is mainly carried out to convert data to appropriate forms for statistical analysis.

8. QUALITY ASSESSMENT OF TIME-SERIES OF OBSERVATIONS

Assessment of variations in trends in hydrological time-series is critical for many climate change and variability studies. Changes that need to be

Table 1.1 Examples of Tools and Their Utilities For Exploratory Data Analysis

Data Type	Tool/Plot/Graph	Utility of the Plot or Graph
Univariate	Summary statistics	Assess central tendency, spread (variability), symmetry, peakedness
	Time-series and run sequence plots	Variations at different temporal scales, trend
	Histogram	Distribution of data, frequency of observations
	Kernel density function	Smoothed distribution of data with no assumptions of distribution
	Cumulative density function	Empirical density function
	Box plots	Visual depiction of entire data (summary statistics, outliers, etc.)
	Probability plots	Assess the distribution of the sample
	Quantile-quantile Plots	Compare empirical and fitted cumulative density functions
Bivariate	Bihistograms	Evaluation of two histograms side by side
	Lag-plot	Plot based on two lagged time-series datasets
	Scatter plots	Understand linear and nonlinear associations, outliers
	Pearson correlation	Quantitative evaluation of strength of correlation
	Spearman rank correlation (ρ)	Evaluation of strength of correlation using rank-ordered data
	Kendall's τ	Alternative to Pearson correlation
	Autocorrelation function	Serial dependence at different temporal lags, assessment of persistence
Multivariate	Scatterplots	Visualize data, Associations between variables
	Star plot	Multidimensional plot explaining variability
	Glyph scatterplot	Glyphs are used to provide additional information in scatter plots
	Correlation matrix	Quantitative evaluation of correlations in matrix form
	Scatterplot matrix	Plots in matrix format to visualize associations
	Correlation maps	Geographical variation of a process

evaluated at different temporal and spatial scales include: (1) gradual and sudden (episodic) changes; (2) changes that happen at specific time intervals or that coincide with the natural cycles of climate variability; (3) changes in the moments or distributional characteristics and (4) changes in indices that define specific extreme properties of time-series. To evaluate all these changes for obtaining clear insights into the variations of time-series, methods ranging from EDA to parametric and nonparametric statistical methods are required. Before any of the methods are adopted for analysis, serially continuous (i.e., gap-free) or chronologically continuous, error-free, homogenous, and quality assured and controlled (QAQC) data are needed. In some cases, several QAQC checks need to be conducted before data can be used for analysis. Dahmen and Hall (1990) recommend four data screening steps and they include: (1) screening of data based on accumulated values at different temporal scales; (2) visual assessment of data using time-series plots to assess any trends or discontinuities; (3) testing for existence of trends using trend tests and (4) checking of stability of variance using F-test, and mean using an unpaired two-sample t-test when using split data that are nonoverlapping and subsets of time-series. If there are missing data, they should be estimated and the method used for imputation must be reported (WMO, 1988).

8.1 Outliers and Anomalies

Hydrometeorological time-series often contain outliers and anomalous observations due to a variety of reasons. Identification of outliers and anomalous observations is an essential task that needs to be carried out to obtain error-free time-series data for any analysis. The words outlier and anomaly are used interchangeably in many instances and studies. An outlier is an unusual observation, numerically different from a set of observations. Anomaly refers to a pattern in a given dataset that does not conform to an established normal behavior. Anomalies are more difficult to detect with statistical methods. Rule-based methods derived from domain expertise (i.e., knowledge of measurement sensors, information about the physical structures, monitoring network and limitations of the sensors) will help immensely in such situations. Outliers are also referred to as distributional monsters by some researchers as they lie in the tails of the probability distributions of the observed values.

8.2 Statistical, Data Mining and Rule-Based Methods

A number of statistical, rule-based and other (clustering, nearest-neighbor, classification) methods for outlier and anomaly detection are available in the literature. Median filters; statistical control charts; moving range control charts; exponentially weighted moving average charts; moving average charts; Grubb's, Rosner, and Dixon tests; Tukey's boxplots, auto-regressive integrated moving average (ARIMA)-based method, and 3-sigma (3-σ) outlier methods are some of the widely used methods for identification of outliers and anomalies. Many of the statistical methods are parametric in nature and require the normality conditions to be satisfied and a minimum number of samples. In many instances, different transformations (e.g., Box-Cox) should be attempted to see if normality of data can be achieved. Rule-based methods with "if-then-else" constructs can be developed as initial screening tools to identify data anomalies. The rules will consider site-specific lower and upper bounds of measurement values of sensors, and other limits associated with field measurements. Rules that will help identify temporary or long-term sensor failures can be also developed. The failure of sensor may be identified based on lack of recorded variations in observations for a long period of time, unexplained spikes at regular intervals, missing observations, and abnormal trends over a period of time that could not be explained by any influencing physical process linked to variable of interest preceding this trend. McCuen (2003) documents multiple methods for outlier detection and they include Chauvenet's method, and Dixon-Thompson and Rosner's tests.

Univariate methods (or site-specific) that employ rule-based or statistical techniques use time-series data at different temporal resolutions to identify outliers and anomalies. Checking of measurements can be carried out by visual inspection with reliably good results. A number of simple checks include detection of (1) gaps in the data; (2) physically impossible values; (3) constant values over several time intervals; (4) values above prespecified thresholds; (5) improbable zero values; (6) unusually low values and (9) unusually high values. Evaluation of time-series data at a single site (i.e., base site) can be carried out using historical data from the same period (month or season) for boundary consistency checks (range checks). Threshold limits need to be obtained from the historical data, and also these limits can be dynamic with spatially and temporally varying values.

8.2.1 Anomalies and Outliers in Precipitation Data

In case of precipitation records, different threshold values (magnitudes of precipitation depths) can be used for different time scales for identification of outliers or anomalies. Physically impossible values can be easily identified. Simple rule-based method (precipitation data-specific rules) can be developed for quality control of the data. Rule-based methods can help identify:

1. possible erroneous values associated with a specific type of rain gauge;
2. values greater than regional record rainfalls for different durations; and
3. repetition of nonzero constant values over several time intervals.

Probability distributions of precipitation data based on historical data for different temporal durations can be used for identifying data issues especially outliers. Once the distributions (e.g., gamma or exponential) are fitted to the precipitation data and validated, outliers can be identified.

Cluster-based methods using k-nearest neighbors (KNN) can also be applied for identification of outliers. All theavailable statistical and cluster-based methods applied to precipitation data can also be applied to other hydrometeorological data for assessment of outliers and anomalies.

Statistical and cluster-based methodologies can be easily applied for univariate time-series data. However, many of the statistical methods require the normality assumption that is difficult to validate as distributions of many hydrometeorological variables are not Gaussian. Univariate methods cannot identify issues with: (1) time shifts in the observations (i.e., observations noted in incorrect time intervals); (2) measured values that do not conform to local rain or no-rain conditions, (3) incorrect observations recorded by a rain gauge when it is malfunctioning that are physically possible values. Neighborhood-based methods used for spatial consistency checks can help in evaluating outliers and anomalies. In case of neighborhood methods, measured values (observed values) and interpolated values (estimated via spatial interpolation) from reference sites (i.e., nearby sites with good quality, gap-free observations) can be used.

9. HOMOGENEITY AND STATIONARITY

Hydrometeorological time-series often exhibit spurious (nonclimatic) jumps and gradual shifts due to changes in station location, environment (exposure), instrumentation, or observing practices (WMO, 2009). Homogeneity means that all the elements of the data series originate from a single population (WMO, 2009). Observation stations are often moved from one location to another, and these locational changes may lead to discontinuities in extremes, trends, and observations influenced by local weather and other regional climatic influences. All these factors affect the homogeneous nature of the long-term time-series of hydrometeorological data and bias studies related to the extremes. Guidelines on analysis of extreme events developed by WMO (2009) provide real-life examples of nonhomogeneities and stress the need to have complete station history metadata (data about data) for resolving these issues. If observations at a station or a set of stations are suspicious, carefully identified reference stations can be used for evaluation. Mass curves can be developed using the observations at the station suspected of problems and any reference station.

Graphical methods of detection of nonhomogeneity include moving average plots using smoothing methods (McCuen, 2003). These methods will be discussed later in the section of this chapter. It is also important to collect causal information in the process of analysis to evaluate the reasons for nonhomogeneity. Often metadata available about observations and information about the stations (or measuring instruments) is extremely helpful to decipher any nonhomogeneity in the data. For example, for rain gauge data, it is important to associate different rainfall-producing mechanisms (slow moving frontal systems, hurricane events, and summer convective storms) to rainfall depths in specific years for specific storm durations. Spatial summary statistics for all stations should be used to assess the regional or global variability of rainfall in a region. This analysis will help to establish or confirm if the storm events produced by meteorological processes are similar. This assessment will also strictly satisfy the homogeneity requirement of statistical analysis of extreme events. Miller (1972) points out that in the case of extreme precipitation, nonhomogeneity tends to be difficult to decipher and nonhomogeneity in yearly precipitation totals is easier to detect. Several other tests are available for assessment of homogeneity of time-series. These tests include Pettitt's test (Pettitt, 1979), Buishand's test (Buishand, 1982), Alexandersson's standard normal homogeneity test (SNHT) (Alexandersson, 1986), and Von Neumann ratio test (Von Neumann, 1941), which can be used for assessment of homogeneity. Buishand and Alexandersson have used the tests for evaluating the homogeneity of rainfall records.

9.1 Stationarity

According to WMO (2009) stationarity means that, excluding random fluctuations, the data series is invariant with respect to time. Hydroclimatic variable extremes are evaluated as block maxima or a single extreme value per year. Annual extremes time-series should be checked for stationarity. This is an important element of the statistical analysis conducted after the initial phase of the data collection. Trend analyses should be conducted for all the available annual extreme data for all durations and tests for statistically significant trends should be based on Mann-Kendall or other tests. According to Ashkar (1996) two important forms of nonstationarity in a time-series (e.g., streamflow time-series) are (1) jumps and (2) trends. Another impact form is the existence of cycles that are associated with long-term climatic oscillations. Statistical tests for detecting stationarity in any hydrologic time-series include the Mann-Whitney test for jumps and Wald-Wolfowitz test for trend (Bobee and Ashkar, 1991). Machiwal and Jha (2012) describe the use of student's t-test and simple t-test for evaluation of stationarity of a time-series. To conduct these tests, the data are divided into a number of subseries and the t-tests are conducted to evaluate statistically significant changes in the mean values of any two subseries.

Betancourt (2009) argues that systems for management of water throughout the developed world have been designed and operated under the paradigm of hydrologic stationarity (HS). The stationarity assumption suggests that hydrologic variables have time-invariant probability density functions whose properties can be estimated from the instrumental record (Betancourt, 2009). Given the magnitude and time lags of climate change associated with the buildup of greenhouse gases, stationarity may indeed be dead (Milly et al., 2008). A viable successor to stationarity must encompass principles and methods for identifying nonstationary probabilistic models of relevant environmental variables and for using such models to optimize water systems (Betancourt, 2009; Milly et al., 2008). Nonstationary hydrologic variables can be handled stochastically to describe the temporal evolution of their means and variances, with estimates of uncertainty.

9.2 Examples of Nonhomogeneity in Time-Series

In a recent study by author, issues related to data homogeneity due to changes in the instrument or the location of measurement were noted in monthly temperature and precipitation data at several sites in Japan provided by japan meteorological agency (JMA). To analyze the data for trends and changes in the observations, two options are available, and they are: (1) eliminate the data from those temporal windows in which homogeneous measurements are not available or (2) adjust the data from suspected nonhomogeneous temporal windows and test for homogeneity. WMO (1988) classifies data in to three different grades referred to as "very good,"

"good," and "acceptable." When station history is known, and the record has been checked for homogeneity with no adjustment required, then the data from that station is graded as very good. Data from a station is considered good when station history is known and data has been homogenized. When station history is not well-known and the data has not been homogenized then the data is considered acceptable. Examples of situations that the author encountered in a recent study with long-term datasets from Japan are shown in Fig. 1.1. The original and filtered (i.e., after inhomogeneous data are removed) datasets for different conditions are shown. In general, it was noted that trend assessments (i.e., increasing or decreasing) were different at some sites based on original and filtered datasets. In the absence of mechanisms to adjust the data, it is recommended that data be removed from those temporal windows to derive any conclusions about the data. It is obvious that removing data results in loss of valuable information from the available time-series. However, the inclusion of the inhomogeneous data will drastically affect the results of the assessments and ultimately conclusions derived based on the analysis.

10. NONPARAMETRIC TESTS FOR INDEPENDENCE

10.1 Serial Correlation Coefficient

Independence of time-series data can be evaluated by using serial correlation coefficient. An autocorrelation plot of the time-series will reveal randomness of the series or existence of persistence. If the autocorrelation at all lags other than at lag zero lies within a critical region defined for a hypothesis test (Anderson, 1942; Dahmen and Hall, 1990), the time-series can be considered independent.

10.2 Runs Test

The runs test also referred to as Wald-Wolfowitz test (Wald and Wolfowitz, 1943) is a nonparametric test that can be used to examine the randomness of a sample. There is no parametric equivalent of this test. The runs test can be used to decide if a dataset is derived based on a random process and to confirm the evidence of nonrandomness in a time-series (Nott, 2006). A run is defined as a series of increasing values or a series of decreasing values. The initial step in the runs test is to list the values in sequential order and count the number of runs. The number of increasing (or decreasing) values is the length of the run. In a random dataset, the probability that the $(i + 1)$th value is larger or smaller than the *ith* value follows a binomial distribution,

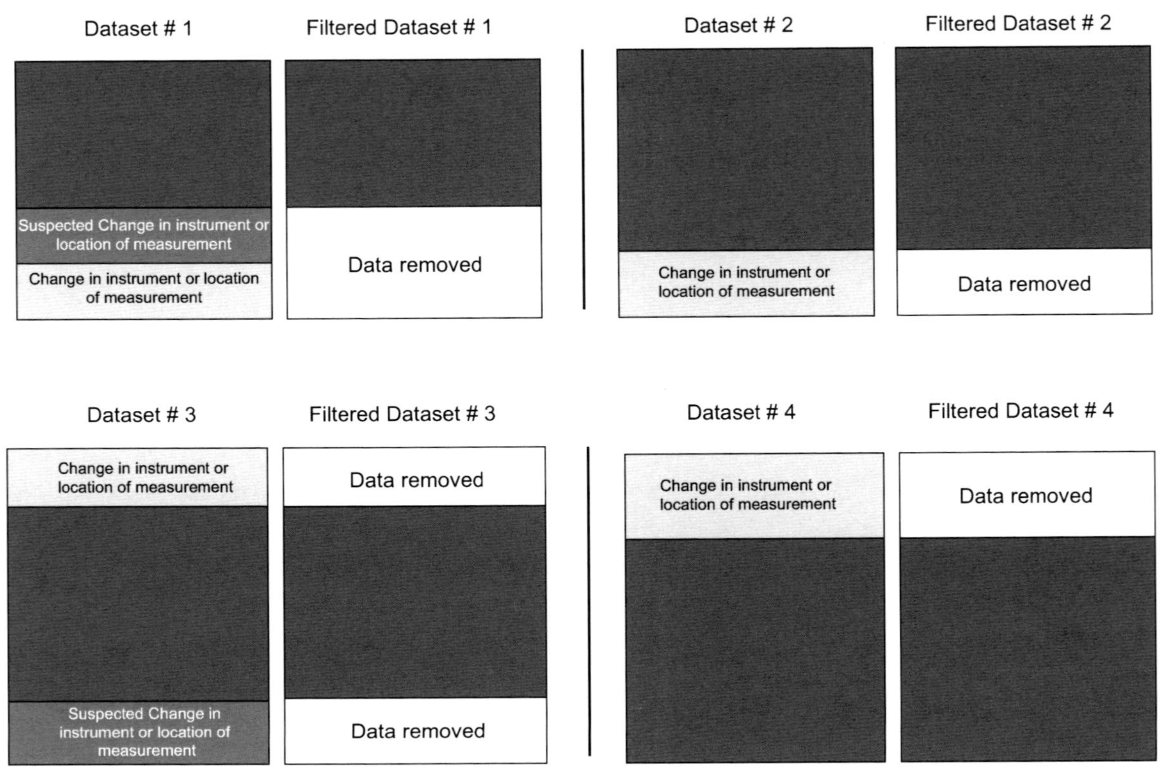

Figure 1.1 Examples of removal of data for basic analysis of trends and changes when suspected or confirmed changes occur in the data measurements.

which forms the basis of the runs test (NIST, 2018). The first step in the runs test is to compute the sequential differences $(Y_i - Y_{i-1})$. Positive values indicate an increasing value, whereas negative values indicate a decreasing value. In other terms, if $Y_i > Y_{i-1}$, an unit value (one) is assigned for an observation and a 0 (zero) otherwise. The series is then transformed to a series of 1s and 0s. To determine if the number of runs is the correct number for a series that is random, let n be the number of observations, n_1 be the number above the mean, n_2 be the number below the mean, and R be the observed number of runs. Then, using combinatorial methods, the probability $P(R)$ can be established and the mean and variance of R can be derived (Cromwell et al., 1994; Gibbons, 1997). When n is relatively large (>20) the distribution of R is approximately Gaussian.

$$\overline{x_R} = \frac{2n_1n_2}{n_1 + n_2} + 1 \tag{1.10}$$

$$S_R = \sqrt{\frac{2n_1n_2(2n_1n_2 - n_1 - n_2)}{(n_1 + n_2)^2(n_1 + n_2 - 1)}} \tag{1.11}$$

$$Z^* = \frac{R + h - \overline{x_R}}{S_R} \tag{1.12}$$

The variable Z^* is the z-score for a normal approximation of the data, R (Corder and Foreman, 2009) is the number of runs, and h is the correction for continuity, ±0.5 where

$$h = +0.5 \quad \text{if} \quad R < \frac{2n_1n_2}{(n_1 + n_2 + 1)} \tag{1.13}$$

$$h = -0.5 \quad \text{if} \quad R > \frac{2n_1n_2}{(n_1 + n_2 + 1)} \tag{1.14}$$

The null hypothesis is rejected if the calculated test statistic is greater than the selected critical value obtained from the standard normal distribution table at a specific significance level. In other words, the time-series is decided to be nonrandom.

10.3 Ranked Von Neumann Test

Ranked Von Neumann test or ranked version of Von Neumann's ratio test, a nonparametric test, can be used for testing the randomness in a sequence of observations. The test statistic is given by Eq. (1.15) and the null and alternative hypotheses are the data are independent, and data are not independent, respectively. The variable n refers to number of sample data and r_i is the rank of the data when the values are arranged from smallest to largest.

$$V_o = \frac{12}{n(n^2 - 1)} \sum_{i=2}^{N} (r_i - r_{i-1})^2 \tag{1.15}$$

The test statistic is compared with a critical value $V_{n,\alpha}$ at a significance level of α from the table provided by Bartels (1982). If the value of V_o is less than $V_{n,\alpha}$, then null hypothesis is rejected. The power of the test is expected to be diminished when ranks are tied.

11. HANDLING MISSING DATA

Missing data that exists in all hydrometeorological time-series creates problems for the analysis of time-series for climate change and variability studies. Evaluation of different time-series characteristics requires chronologically continuous gap-free data. Spatial and temporal interpolation methods can be used for filling missing data. The next two sections describe spatial and temporal interpolation methods and their applicability in estimating missing precipitation data.

Precipitation is a vital input for many hydrological modeling studies and has a direct bearing on the hydrological modeling and water resources management at different spatial and temporal scales. The importance of rainfall as being the most sensitive input in the simulation models is stressed by many researchers in their studies (e.g., Larson and Peck, 1974; Vieux, 2001; Xu and Singh, 1998). Xu and Singh (1998) indicates that the accuracy of a streamflow simulation model primarily depends on how well the variability of the rainfall is defined. In-filling of historical precipitation data that is missing due to a variety of reasons is an important task that relies on availability of data from monitoring networks and computationally tractable interpolation methods that are conceptually sound and robust.

11.1 Spatial and Temporal Interpolation

Spatial interpolation or temporal interpolation methods can be used for infilling missing data in any time-series. The former method uses observations available at different sites in a region for infilling the data at a site with missing data (i.e., base site), while the latter method employs only data from the site (base site) itself to infill data. The success of any spatial interpolation is attributed to the existence of a strong spatial correlation between any site and base site. Temporal interpolation depends on the existence of serial autocorrelation in time-series for developing linear or nonlinear interpolation models or autoregressive models. In some cases, spatial interpolation can only be used as temporal interpolation fails due to lack of high serial correlation at several lags in hydroclimatic variables.

11.2 Estimation of Missing Precipitation Data

Long-term precipitation datasets contain gaps due to instrument malfunctioning, the inability of the operator to collect data, discontinuity, and a variety of other reasons. Statistical analyses of long-term precipitation data require filling of these data gaps with appropriate values to obtain a serially complete dataset. Data filling techniques are normally selected based on location and desired accuracy required for the hydrologic analyses. Missing data filling methods are also based on consideration given to the nature of missing data defined by one of the three conditions (Rubin, 1987, 1996; Little and Rubin, 2002; Graham, 2012): (1) missing at random (MAR); (2) missing completely at random (MCAR); and (3) missing not at random (MNAR). The conditions MAR and MCAR indicate that the probability of having a missing observation does not depend on observed values or unobserved values. MNAR is opposite to the two conditions (i.e., MCAR and MAR). Mapping the mechanisms for missing data is difficult. One of the most straightforward definitions for MCAR is … missingness does not depend on the values of the data, missing or observed … (Little and Rubin, 2002). The need for serially complete time-series of hydrological data, especially continuous precipitation records, is becoming increasingly evident from the studies that require reconstruction of past climate and hydrological history of a site for design, modeling, and management of water resources systems. The recent availability of radar- and satellite-based rainfall estimates is also providing impetus to many studies that involve in-filling missing precipitation records.

11.3 Deterministic and Stochastic Methods

Spatial interpolation methods ranging from conceptually simple weighting techniques to methods using artificial intelligence (AI) paradigms (Teegavarapu, 2008) are now available for estimation of missing precipitation data. In this section missing rainfall data and techniques for its estimation are discussed. The main Achilles' heel of the many interpolation methods is the lack of guidance on the objective selection of the control points needed for interpolation. Often these points for interpolation are selected based on meteorological expertise and judgment. Improvements in interpolated estimates can only be evaluated by exhaustive experiments with control points (observation points) by moving from local to global interpolation schemes. This evaluation is a subjective and time-consuming exercise.

Estimating missing precipitation data is generally accomplished by traditional weighting methods (Smith, 1993), distance-based weighting methods (Simanton and Osborn, 1980; Wei and McGuinness, 1973), nonlinear deterministic and stochastic variance—dependent interpolation methods (e.g., kriging) (Teegavarapu, 2007; Zimmerman et al., 1999), and regression and time-series analysis methods (Salas, 1993). Comparative studies of rainfall estimation using these methods can be found in Singh and Chowdhury (1986), Tabios and Salas (1985), and Tung (1983). Variants of regression models incorporating meteorological variables that change in spatial domain (i.e., elevation) were proposed by Daly et al. (1994), Daly et al. (2002), and Tsanis (2004). Use of elevation information for interpolation of monthly precipitation data was investigated by Llyod (2005). Goovaerts (2000) used geostatistical approaches for evaluating the use of elevation information for spatial interpolation of rainfall. Models incorporating locally weighted polynomials reported by Loader (1999) and Regonda et al. (2006) are conceptual improvements over traditional weighting methods in which the number of neighbors and polynomial functions are objectively chosen. Inverse distance weighting method (IDWM) recommended by the *Handbook of Hydrology* (ASCE, 2001) is the most commonly used method for estimating missing data in the fields of hydrology and geographical sciences. Recent studies of Teegavarapu and Chandramouli (2005) and Tomczak (1998) provided several conceptual variants of IDWM. Lu and Wong (2008) modified IDWM weights by a distance-based decay parameter to adjust the diminishing strength in relationship with increasing distance. Teegavarapu (2009) used an association rule mining (ARM) approach to improve estimates of missing precipitation data. Global interpolation methods that use trend surface analysis with polynomial equations of spatial coordinates (Wang, 2006) and regression are also applicable for spatial interpolation. However, selection of the appropriate functional form to model the trend poses a major problem due to large range of candidate functions (O'Sullivan and Unwin, 2010). Also, thin-plate spline methods tend to generate steep gradients in data-poor areas leading to compounded errors in the estimation process (Chang, 2009). Improvement in interpolation can be achieved with a variant of the above method referred to as thin-spline with tension. Xia et al. (1999) used inverse distance, normal ratio, single best estimator, multiple regression methods for estimation of missing climatological data. Regression method was proved to be the best method of all the methods investigated in their study. In another study, Xia et al. (2001) reported the use of thin-splines, closest station, multiple linear regression, and Shepards' method (Shepard, 1968) for estimation of daily climatalogical data. They indicated that thin-splines method was the best among all the methods investigated. Hutchinson (1995) used thin-splines for interpolation of precipitation data. Ramos-Calzado et al. (2008) proposed a new approach for estimating missing precipitation data considering the rainfall measurement uncertainty. Improvements were achieved in the estimation of precipitation data when the stations with lowest measurement uncertainty were selected in the interpolation process. Teegavarapu et al. (2012) evaluated spatial interpolation methods for transformation of precipitation estimates from one grid to another. These methods use extents of spatial overlays as weights in nearest neighbor interpolation.

Variance-dependent stochastic surface interpolation methods, belonging to the general family of kriging, have been applied to hydrological spatial interpolation problems (Vieux, 2001; Grayson and Gunter, 2001). Kriging in various forms has been used to estimate missing precipitation data as well as to interpolate precipitation from point measurements (Dingman, 2002; Vieux, 2001; Ashraf et al., 1997). Cokriging of radar and rain gauge data has been employed by Krajewski (1987) to estimate mean aerial precipitation. Seo et al. (1990a, 1990b) and Seo (1996) described the use of cokriging and indicator kriging for interpolating rainfall data. Seo and Smith (1993) employed a Bayesian approach for short-term rainfall prediction using radar data in conjunction with rain gauge data. Real-time estimation of rainfall fields using radar and rain gauge data was discussed by Seo (1998). The use of ordinary kriging (OK) along with universal function approximation–based kriging for estimating missing daily precipitation data was reported by Teegavarapu (2007).

Regression and time-series models used in the past for estimating missing rainfall data require the definition of functional forms of the relationships a priori. Empirical models derived using Genetic algorithms (GAs), artificial neural networks (ANNs), and genetic programming (GP) have found numerous applications in the development and application of inductive models. Teegavarapu and Chandramouli (2005) used ANNs for estimation of missing precipitation data. Also, Teegavarapu (2007) demonstrated the use of universal functional approximation within a stochastic variance–dependent interpolation technique for estimation of missing precipitation data. Recent work of Teegavarapu et al. (2009) focused on the development of optimal functional forms for estimating missing precipitation data. The methods using optimal forms provided better estimates compared to those by traditional distance-based methods. In the current work, the functional forms are fixed, and the parameters are optimized. Optimal functional forms were developed by Teegavarapu et al. (2008) to characterize relationships between rain gauge and radar observations to estimate missing data at rain gauges. These function forms have a set of fixed algebraic relationships that can be specified by the users.

In spite of several improvements in traditional methods, limitations of spatial interpolation methods continue to exist. Vieux (2001); Grayson and Gunter (2001); O'Sullivan, and Unwin, (2010); Teegavarapu (2007, 2008, 2009); Teegavarapu et al. (2009) and Brimicombe (2003) discussed numerous limitations of the IDWM and other spatial interpolation methods. Eischeid et al. (2000) also report several interpolation methods for estimation of missing daily temperature and precipitation records and discuss their limitations. The main motivating factors to revisit the IDWM and other spatial interpolation methods and to develop new methods are described in the arguments here. The success of IDWM and many other interpolation methods depend primarily on the existence of positive spatial autocorrelation (Vasiliev, 1996), which is summarized by the following statement. Data from locations near one another in space are more likely to be similar than data from a location remote from one another (O'Sullivan and Unwin, 2010). However, it can be easily argued that distance alone is not a surrogate measure of spatial autocorrelation, and also the existence of negative autocorrelation may become a major limitation in the application of IDWM for estimation of missing data. Another important issue relevant to IDWM is the selection of neighborhood points of observations for estimation of missing data at a point of interest. Selection of locations in space for the use of observations in arriving at the distances is arbitrary. Many researchers recommend the use of three or four closest stations for application of IDWM. The arbitrariness in the choice of weighting parameter and the definition of the neighborhood are two obvious and major limitations of IDWM. Some of these limitations are common to all deterministic and stochastic interpolation methods except locally weighted polynomial-based methods. A more extensive review of methods for estimation of missing precipitation is provided in Chapter 2.

12. EXTRACTING EXTREME VALUES FOR ANALYSIS

Annual extreme value series often referred to as block maxima series are used for frequency analysis and in many other applications in the field of hydrology. In case of annual extreme series, the maximum value can be obtained from a calendar year or a water-year (or a hydrological year). The definition of water year changes from region to region. For example, in the United States, the United States Geological Survey (USGS) defines the water-year as a period of 12 months from October 1st of the current year to September 30th of the following year. However, extreme values over a prespecified threshold are also used for analysis. These series are referred to as partial duration series (PDS) or peak-over-threshold (POT) series. One limitation of annual series is that it may omit a value that is lower than the maximum value in a specific year but is higher than all other values in a series. The partial duration series overcomes this limitation by selecting all extreme values above a threshold. In some instances, the number of extreme values adopted from a PDS is the same as the number of years of data available. Analysis of annual maximum series provides estimates of the average period between years when a particular value is exceeded. This is referred to as average recurrence interval (ARI). The PDS requires the definition of a threshold value of the variable under consideration and independence between any two events that are selected.

13. STEPS FOR EVALUATION OF HYDROCLIMATIC TIME-SERIES

A number of steps to be followed for evaluation of hydroclimatic variable time-series for any trends and variations are shown using two phases in Fig. 1.2 (phase I on the left and phase II on the right). Phase I involves a quality assessment of the data and phase II include steps for assessment of time-series for climate variability and change considerations.

Quality assessment needs to be conducted in phase I considering all the issues discussed in the previous sections. In case of phase II for assessment of time-series for climate variability, data is split into multiple temporal windows that coincide with the natural cycles of climate (e.g., two or more phases of coupled oceanic and atmospheric oscillations). Evaluation of changes in hydroclimatic variables due to climate change can be carried out using constant (i.e., fixed) and moving temporal windows. Variations in the time-series from one temporal window to another can be assessed and confirmed using a number of parametric and nonparametric hypothesis tests.

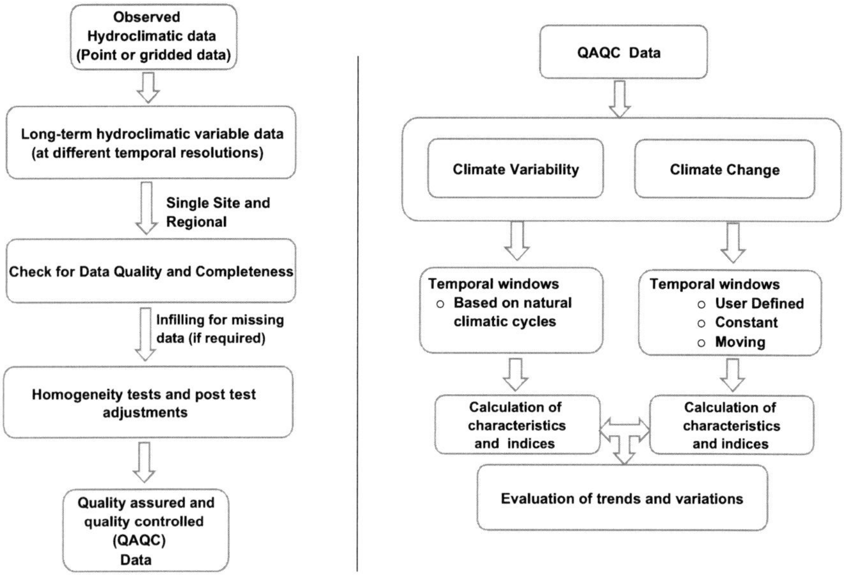

Figure 1.2 Evaluation of hydroclimatic data for changes due to climate variability and change in two phases (phase I and II).

14. TREND ANALYSIS: NONPARAMETRIC APPROACHES

Trend analysis of any hydroclimatic variable can be helpful to understand and confirm the existence (or nonexistence) of statistically significant changes in extremes or characteristics. Parametric and nonparametric methods can be used for evaluation of changes in space and time. New methods or variants of existing trend evaluation methods can be developed considering issues related to homogeneity, data length, and others factors. Trend analysis combined with evaluation of changes in hydroclimatology, land use and physiography can also help to ascertain and confirm any attributable reasons (natural climate variability influenced or anthropogenic changes) to the changes or trends. The number of sites selected for any study will depend on several factors: (1) long historical record length; (2) availability of error and gap-free data records and (3) availability of sites in watersheds with exhaustive hydrometeorological data. The following are the tasks needed to be completed for evaluation of trends:

1. Collect and evaluate the time-series data at several sites in the region of interest and check for homogeneity issues.
2. Extract extremes and conduct analysis to check for existence of change points in the time-series.

3. Develop and test nonparametric and parametric statistical approaches for trend assessment considering issues such as serial autocorrelation, length of the data, temporal windows (moving and constant) and temporal windows linked to different phases of natural climate variability or anthropogenic activity.
4. Assess natural or anthropogenic-based variations in extremes using detection and attribution methods.

Parametric approaches require the sample distribution to be completely described before any statistical inference tests can be undertaken. Some tests require normality assumptions to be fulfilled. Parametric methods are also very sensitive to outliers. Nonparametric approaches do not require any prior information about the distributional characteristics of the samples. Inference tests are typically based on medians or ranked data. Trend analysis of time-series is generally carried out using two well-known and widely accepted nonparametric tests, and they are (1) Spearman's Rho (ρ) test and (2) Mann-Kendall's test and its variants. Applications of Mann-Kendall's test are more documented in the literature compared to Spearman's test. However, it was not in many cases the power of both the tests is same. The nonparametric Mann-Kendall test with Sen's slope information can be used to assess the existence of trends in time-series data. For both methods, the datasets are not required to be normally distributed as they are distribution-free tests. The methods are less sensitive to extreme events and missing data points (Partal and Kahya, 2006; Yilmaz and Perera, 2015). Spatial evaluation of trends can be carried out either by grouping the data into pooled datasets considering homogeneous subregions defined by hydroclimatology and other physical features of the watersheds or basins in the region of interest or by using site-based trend assessment results.

14.1 Spearman's Rank Correlation Coefficient (ρ) Test

Spearman's Rho (ρ) test is a nonparametric test that can be used to evaluate the association between two ranked variables. The test can be used as a statistical method that assesses a possible linear association between the time and the variable of interest. The null hypothesis (H_o): $\rho = 0$ and the alternative hypothesis (H_a): $\rho \neq 0$ are defined. The Spearman's rank correlation coefficient is calculated using Eq. (1.16).

$$R_s = 1 - \frac{6\sum_{i=1}^{n}(D_i - i)^2}{n(n^2 - 1)} \tag{1.16}$$

R_s is the Spearman's Rank Correlation Coefficient, D_i is the rank of ith observation, n is the total length of the time-series data. The significance of the correlation can be tested using a table of critical values (one-tailed or two-tailed test values) when the number of data pairs are less than 30. The null hypothesis is rejected if the absolute value of R_s is greater than the critical value given in the table. If the number of data pairs are greater than 30 (Gauthier, 2001), a statistic (Z_s) which is also referred to as t-statistic is calculated using Eq. (1.17).

$$Z_s = R_s\sqrt{\frac{n-2}{1-R_s^2}} \tag{1.17}$$

Z_s is student's t-distribution-based statistic with $(n-2)$ degrees of freedom. A critical value for Z_s can be obtained at a 5% significance level of student's t-distribution table, defined as $t_{(n-2,1-\alpha/2)}$. If $|Z_s| > t_{(n-2,1-\alpha/2)}$, ($H_o$) is rejected and a significant trend exists in the hydrologic time-series (Ahmad et al., 2015).

14.2 Mann-Kendall Test

Mann-Kendall test (Mann, 1945; Kendall, 1975) is a commonly used nonparametric test for assessment of trends in hydrological time-series. It can be viewed as a nonparametric test for zero slope of the linear regression of time-ordered data with time (Gilbert, 1987, Hollander and Wolfe, 1973). In case of Mann-Kendall test, the null hypothesis (H_o) represents the case when there is no trend in the record, while the alternative hypothesis (H_a) indicates the existence of an upward or downward monotonic trend in the data. The test statistic (S) required for the Mann-Kendall test is obtained using Eq. (1.18).

$$S = \sum_{k=1}^{n-1}\sum_{j=k+1}^{n} sgn(x_j - x_k) \tag{1.18}$$

where x_j and x_k are measurements obtained at time j and k. n is the number of observed events. The function sgn () is described using Eq. (1.19).

$$sgn(x_j - x_k) = \begin{cases} +1 & \text{if } (x_j - x_k) > 0 \\ 0 & \text{if } (x_j - x_k) = 0 \\ -1 & \text{if } (x_j - x_k) < 0 \end{cases} \tag{1.19}$$

The variance of the test statistics (S) is calculated using Eq. (1.20).

$$Var(S) = \frac{\left[n(n-1)(2n+5) - \sum_{i=1}^{m} t_i(t_i-1)(2t_i+5)\right]}{18} \tag{1.20}$$

where n is the number of observations, m is the total number of tied groups, and t_i denotes the number of ties of extent i. The standardized test statistics (z) is computed using Eq. (1.21).

$$z = \begin{cases} \frac{S-1}{\sqrt{Var(S)}} & \text{if } S > 0 \\ 0 & \text{if } S = 0 \\ \frac{S+1}{\sqrt{Var(S)}} & \text{if } S < 0 \end{cases} \tag{1.21}$$

The null hypothesis (H_o) is accepted if $|z| \le z_{\alpha/2}$ where α is user-specified level of significance (Partal and Kahya, 2006; Yilmaz and Perera, 2015). To quantify the magnitude of the trend detected by the Mann-Kendall test, the Theil-Sen's slope estimate can be used. According to Sen (1968), the slope of the lines (Q_i) connecting N distinct pairs can be computed using Eq. (1.22).

$$Q_i = \frac{Y_j - Y_k}{j - k} \quad \text{for } i = 1, \ldots, N \tag{1.22}$$

where Y_j and Y_k are data values corresponding to j and k ($j > k$), respectively. After arranging the values of $Q_i^{'}s$ in ascending order of magnitude, the median of the slopes (Sen's slope estimator) (Q_{med}) is computed using Eq. (1.23). In case of Theil-Sen slope estimation, the original observations are used in obtaining the slope estimate. The use of median of pairwise slope values in the Theil-Sen procedure makes it a nonparametric.

$$Q_{med} = \begin{cases} Q_{(N+1)/2} & \text{if } N \text{ is odd} \\ \frac{Q_{(N/2)} + Q_{(N+2)/2}}{2} & \text{if } N \text{ is even} \end{cases} \tag{1.23}$$

The Theil-Sen slope helps to evaluate the magnitude of the trend through a robust linear regression with minimal affect by outliers. The Theil-Sen slope procedure does not require normality assumption for residuals and is not strict about homoscedasticity (EPA, 2000).

14.2.1 Influence of Serial Autocorrelation

One of the problems in detecting trends in hydroclimatological data is the presence of serial correlation in the time-series which is also referred to as persistence. Zhang et al. (2001) suggested a prewhitening (PW) procedure to eliminate the effects of the serial correlation on the statistical testing for the Mann-Kendall test. Yue et al. (2002) found that PW approach affects the magnitude of slope present in original data series and reduces the significance of trend present in original data. They have proposed a trend-free prewhitening (TFPW) approach to address this problem. If no autocorrelation exists as confirmed by an auto correlogram, trend evaluation using the MK test without PW or TFPW can be carried out.

14.2.2 Mann-Kendall Test With Data Prewhitening

One of the problems in detecting trends in hydroclimate data is the presence of serial correlation (i.e., persistence). Zhang et al. (2001) suggested a pre-whitening (PW) procedure to eliminate the effects of the serial correlation on the statistical testing for the Mann-Kendall test. Statistical significance of the trends in time-series of hydroclimate data was examined by Mann-Kendall test with PW using the following steps (Zhang et al., 2001; Teegavarapu, 2013):

1. Compute lag-1 serial correlation coefficient (r_1).
2. If $r_1 < 0.1$ then Mann-Kendall test was applied to the original time-series.
3. Otherwise, the Mann-Kendall test was applied on the PW series ($x_2 - r_1 x_1$, $x_3 - r_1 x_2, \ldots x_n - r_1 x_{n-1}$).

In a recent study by Teegavarapu and Nayak (2017), trends in precipitation extremes were evaluated using PW procedure.

14.2.3 Mann-Kendall Test with Trend-Free Prewhitening

Yue et al. (2002) found that PW approach affects the magnitude of slope present in original data series and reduces the significance of trend present in original data. They proposed following TFPW approach to address this problem. The steps to implement MK-TFPW (Teegavarapu, 2013) to detect significance of trends are given below:

1. Estimate Sen's slope (*SS*) of the data series
2. Detrend the data using following Eq. (1.24)

$$X_t' = X_t - SS(t) \tag{1.24}$$

Compute lag-1 serial correlation coefficient (r_1) for the detrended data series

3. If detrended data series does not show serial correlation, then MK test was applied to the original time-series
4. Otherwise, MK test was applied on TFPW data series (Y_t) computed as follows

$$Y_t^{'} = X_t^{'} - (r_1 X_{t-1}) \tag{1.25}$$

$$Y_t = Y_t^{'} - SS(t) \tag{1.26}$$

14.2.4 Seasonal Mann-Kendall Test

A variant of Mann-Kendall test considering the seasonality in time-series data was proposed by Hirsch et al. (1982). A seasonal Kendall statistic given by Eq. (1.27) is calculated by aggregating Kendall statistic value for each season. The variable k is the index for the season and ns is the total number of seasons. The z-statistic can be estimated using Eq. (1.27), by using the Var(S^o) in place of Var(S). The value of Var(S^o) is calculated using Eq. (1.28).

$$S^o = \sum_{k=1}^{ns} S_k \tag{1.27}$$

$$Var(S^o) = \sum_{k=1}^{ns} Var(S_k) \tag{1.28}$$

15. STUDIES USING MANN-KENDALL AND SPEARMAN'S RHO TESTS

A number of studies carried out in different regions of the world have used original and variants of Mann-Kendall and Spearman's Rho tests for assessment of trends in hydroclimatic variables for different temporal scales. Tables 1.2–1.6 provide a list of these studies along with the information about the type of hydroclimatic variable and temporal scale at which the assessment was conducted. No assertions are made about the exhaustiveness of the reported list of studies. It can be inferred from the list that MK test is the most used test in the literature followed by SR test.

While Mann-Kendall and Spearman's Rho tests are most commonly applied trend assessment tests, other tests such as Cox-Stuart test and Noether's binomial test for cyclical trend documented by McCuen (2003) are also useful.

Table 1.2 List of Studies Using Spearman's Rho and Mann-Kendall Tests for Different Hydroclimatic Variables

Hydroclimatic Variable	Spatial/Temporal Scale/Extremes	Name of the Test	Reference
Temperature	Monthly	TSS	Abdou (2014)
River flow	Annual and monthly	SR, MK, TSS	Abghari et al. (2013)
Precipitation	Monthly	MK, SR	Ahmad et al. (2015)
Temperature	Annual and seasonal	MK, TSS	Amirabadizadeh et al. (2015)
Streamflow, precipitation, and evaporation	Daily	MK, TSS	Amo-Boateng et al. (2014)
Temperature	Annual average and seasonal	MK	Arora et al. (2005)
Maximum temperatures and heat waves	Monthly	MK, SR	Austria and Bandala (2016)
Evapotranspiration	Monthly maximum and minimum	MK	Bandyopadhyay et al. (2009)
Precipitation	Seasonal and annual	MK, SR	Bartolini et al. (2014)
Rainfall	Annual	MK	Basistha et al. (2009)
Streamflow	Annual	MK	Onoz and Bayazit (2003)
Streamflow	Seasonal and annual	MK	Birsan et al. (2005)
Temperatures	Minimum and maximum daily	MK	Brunetti et al. (2000)
Precipitation	Seasonal and annual	MK	Brunetti et al. (2001)
Rainfall	Annual and seasonal	MK	Caloiero et al. (2011)
Rainfall	Annual, seasonal, and monthly	MK	Cannarozzo et al. (2006)
Rainfall	Annual and seasonal	MK, SR	Chakraborty et al. (2013b)
Evapotranspiration	Monthly	MK	Chakraborty et al. (2013a)
Water quality	Annual	MK	Chebana et al. (2013)

Table 1.3 List of Studies Using Spearman's Rho and Mann-Kendall Tests for Different Hydroclimatic Variables

Hydroclimatic Variable	Spatial/Temporal Scale/Extremes	Name of the Test	Reference
Runoff, precipitation, and temperature	Seasonal and annual	MK	Chen et al. (2007)
Rainfall	Annual	MK	Cislaghi et al. (2005)
Streamflow	Monthly	MK	Cristea et al. (2014)
Evapotranspiration	Seasons	MK, TSS	Croitoru et al. (2013)
Evapotranspiration and weather variables	Annual and seasonal	MK	Darshana et al. (2013)
Streamflow	Annual	MK	Detzel and Mine (2014)
Evapotranspiration	Monthly and annual	MK, TSS	Dinpashoh et al. (2011)
Precipitation	Annual and seasonal	MK, TSS	Duhan and Pandey (2013)
Precipitation, temperature, and evaporation	Annual	MK	Ejder et al. (2016)
Evapotranspiration	Annual and seasonal	MK	Espadafor et al. (2011)
Water level	Annual and monthly	MK, SR, TSS	Fathian et al. (2015)
Precipitation	Annual and monthly	MK	Fathian et al. (2016)
Temperatures	Mean annual and mean maximum	MK	Gadgil and Dhorde (2005)
Rainfall	Annual, seasonal, and monthly	MK, TSS	Gajbhiye et al. (2016)
Precipitation	Daily	MK	Gallego et al. (2006)
Precipitation, Streamflow, and Evapotranspiration	Annual and seasonal	MK	Garbrecht et al. (2004)
Streamflow and precipitation	Yearly monsoon and premonsoon	MK, SR	Gautam et al. (2010)
Streamflow	Seasonal	MK, TSS	Gautam and Acharya (2012)
Precipitation	Monthly	MK	Gemmer et al. (2004)

Continued

Table 1.3 List of Studies Using Spearman's Rho and Mann-Kendall Tests for Different Hydroclimatic Variables—cont'd

Hydroclimatic Variable	Spatial/Temporal Scale/Extremes	Name of the Test	Reference
Water quality	Monthly, seasonal, and annual	MK, SR	Gocic and Trajkovic (2013)
Precipitation	Monthly	MK	Gonzalez-Hidalgo et al. (2009)
Precipitation	Daily	MK	Guo and Li (2015)
Temperature and precipitation	Annual	MK, TSS	Guo and Xia (2014)
Rainfall	Annual extreme	MK, SR	Hajani et al. (2014)

Table 1.4 List of Studies Using Spearman's Rho and Mann-Kendall Tests for Different Hydroclimatic Variables

Hydroclimatic Variable	Spatial/Temporal Scale/Extremes	Name of the Test	Reference
River flow	Annual	MK	Hamed (2008)
River flow	Annual	MK, SR	Hamed (2016)
Rainfall and streamflow	Annual	MK	Hamed and Rao (1998)
Heavy precipitation and rainfall	Sub-daily	TSS	Hanel et al. (2016)
Water quality and discharge	Seasonal	MK	Hirsch et al. (1991)
River flow	Annual	MK	Hodgkins et al. (2003)
Water level	Annual	MK, SR	Hu and Wang (2009)
Temperature and rainfall	Daily	MK	Hu et al. (2012)
Temperature and precipitation	Interannual	TSS	Irizarry-Ortiz et al. (2012)
River discharge	Annual	MK	Dery and Wood (2005)
Temperature	Monthly, seasonal, and annual	MK	Jhajharia and Singh (2011)
Pan evaporation	Seasonal	MK	Jhajharia et al. (2009)
Evapotranspiration	Annual and seasonal	MK, TSS	Jhajharia et al. (2012)
Temperature	Monthly	MK	Jhajharia et al. (2014)

Table 1.4 List of Studies Using Spearman's Rho and Mann-Kendall Tests for Different Hydroclimatic Variables—cont'd

Hydroclimatic Variable	Spatial/Temporal Scale/Extremes	Name of the Test	Reference
Precipitation and river flow	Monthly	MK	Jiang et al. (2007)
Precipitation	Seasonal	MK	Jones et al. (2015)
Streamflow	Monthly	MK, SR	Kahya and Kalayci (2004)
Precipitation	Seasonal	MK, SR	Kalra and Ahmad (2011)
Rainfall	Annual	MK	Kampata et al. (2008)
Precipitation	Annual, monthly, and seasonal	MK, TSS	Karpouzos et al. (2010)
Water temperatures	Annual mean	MK	Kaushal et al. (2010)
Pan evaporation	Monthly	MK	Kisi (2015)
Streamflow	Annual and seasonal	MK	Kumar et al. (2009)
Rainfall	Monthly	MK, TSS	Kundu et al. (2015)
Stream quality	Seasonal	MK	Lettenmaier (1988)
Rainfall	Annual	MK	Li and Sun (2015)
Runoff	Annual and monthly	MK, TSS	Li et al. (2008)
Evapotranspiration	Annual	MK	Li et al. (2010)
Water quality and discharge	Seasonal	MK	Libiseller and Grimvall (1999)
Streamflow	Annual minimum and maximum	MK	Lins and Slack (1999)
Rainfall	Annual and seasonal	MK	Longobardi and Villani (2010)
Streamflow and rainfall	Monthly	MK	Marengo et al. (1998)
Temperature and precipitation	Annual	MK	Marin et al. (2014)
Precipitation	Seasonal	MK, SR, TSS	Martino et al. (2013)
Precipitation	Monthly	TSS	Masson and Frei (2016)

Table 1.5 List of Studies Using Spearman's Rho and Mann-Kendall Tests for Different Hydroclimatic Variables

Hydroclimatic Variable	Spatial/Temporal Scale/Extremes	Name of the Test	Reference
Climate	Annual	SR	McCabe and Wolock (1997)
Water quality	Daily	MK	McLeod et al. (1991)
Water discharge and sediment load	Annual	MK	Memarian et al. (2012)
Rainfall, heavy rain, and dry days	Annual and monthly	MK	Méndez-Lázaro et al. (2014)
Precipitation	Seasonal	MK	Miao et al. (2012)
Evaporation	Monthly maximum and minimum	MK	Min et al. (2010)
Rainfall	Annual and monthly	MK	Modarres and Silva (2007)
Temperature	Annual and seasonal	MK, TSS	Mohsin and Gough (2010)
Rainfall and temperature	Annual, seasonal, and monthly	MK,TSS	Mondal et al. (2015)
Evapotranspiration	Annual and seasonal	MK,TSS	Nafarzadegan et al. (2013)
Rainfall and relative humidity	Monthly	MK,TSS	Ogolo and Adeyemi (2009)
Rainfall	Annual, seasonal, and monthly maximum	MK, TSS	Ouarda et al. (2014)
Precipitation	Annual mean and monthly total	MK, TSS	Partal and Kahya (2006)
Streamflow	Mean daily	MK	Petrow and Merz (2009)
Precipitation	Annual, monthly, and seasonal	MK	Piccarreta et al. (2004)
Rainfall	Monthly, seasonal, and annual	MK	Pranuthi et al. (2014)
Precipitation	Annual	MK	Rahmat et al. (2016)
Rainfall and air temperature	Seasonal and annual	MK, SR	Reza et al. (2011)
Precipitation and river discharge	Annual and monthly	MK, SR	Sabzevari et al. (2015)
Extreme precipitation	Annual and seasonal	MK	Santos and Fragoso (2013)

Table 1.5 List of Studies Using Spearman's Rho and Mann-Kendall Tests for Different Hydroclimatic Variables—cont'd

Hydroclimatic Variable	Spatial/Temporal Scale/Extremes	Name of the Test	Reference
Temperature and evapotranspiration	Daily	MK	Saptomo et al. (2009)
Baseflow	Annual	MK	Sawaske and Freyberg (2014)
Precipitation	Seasonal and annual	MK, TSS	Sayemuzzaman and Jha (2014)
Temperature and precipitation	Subseasonal	MK, TSS	Sayemuzzaman et al. (2014)
Surface temperature range	Seasonal and annual	MK, TSS	Sayemuzzaman et al. (2015a)
Temperature	Annual and seasonal	MK, TSS	Sayemuzzaman et al. (2015b)
Rainfall	Annual, June–September and March–May	MK	Seleshi and Zanke (2004)
Temperature	Annual mean, minimum, and maximum	MK, SR	Şen (2014)
Precipitation	Monthly	MK	Serrano et al. (1999)
Precipitation	Monthly, seasonal and annual	MK, SR	Shadmani et al. (2012)
Evapotranspiration	Annual and seasonal	MK, TSS	Shan et al. (2015)
River-ice observations	Annual	MK	Smith (2000)
Precipitation	Annual and seasonal	MK	Some'e et al. (2012)
Extreme temperature and precipitation	Annual and seasonal	MK	Su et al. (2006)
Precipitation and evapotranspiration	Monthly and annual	MK, SR, TSS	Tabari and Aghajanloo (2013)
Precipitation	Annual and seasonal	MK	Tabari and Talaee (2011a)
Air temperatures	Annual, monthly, and seasonal	MK, TSS	Tabari and Talaee (2011b)
Temperature	Annual, monthly, and seasonal	MK	Tabari and Talaee (2011c)
Evapotranspiration	Annual	MK, SR	Tabari et al. (2012)

Table 1.6 List of Studies Using Spearman's Rho and Mann-Kendall Tests for Different Hydroclimatic Variables

Hydroclimatic Variable	Spatial/Temporal Scale/Extremes	Name of the Test	Reference
Rainfall	Annual, seasonal, and monthly	MK, TSS	Talaee (2014)
Temperature and precipitation	Annual mean	MK, TSS	Tanarhte et al. (2012)
Rainfall	Annual and seasonal	MK, TSS	Taxak et al. (2014)
Rainfall	Annual maximum, extreme precipitation indices	MK	Teegavarapu and Nayak (2017)
Pan evaporation	Monthly	MK	Topaloğlu et al. (2012)
Precipitation, temperature	Daily	MK	Vergni and Todisco (2011)
Flood	Annual	MK, SR	Villarini et al. (2009)
Rainfall	Annual and seasonal	MK	Wagesho et al. (2013)
Groundwater quality	Annual	MK	Wahlin and Grimvall (2010)
Evaporation, streamflow, temperature, rainfall	Monthly and daily	SR	Wang et al. (2015)
Precipitation	Annual maximum daily	MK	Xu et al. (2003)
Evapotranspiration and pan evaporation	Seasonal and annual	MK	Xu et al. (2006)
Temperature, relative humidity, and net solar radiation	Annual minimum and maximum	MK	Xu et al. (2010)
Streamflow	Seasonal	TSS	Yeh et al. (2015)
Extreme Rainfall	Monthly	MK, SR, TSS	Yilmaz and Perera et al. (2015)
Precipitation	Monthly	MK	Yue and Hashino (2003)
Streamflow	Annual minimum, mean, and maximum daily	MK	Yue and Pilon (2003)
River flow	Annual maximum daily	MK	Yue and Pilon (2004)

Table 1.6 List of Studies Using Spearman's Rho and Mann-Kendall Tests for Different Hydroclimatic Variables—cont'd

Hydroclimatic Variable	Spatial/Temporal Scale/Extremes	Name of the Test	Reference
Streamflow	Annual mean, minimum, and maximum daily	MK	Yue and Wang (2002)
Water quality, streamflow, and precipitation	Annual	MK	Yue and Wang (2004)
Streamflow	Annual maximum daily	MK, SR	Yue et al. (2002)
Precipitation	Annual and seasonal	MK, TSS	Yurekli (2015)
Streamflow	Annual and monthly	MK	Zhang et al. (2001)
Water level and streamflow	Annual	MK	Zhang et al. (2006)
Flow regimes and precipitation	Monthly, seasonal, and annual	MK	Zhang et al. (2011)
Precipitation	Monthly	MK, SR	Zhao et al. (2015)
Evapotranspiration	Annual	MK	Zheng and Wang (2014)

16. APPLICATION OF SPEARMAN'S RHO AND MANN-KENDALL TESTS

The two nonparametric trend tests (viz., Spearman's Rho and Mann-Kendall) were applied to 24-h duration precipitation annual extreme time-series. The observed precipitation extremes are provided in Table 1.7. The independence of the observations is initially tested using a runs test.

The runs test is performed on the sequence of precipitation values provided in Table 1.7. The null hypothesis, H_o, is that the series of precipitation values come in a random order and the alternative hypothesis is the order of the values is not random. As discussed in Section 1.9.2, the test is carried out based on the number of runs of consecutive values above or below the mean value of the time-series. Too few runs suggest clustering

Table 1.7 Precipitation Data

Year	Precipitation Depth (mm)	Year	Precipitation Depth (mm)
1960	61	1987	65
1961	53	1988	47
1962	58	1989	59
1963	75	1990	62
1964	99	1991	87
1965	50	1992	77
1966	79	1993	195
1967	89	1994	150
1970	138	1995	154
1971	80	1996	54
1973	88	1997	77
1974	49	1998	117
1975	87	1999	167
1976	48	2000	106
1978	78	2001	117
1979	214	2002	68
1980	95	2003	121
1981	198	2004	69
1982	67	2005	60
1984	130	2006	65
1985	82	2007	116
1986	66		

of high values together and low values together. Too many runs indicate a tendency for high values and low values to alternate. The mean value of the series is 93.4 mm. The numbers of values above and below mean are 15 and 28, respectively. The Z statistic value is -1.3742 and the number of runs is 16. Based on the test results at 5% significance level and a P-value of 0.170, there is no evidence to reject the null hypothesis that the values come in a random order. Once it is determined that the time-series observations are independent, Spearman's Rho test is conducted. A plot of observations and results from Spearman's Rho that uses t-test are shown in Figs. 1.3 and 1.4, respectively.

The number of data pairs used for this example is more than 30. Hence a t-statistic is calculated and conclusions are drawn. The value of t-statistic and critical values are provided in Fig. 1.3. The dark colored areas shown in

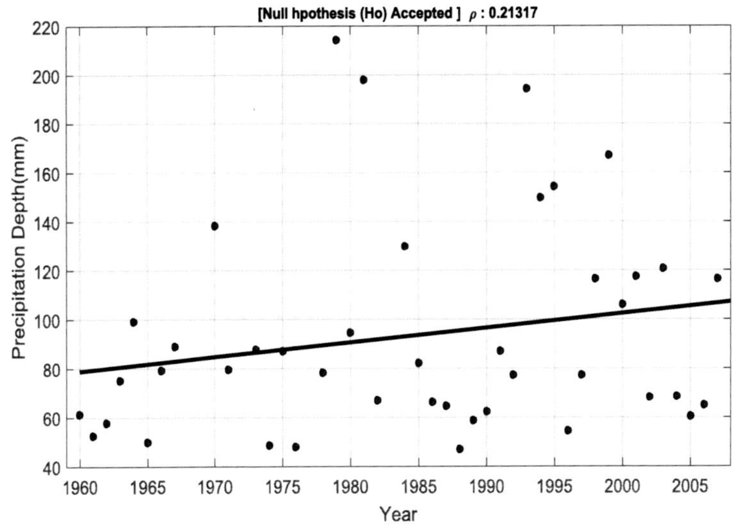

Figure 1.3 Time-series plot of precipitation depths with a fitted trend line.

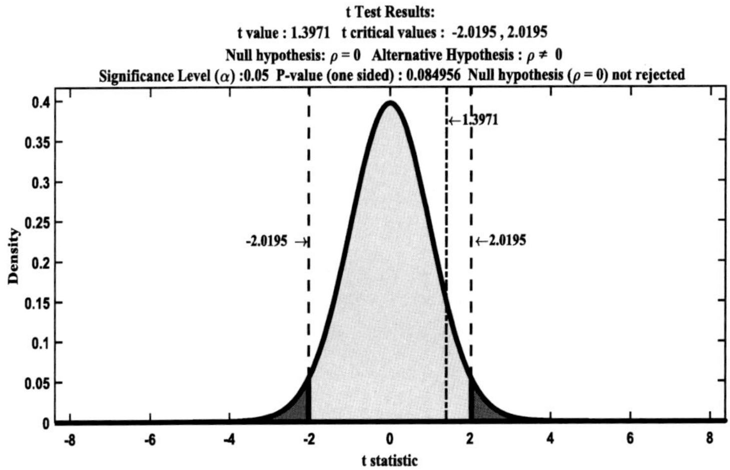

Figure 1.4 Results for the Spearman rho test including details of t-statistic and critical values.

Fig. 1.3 relate to rejection regions for the two-sided t-test for a significance level of 5%. The autocorrelogram constructed using precipitation data is shown in Fig. 1.5. The figure shows no statistically significant correlation at different lags as the serial correlation values are in within the bounds of critical region (Anderson, 1942) defined by two horizontal lines.

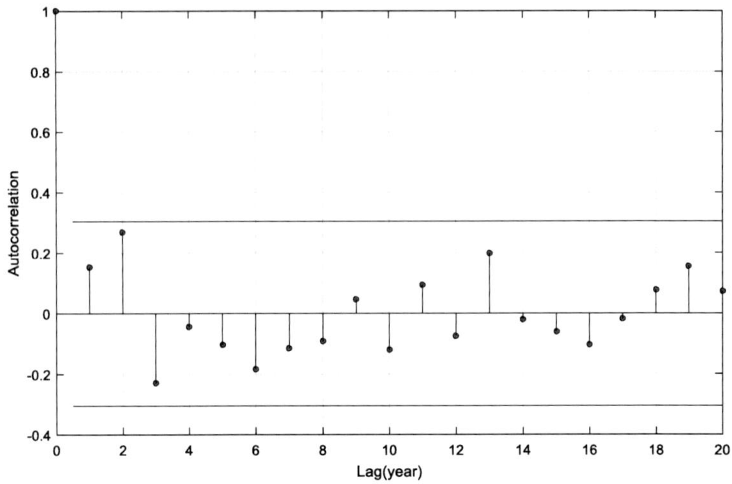

Figure 1.5 Autocorrelogram based on the precipitation data provided in Table 1.1.

The critical region (U) (Anderson, 1942; Dahmen and Hall, 1990) is defined at 5% significance level as

$$\left\{-1,\frac{\left(-1-1.96(n-2)^{0.5}\right)}{(n-1)}\right\}U\left\{\frac{\left(-1+1.96(n-2)^{0.5}\right)}{(n-1)},+1\right\} \tag{1.29}$$

Trend analysis using the Mann-Kendall test as well as estimation of Theil-Sen slope is shown in Fig. 1.6. The dotted line is the one based on Theil-Sen slope calculation concept. The S value is 107, and the Z statistic is 1.1096, and Kendall's tau is 0.118.

Monthly precipitation data available from 1987 to 2012 at one site in Iran are used to conduct the seasonal Mann-Kendall test as the data show seasonality which can be concluded from autocorrelogram shown in Fig. 1.7. Results from seasonal MK test point to null hypothesis being true with a P-value of 0.373 and Sen slope is equal to zero as shown in Fig. 1.8. The test is conducted at 5% significance level. To address the issue of serial autocorrelation PW procedure previously discussed can be adopted.

17. PARAMETRIC TREND ANALYSIS: REGRESSION-BASED METHOD

Linear regression as one of the parametric procedures can be used for assessing linear trends. If observations are available with respect to time, a

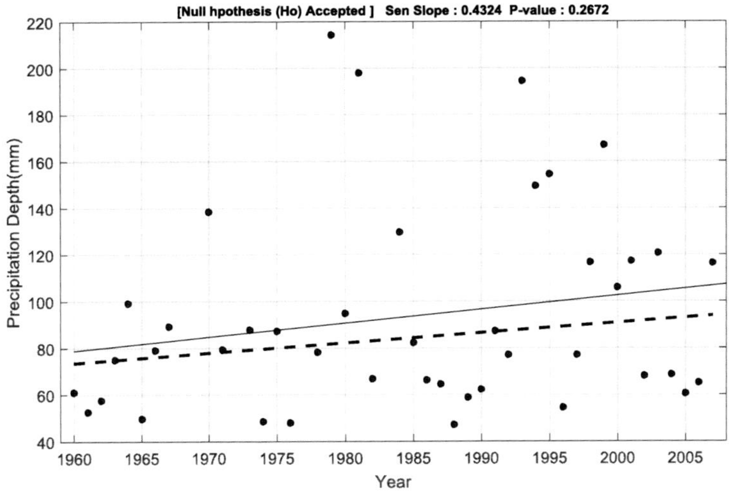

Figure 1.6 Trend analysis results using nonparametric Mann-Kendall test with Theil-Sen slope-based line (showed as a *dotted line*).

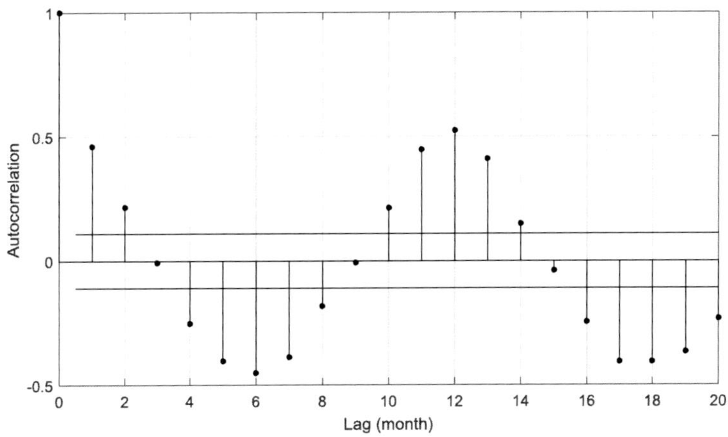

Figure 1.7 Autocorrelogram of monthly precipitation data.

linear regression given by Eq. (1.30) can be developed. An ordinary least squares (OLS) method can be used to estimate the parameters (slope: β, intercept: α). The variable $y(t)$ is the estimate value at time t and $\varepsilon(t)$ is the error. The null hypothesis H_o: $\beta = 0$ and alternative hypothesis H_a: $\beta \neq 0$ are evaluated using t-statistic and t-test. Null hypothesis suggests that data are not linearly dependent on time. A confidence interval can be constructed for the trend which is same as that for slope.

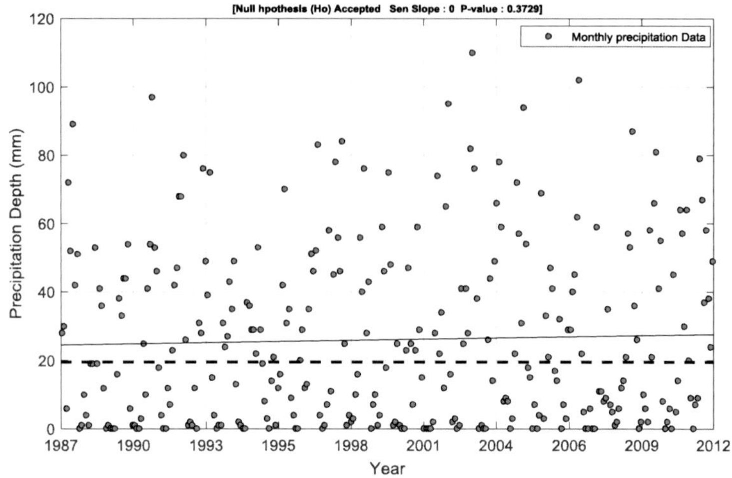

Figure 1.8 Trend analysis results using nonparametric Mann-Kendall test with Theil-Sen slope-based line (showed as a *dotted line*) for monthly precipitation data.

$$y(t) = \beta \cdot t + \alpha + \varepsilon(t) \tag{1.30}$$

A number of assumptions need to be met before the results from the parametric model can be used for making any conclusions about the existence or nonexistence of trend in the data. The observations themselves are assumed to be independent, with no heteroscedasticity in the data. Additional assumptions include: (1) residuals are independent; (2) residuals are normally distributed, and (3) residuals are identically distributed. Residuals based on observed and estimated data are estimated for each time interval using the regression equation. They are evaluated for any systematic bias, time-dependent bias, any signs of heteroscedasticity and normality of residuals. A residual (e_i) is calculated for each time interval i and is given by Eq. (1.31). In calculating the residuals, estimated values ($\theta_{o,i}$) using Eq. (1.31) and observed values ($\theta_{e,i}$) are used. Visual evaluation will reveal many of the issues associated with residuals. Probability plots of residuals can be evaluated for normality. A summary of few visual and statistical tests for evaluation of residuals is provided in Table 1.8.

$$e_i = \theta_{o,i} - \theta_{e,i} \quad \forall i \tag{1.31}$$

Table 1.8 Tests for Evaluation of Residuals and Requirements

Test or Plot	Evaluation	Requirement
Time-series plot of residuals	Time dependency	Random variation and no structured pattern in the variation of residuals with time
Autocorrelation plot (autocorrelogram)	Check for serial correlation at different temporal lags	Low or negligible correlation at all lags
Probability plot (Normal)	Normality of residuals	Linear plot of residuals
Histogram	Visual check for Normality	Gaussian distribution
Durbin-Watson test	Test for serial correlation	No serial correlation

Normality of residuals can be evaluated visually using: (1) histogram; (2) normal probability plot; (3) quantile-quantile plot. Goodness-of-fit hypothesis tests can also be used to evaluate the normality. Independence of residuals can be evaluated using an autocorrelation function, and constant variance (homoscedasticity) of residuals can be checked using Tukey-Anscombe plot which shows the variation of residual values with respect to estimated (fitted) values from the regression. The Durbin-Watson test (DW test) (Durbin and Watson, 1950, 1951) statistic, *dw*, can be calculated based on Eq. (1.32) and is compared with a critical value which in turn is based on sample size.

$$dw = \frac{\sum_{i=2}^{N} (e_i - e_{i-1})^2}{\sum_{i=1}^{N} e_i^2} \tag{1.32}$$

The Ljung–Box Q test (Ljung and Box, 1978), a portmanteau test, can be used to check the null hypothesis that no autocorrelation exists in a series of residuals for a prespecified number of lags (L) against the alternative that serial correlation coefficient, $\rho(k)$, for lags up to L is nonzero. The variable n is the length of time-series. In case of a portmanteau test, the null hypothesis is well defined while the alternative hypothesis is not clearly specified. The test statistic Q is calculated using Eq. (1.33).

$$Q = n(n+2)\sum_{k=1}^{L}\left(\frac{\rho(k)^2}{(n-k)}\right) \tag{1.33}$$

The number of lags used for the test will affect the power of the test and appropriate value recommended by Box et al. (1994) can be used.

17.1 Illustrative Example

The dataset of precipitation depths provided in Table 1.9 is used to develop a regression model with time (i.e., year) as the independent variable and precipitation depth as the dependent variable. The results of the regression analysis and the ANalysis Of Variance (ANOVA) results are provided in Table 1.9. The coefficient of determination (R^2) which is the ratio of explained variation (sum of squares calculated based on regression) and total variation (sum of squares of regression and residuals) is also provided in Table 1.9.

The regression parameters estimated are β. The regression statistics point to lack of evidence in support of null hypothesis (H_o: $\beta = 0$) for parameter

Table 1.9 Regression Diagnostics for Linear Model Fitted to Precipitation Depths

Regression Statistics			**Value**		
Multiple R	0.196				
R Square	0.038				
Adjusted R Square	0.015				
Standard Error	42.475				
Observations	43				
ANOVA					
	df (Degrees of Freedom)	*Sum of Squares*	*Mean Sum of Squares*	*F stat*	*P-value*
Regression	1	2943.856	2943.856	1.632	0.209
Residual	41	73,968.609	1804.112		
Total	42	76,912.465			
Coefficient	*Coefficient Value*	*t Stat*	*P-value*		
α	−1072.35	−1.1750	0.2468		
β	0.59	1.2774	0.2086		

β with a probability value of 0.2086. The t-statistic is calculated with degrees of freedom equal to N−2 (i.e., 43−2, 41). Visual evaluations of different assessments are shown in Figs. 1.9 and 1.10.

The histogram of residuals shows a positively skewed distribution and the residuals are not normally distributed. The Chi-square and KS goodness-of-fit hypothesis tests for normality suggest otherwise with null hypothesis being not true. However, Lilliefors's' and Anderson Darling tests point to null hypothesis not being true which seems correct based on the visual assessment of histogram of residuals. Regression is carried out using log-transformed ($\log_e$) values. The regression statistics point to evidence in support of null hypothesis (H_o: $\beta = 0$) with a probability value of 0.147 with test carried at 5% significance level. The results of the regression analysis with log-transformed precipitation depths are shown in Table 1.10. Different plots to help evaluate the appropriateness of linear model are shown in Figs. 1.11 and 1.12.

The histogram of residuals shows a distribution resembling Gaussian and the residuals are somewhat normally distributed. The Chi-square and kolmogorov-smirnov (KS) goodness-of-fit hypothesis tests for normality suggest that null hypothesis is true with probability values of 0.415 and

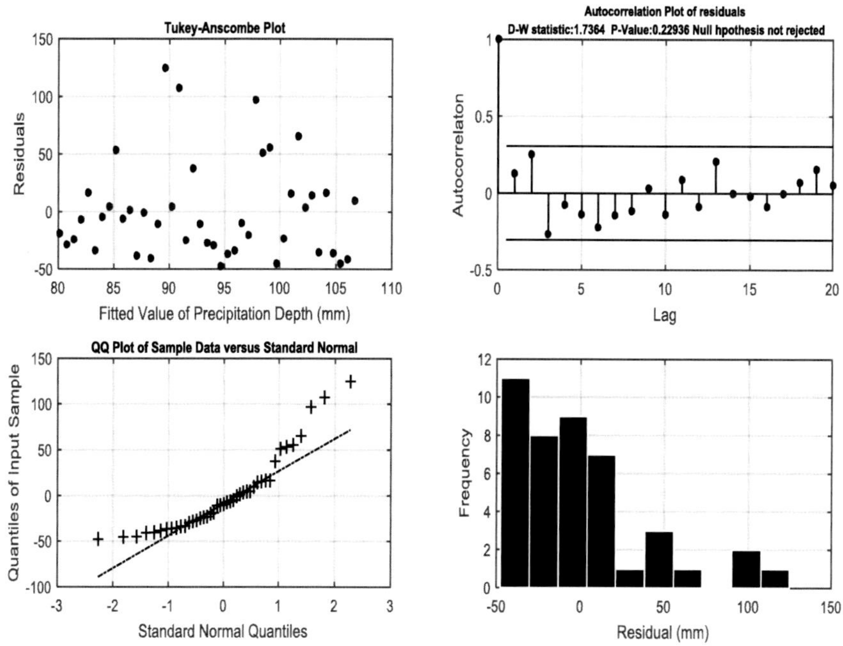

Figure 1.9 Visual evaluation of parametric trend analysis using different plots.

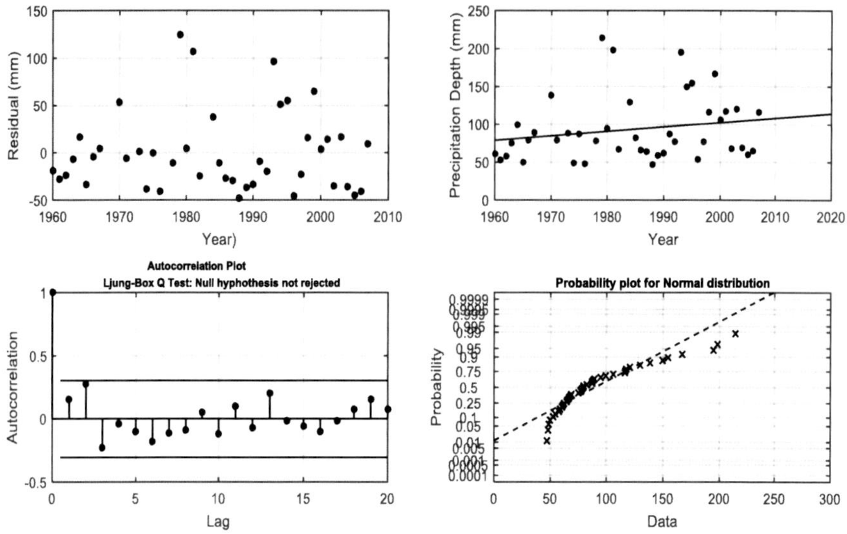

Figure 1.10 Visual evaluation of parametric trend analysis using different plots.

Table 1.10 Regression Diagnostics for Linear Model Fitted to Log-Transformed Precipitation Depths

Regression Statistics

Multiple R	0.225				
R Square	0.051				
Adjusted R Square	0.027				
Standard Error	0.403				
Observations	43				
ANOVA					
	df (Degrees of Freedom)	*Sum of Squares*	*Mean Sum of Squares*	*F stat*	*P-value*
Regression	1	0.355	0.355	2.182	0.147
Residual	41	6.674	0.163		
Total	42	7.029			
Coefficient	*Coefficient Value*	*t Stat*	*P-value*		
A	−8.355	−0.964	0.341		
B	0.006	1.477	0.147		

0.819, respectively. The Lilliefors's' and Anderson Darling tests point to null hypothesis being true with probability values of 0.439 and 0.149, respectively. Results from the Ljung–Box Q test are shown in Fig. 1.11 for the time-series.

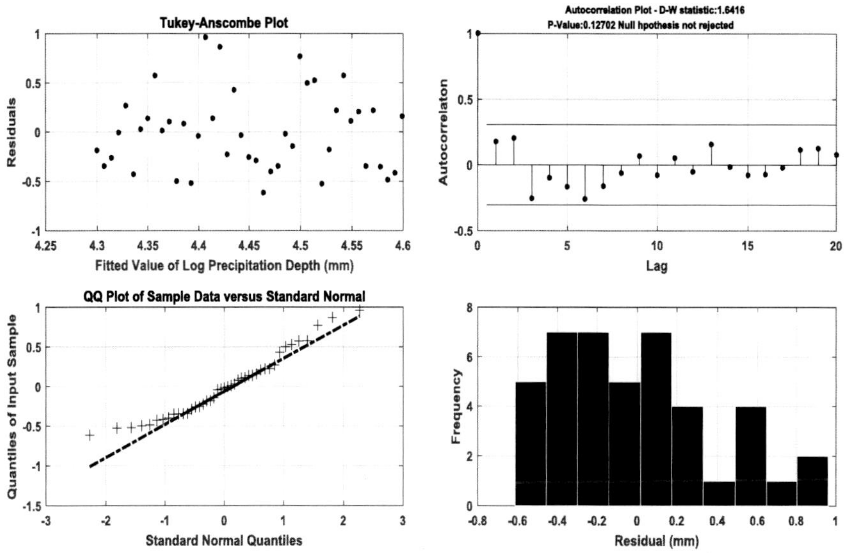

Figure 1.11 Visual evaluation of parametric trend analysis using different plots with log-transformed data.

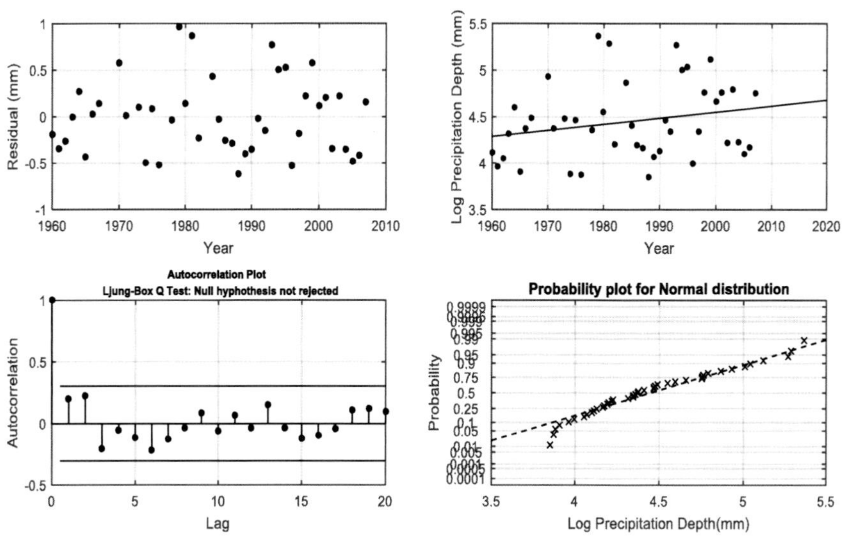

Figure 1.12 Visual evaluation of parametric trend analysis using different plots with log-transformed data.

The Tukey-Anscombe plots can help in the determination of homoscedasticity or heteroscedasticity. Homoscedasticity (i.e., constant error variance) or heteroscedasticity (i.e., nonconstant error variance) can be assessed by plotting standardized residuals and estimated values. Checking for heteroscedasticity of residuals is essential and can be done using different tests: Breush-Pagan test and White Test (Woodridge, 2003). Visual observations of residuals as a first step, can be used for matching with the patterns that identify with those with heteroscedasticity (Hair et al., 2010). Also, it is important to note that heteroscedasticity has to be severe before it can lead to serious bias in the standard errors (Allison, 1998).

18. SMOOTHING METHODS

Smoothing methods belong to the class of approaches that use average values of observations calculated over specific temporal windows to reduce noise or variations in the time-series. Moving averages for precipitation depths from Table 1.7 calculated based on 5-, 10-, and 15-year temporal windows are shown along with original data in Fig. 1.13. It is evident from the plots that as the temporal window for averaging increases, the curves representing the data become smoother and provide a better understanding of the data by reducing the variability. Smoothing by moving average process helps in clearly identifying any long-term trends by reducing the noise introduced by short-term temporal fluctuations.

Apart from moving average smoothing filters, different types of filters are used in assessment of many time-series data, and they include: (1) locally weighted scatter plot smooth (LOWESS) interpolation with variants of linear and quadratic fits; (2) Robust LOWESS (with variants of linear and quadratic fits) and (3) Savitzky and Golay. Results from these smoothing filters are shown in Figs. 1.14 and 1.15. Rodda and Little (2015) discuss the

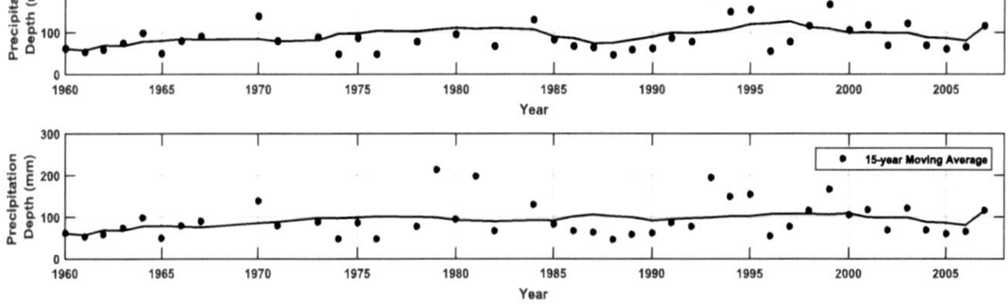

Figure 1.13 Moving average plots using different temporal windows.

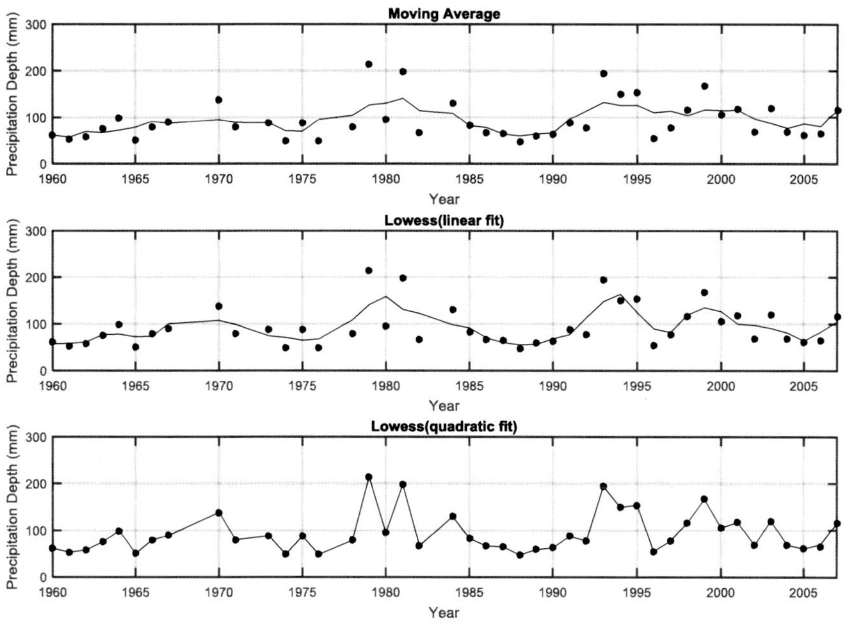

Figure 1.14 Moving average, LOWESS plots of the precipitation time-series.

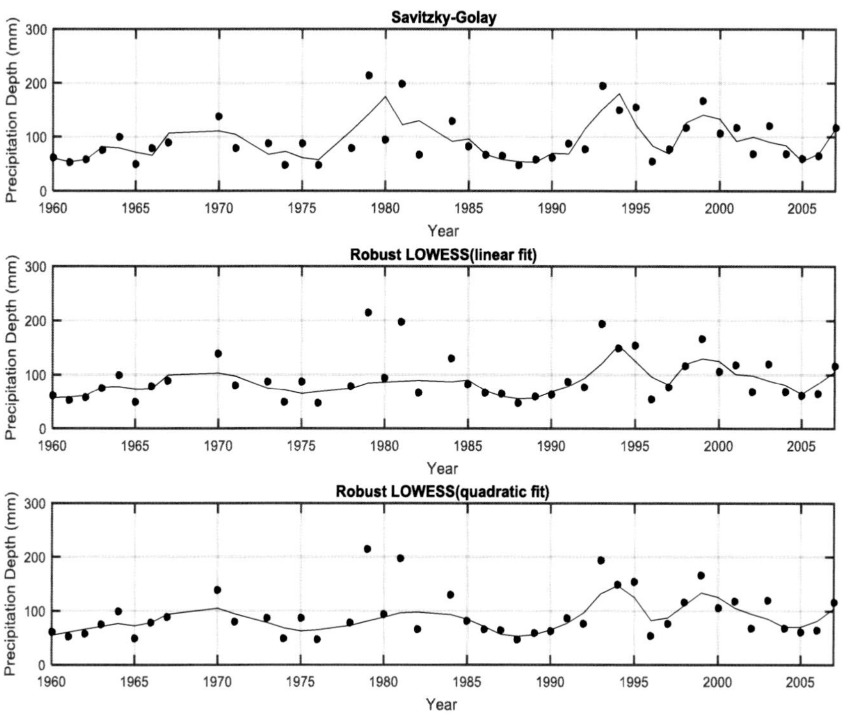

Figure 1.15 Smoothened plots of precipitation time-series using different filters.

advantages and disadvantages of linear smoothing and different filters. They provide details about running mean (average), exponential, low-pass and frequency domain filtering methods. In case of moving average filters, the temporal window (i.e., the length of period) over which the mean time-series values are calculated will affect the nature of smoothing. The larger the temporal window, the smoother the timer-series and if the temporal size is large, information from long history of time-series can be included (Rodda and Little, 2015).

19. ASSESSMENT OF CHANGES IN STATISTICAL MOMENTS OF DATA

19.1 Parametric and Nonparametric Methods

Statistical analyses can be conducted to assess the changes in moments based on historical data. Any hypothesis test will generally involve a sequence of steps, and they include: (1) collection of sample data with confirmation of homogeneity and independence; (2) check and validate the assumptions that are essential for the datasets that are related to a specific hypothesis test planned; (3) formulate null and alternative hypotheses referred to as H_o and H_a respectively (4) specify a significance level (e.g., 1%, 5%); at which the test needs to be conducted; (5) identify the probability distribution that relates to the sampling distribution; (6) derive the values of test statistic, critical value of null hypothesis and establish acceptance and the rejection regions (7) make a decision to accept or reject the null hypothesis if the test statistic falls within the acceptance or within the rejection region, respectively.

The data is split into two datasets (samples) and compared. Initially the samples are checked to see if they fit a specific probability distribution (e.g., normal or log-normal distribution), and then the sample variances (S_1^2, S_2^2) are also checked to see if they were equal or unequal. Probability plots on normal or log-normal paper are generally used to visually check the hypothesized distributions. Goodness-of-fit tests can also be used for more comprehensive checks. In few studies, annual precipitation totals were characterized using normal and gamma distributions. Thompson (1999) indicates that annual precipitation tends to follow a normal distribution, particularly in humid climates and existence of low values in arid and semiarid climates can be characterized by positively skewed distributions. However, time dependence of these values may violate the assumption of independence due to persistence at interannual time scales. Droughts and wet-years are examples of multiyear persistence (Thompson, 1999). Naoum and Tsanis (2003) report that annual rainfall amounts can be characterized using normal distribution for the island of Crete in Greece. When data seem to follow normal distributions, a few goodness-of-fit tests are applicable. The Kolmogorov–Smirnov test (KS test), can be used to ensure the validity of hypothesized data distributions. The Anderson Darling test (AD test) that is sensitive to the tails of the distribution can also be used. For datasets with equal variances, a comparison can be made using a t-test. For datasets with unequal variances, a comparison can be made using Satterthwaite's modified t-test (Satterthwaite, 1946; McBean and Rovers, 1998).

Sample means $(\overline{x}_1, \overline{x}_2)$ can then be calculated for the datasets and checked to see if they were statistically different at a specified significance level using hypothesis tests. A significance level of 5% or 1% is generally used. F-tests are required to ascertain if the variances were equal or not equal. An unpaired two-sample t-test can then be used to make inference about differences in sample mean values. The procedures for these tests are discussed in detail by Gilbert (1987) and Ott (1995). Summary of statistical hypothesis tests used for the assessment of changes in moments for normally distributed data are provided in Table 1.11.

Table 1.11 Hypothesis Tests Used Assessment of Changes in Moments for Normally Distributed Data

Statistical Hypothesis Test	Purpose	Null Hypothesis (H_o)	Alternative Hypothesis (H_a)
Kolmogorov-Smirnov(KS) Lilliefors Chi-Square Anderson Darling Jarque-Bera D'Agostino-Pearson Shapiro Wilk Cramer-von Mises	Test for goodness-of-fit	Data follows the hypothesized distribution (i.e., normal)	Data does not follow the hypothesized distribution
F-test	Test for equality of population variances	$\sigma_1^2 = \sigma_2^2$	$\sigma_1^2 \neq \sigma_2^2$
Two-sample unpaired t-test [Equal variance—student's two-sample t-test] (Unequal variance—Satterthwaite's two-sample unpaired t-test)	Tests for equality of hypothesized population means	$\mu_1 = \mu_2$	$\mu_1 \neq \mu_2$

19.2 Inferences About Changes in Mean Values

The use of two-sample unpaired t-tests requires several conditions that need to be met before any inference can be made about population means. The assumptions for a two-sample unpaired t-test are (1) normality; (2) independence of observations; and (3) equal variances. Two different types of t-tests are used based on the knowledge about sample variances. The t-test statistic calculations are identified by Eqs. (1.34) and (1.35) when the sample variances based on two different sampling periods are equal.

$$t = \frac{\overline{x}_1 - \overline{x}_2}{\sqrt{\frac{n_1 + n_2}{n_1 n_2}\left(\frac{(n_1 - 1)S_1^2 + (n_2 - 1)S_2^2}{n_1 + n_2 - 2}\right)}} \tag{1.34}$$

The degrees of freedom (*df*) is defined by the Eq. (1.35).

$$df = n_1 + n_2 - 2 \tag{1.35}$$

where n_1 is the number of samples in dataset #1, n_2: number of samples in dataset 2. S_1^2 and S_2^2 are variances, and $\overline{x}_1$ and $\overline{x}_2$ are mean values of datasets 1 and 2, respectively. The t-test statistic used for unequal variances is defined by Eqs. (1.36) and (1.37). The test is referred to as Satterthwaite's modified t-test (Satterthwaite, 1946). The degrees of freedom (*df*) in case of this t-test is given in Eq. (1.37).

$$t = \frac{\overline{x}_1 - \overline{x}_2}{\sqrt{\frac{S_1^2}{n_1} + \frac{S_2^2}{n_2}}} \tag{1.36}$$

$$df = \frac{\left[\frac{S_1^2}{n_1} + \frac{S_2^2}{n_2}\right]^2}{\frac{S_1^4}{n_1^2(n_1 - 1)} + \frac{S_2^4}{n_2^2(n_2 - 1)}} \tag{1.37}$$

An illustrative example of the application of two-sample t-test using the two datasets is provided. The datasets provided in Table 1.12 are annual 24-h duration extreme values from observations taken from rain gauge in the state of Florida, USA.

The sample datasets are initially evaluated for normality using Chi-square, goodness-of-fit, and KS tests at 5% significance level. For the Chi-square test, the number of bins used for this particular analysis is 10. The *P*-value of KS test was found to be 0.406. Both the tests indicate

Table 1.12 Annual 24-H Precipitation Extreme Datasets

Dataset# 1 (Precipitation Depth (mm))	Dataset# 2 (Precipitation Depth (mm))
153	94
93	122
84	91
77	86
103	69
212	76
163	157
66	97
274	79
132	66
98	114
108	130
90	99
96	224
117	91
177	56
63	86
74	104
139	79
140	97
98	124
90	86
63	150
75	117
112	97
99	86
144	152
137	130
86	89
	175

normality of the sample datasets. The F-test is conducted to evaluate if the samples are drawn from two populations with equal or unequal variances. The F-test confirmed that population variances are equal. Using the information from the F-test, the two-sample t-test is conducted, and the results of this test are shown in Fig. 1.16. The critical values, *P*-value, and t-statistic are also provided in Fig. 1.16. All the tests are conducted at 5% significance level. Results from the two-sample t-test show that null hypothesis is not rejected suggesting that there is no statistically significant difference in the two population means from which the two samples of

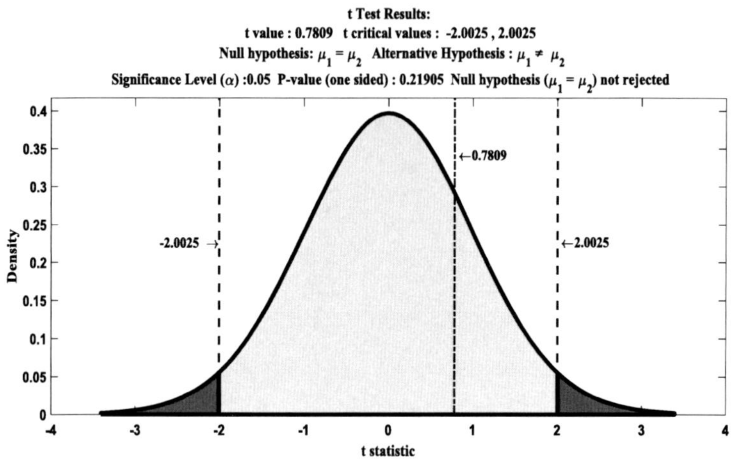

Figure 1.16 Two-sample unpaired t-test for two samples of annual 24-h extreme precipitation depths.

extreme precipitation depths are drawn. The graphic shown in Fig. 1.16 is generated using MATLAB code developed by the author.

19.3 Methods for Evaluating Changes in Variances

In the previous section, results of the two-sample unpaired t-test were reported. However, to conduct the t-test, an assessment of variances of two datasets needs to be conducted. The assessment can be carried out using an F-test to confirm the equality or inequality of variances of the populations from which the data samples are drawn. The F statistic is calculated using Eq. (1.38).

$$F = \frac{\sigma_1^2}{\sigma_2^2} \tag{1.38}$$

The test statistic F is the ratio of two population variances and this ratio follows a Fisher distribution which is also used in hypothesis testing associated with analysis of variance and linear and multiple regression. The F statistic can be estimated using sample variances $\frac{s_1^2}{s_2^2}$. Following the notation of Dahmen and Hall (1990), the rejection region, U, is bounded by Eq. (1.39).

$$\left\{0,\ \left\{F_{\nu_1,\nu_2,\alpha}\right\}\right\}U\left\{\left\{F_{\nu_1,\nu_2,1-\alpha/2}\right\},\ +\infty\ n_1\right\} \tag{1.39}$$

The variables ν_1, ν_2 are degrees of freedom for the datasets equal to $n_1 - 1$ and $n_2 - 1$, respectively and α is the significance level. The null and alternative hypothesis are H_o: $\sigma_1^2 = \sigma_2^2$ and H_a: $\sigma_1^2 \neq \sigma_2^2$, respectively.

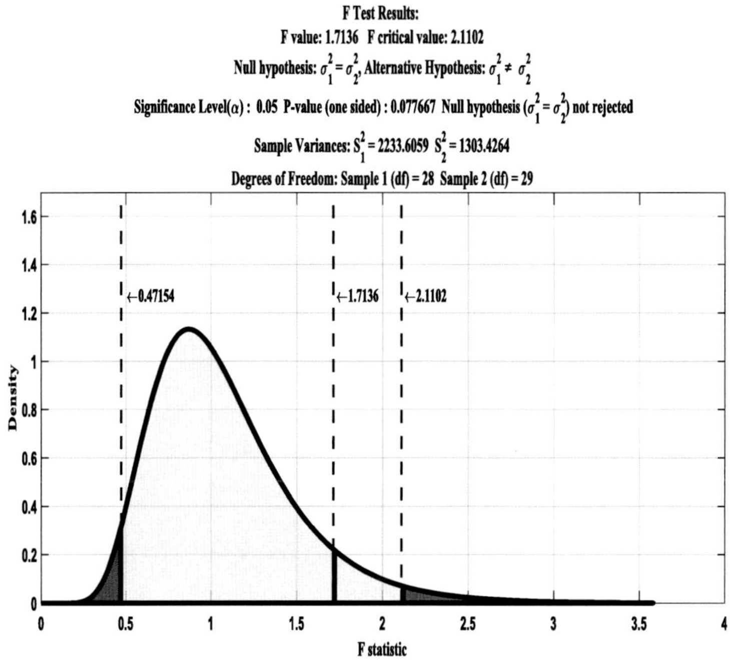

Figure 1.17 Results of F-test for two datasets.

Annual precipitation extreme values available at a rain gauge in the state of Florida, USA, for 24-h duration provided in Table 1.12 are used to carry out the F test. The results for the F test, with F statistic, acceptance region and *P*-value are shown in Fig. 1.17. It can be concluded from the test result, that there is no evidence to reject null hypothesis (H_o: $\sigma_1^2 = \sigma_2^2$) at a significance level of 5%.

20. NONPARAMETRIC METHODS

Several nonparametric techniques can be used to characterize and analyze hydroclimatic variable extremes. The following section discusses a technique that can be used to evaluate data characteristics with no assumptions made about the distribution of the data.

20.1 Kernel Density Estimation

Kernel density estimation (KDE) is a non-parametric alternative to the use of a histogram to characterize the distribution of sample data. In comparison to parametric estimators, the parameters to be estimated from data non-parametric estimators have no fixed structure and depend on all the data points to reach an estimate. The main reasons for using a kernel density estimator in place of a histogram are: (1) histograms do not provide a smooth representation of the data; (2) the shapes of the histograms depend on end points of bins. The selection of the number of bins is subjective even though few rules of thumb are available and (3) the visual representation of the data distribution also depends on the width of the bins. There are no clear guidelines for selection of the widths. In two recent studies, Shimazaki and Shinomoto (2007, 2010) provide approaches for optimum bin width in case of histograms and band width for kernel density estimation functions. The use of optimum bin width eliminates the limitations associated with subjective bin selection. Kernel estimators center a kernel function (K) at each data point. Adoption of smooth kernel function can overcome the limitations of the histograms. The contribution of data point "x_i" to the estimate at some point "x" depends on how apart "x_i" and "x" are from each other. The extent of this contribution is dependent upon the shape of the kernel function adopted and the width (bandwidth (h)) assigned.

The estimated density at any point "x" is

$$\widehat{f}_h(x) = \frac{1}{nh}\sum_{i=1}^{n} K\left(\frac{x - x_i}{h}\right) \tag{1.40}$$

A number of kernel functions are available for use with KDE and they include uniform, triangular, biweight, triweight, Epanechnikov, normal, Cauchy, and others. Teegavarapu et al. (2013) have used kernel density estimates to evaluate precipitation extremes at different durations in two Atlantic multidecadal oscillation (AMO) phases. In their study, a Gaussian kernel was used for developing KDE for assessing the extreme precipitation events in AMO phases. The Gaussian kernel function is given by Eq. (1.41).

$$K\left(\frac{x - x_{ip}}{h}\right) = \frac{1}{\sqrt{2\pi}}\exp\left(-\frac{1}{2}\left(\frac{x - x_{ip}}{h}\right)^2\right) \quad \forall\, ip \tag{1.41}$$

Kernel density estimates using Gaussian smoothing function and optimal bandwidth (h) can be used to characterize the precipitation extremes and their temporal occurrences in two temporal windows. The bandwidth parameter (h) controls the smoothness of the probability density curve, and different bandwidths are experimented before an optimal one is selected for representation of kernel density estimates. The histograms of precipitation data for different durations are superimposed on kernel density estimates

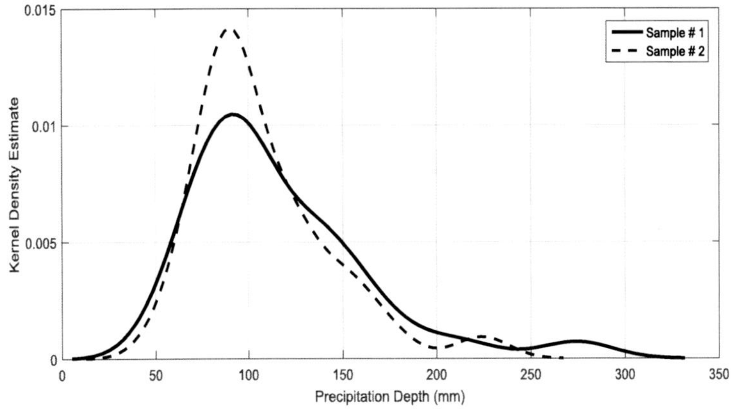

Figure 1.18 Kernel density estimate plots for two samples using a Gaussian kernel.

with appropriate scaling to evaluate the match between the two and also to confirm the use of correct bandwidth. The bandwidth parameter can be optimized based on the sample data. An illustration of two KDE plots using two sample datasets is shown in the Fig. 1.18. It is easy to evaluate the distributions of the data and their characteristics using KDE plots when compared to using two overlapping histograms.

20.2 Resampling Methods

Bootstrap sampling methods (Efron, 1979; Efron and Gong, 1983) and finally the generation of confidence intervals can help make inferences about the sample statistics when limited numbers of datasets related to hydroclimate variable extremes exist due to missing data or any other reasons. Bootstrap sampling method (Efron and Tibshirani, 1993) can be used to obtain samples from data in a specific temporal window and confidence intervals for sample statistic can be assessed. A general procedure of bootstrap methodology is explained and notation used by Davison and Hinkley (1997) and reported by Teegavarapu et al. (2013) is described in this section. The sample values $y_1, y_2, \ldots, y_n$ are thought of as the outcomes of independent and identically distributed ($\widetilde{Ud}$) random variables $Y_1, Y_2, \ldots, Y_n$ whose cumulative distribution function (CDF) is denoted by F. The estimate of F denoted by $\widehat{F}$ is obtained using data $y_1, y_2, \ldots, y_n$. The following steps (Davison and Hinkley, 1997; Teegavarapu et al., 2013) are used to obtain bootstrap sampling confidence intervals:

- Bootstrap (re) sample $y_1^*, y_2^*, \ldots, y_n^* \; \widetilde{ud} \; \widehat{F}$ are obtained from the original samples allowing repetitions.

- $\widehat{F}$, an estimator of F is obtained nonparametrically using empirical distribution function (EDF) of the original data, that is, by placing a probability of "$1/n$" at each data value from sample $y_1, y_2, \ldots, y_n$.
- Sample mean statistic θ^* is computed from bootstrap sample $y_1^*, y_2^*, \ldots, y_n^*$.
- The above steps are repeated "N" times, to obtain N sample means $\widehat{\theta_1^*}, \widehat{\theta_2^*}, \ldots, \widehat{\theta_N^*}$. The practical size of "N" depends on the tests to be run on the data.

The size of N recommended by Chernick (2007) is 1000 and 10,000 for evaluating the sample statistic and confidence intervals respectively. In the current study, these values are used. After N samples are obtained, normally approximated confidence intervals are computed for the uncertainty assessment. If $\widehat{\theta}$ (estimated mean of original data) is approximately normal, then $\widehat{\theta} \sim N(\theta + \beta, \nu)$. The confidence interval (CI) of θ for known bias ($\beta = \beta(F)$) and variance ($\nu = \nu(F)$) (Davison and Hinkley, 1997) is given by:

$$CI = \widehat{\theta} - \beta \pm Z_\alpha \cdot \nu^{1/2} \tag{1.42}$$

$$\text{where, } \beta(F) \doteq \beta(\widehat{F}) \doteq b = \overline{\widehat{\theta^*}} - \widehat{\theta}, \tag{1.43}$$

$$\nu(F) \doteq \nu(\widehat{F}) \doteq \nu = \frac{1}{N-1}\sum_{i=1}^{N}\left(\widehat{\theta_i^*} - \overline{\widehat{\theta^*}}\right)^2 \text{ at 95\% confidence interval,}$$
$$\alpha = 0.025, \; Z_\alpha = -1.96 \tag{1.44}$$

The variable $\overline{\widehat{\theta^*}}$ is the mean of $\widehat{\theta_1^*}, \widehat{\theta_2^*}, \ldots, \widehat{\theta_N^*}$ and Z_α is the α quantile of the standard normal distribution.

In many evaluations, time-series characteristics, indices, or extremes from one temporal window are compared with another one to assess changes. A comparative analysis of extremes or any time-series characteristics in two temporal windows may be biased if the number of observations is larger in one temporal compared to the other. To avoid bias due to this, a resampling technique such as bootstrap sampling method can be used to obtain samples from data. In one case, the number of samples drawn is restricted to a lower number of the datasets from two temporal windows. Given two datasets: temporal window # 1 (sample size, n) and temporal window # 2 (sample size, m), simple random sampling without replacement can be performed on two datasets to obtain equal datasets of the length, x (x = minimum (n, m)). Once datasets of equal length of x are obtained, bootstrap sampling is then used to obtain sample distribution. Random sampling without replacement is repeated several times along with bootstrap sampling in the second step. An approach combining random sampling without replacement and with replacement is illustrated by a series of steps in Fig. 1.19. The distributions of sample statistic (e.g., mean) developed from

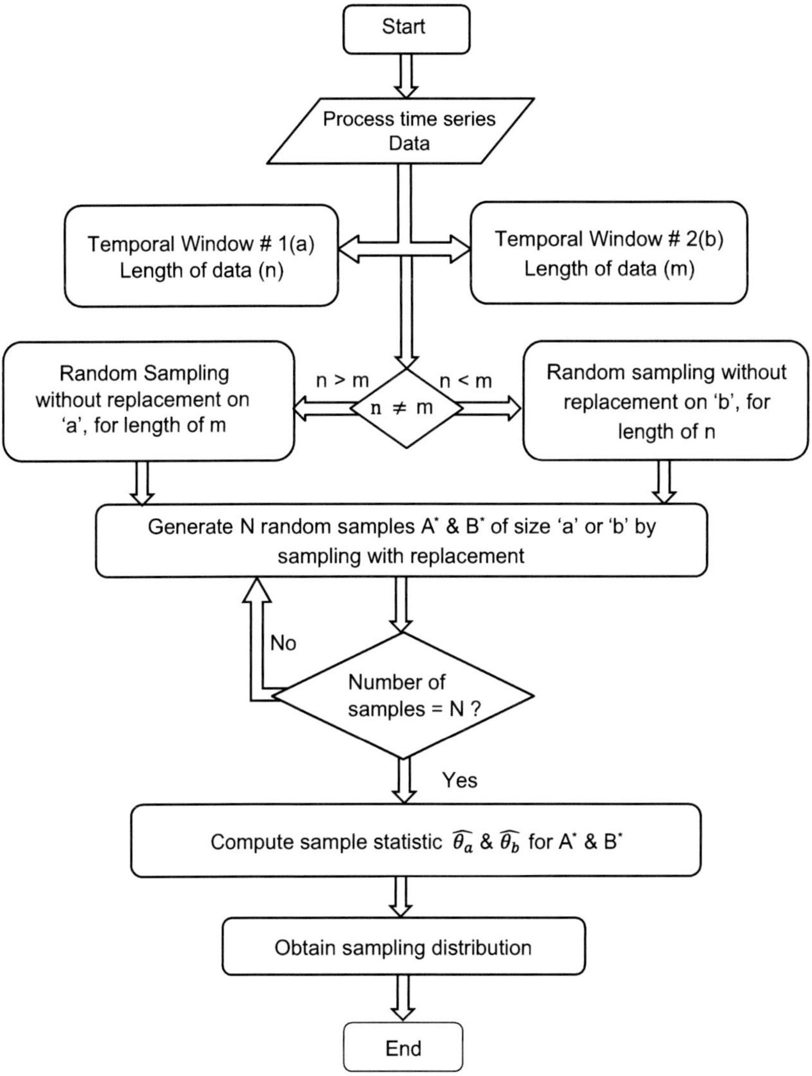

Figure 1.19 Schematic of random sampling without replacement and bootstrap sampling. *(Modified from Teegavarapu, R.S.V., Goly, A., Obeysekera, J., 2013. Influences of Atlantic multidecadal oscillation phases on spatial and temporal variability of regional precipitation extremes. J. Hydrol. https://doi.org/10.1016/j.jhydrol.2013.05.003.)*

the final step can be evaluated using kernel density estimates. The bootstrap sampling method used for obtaining confidence intervals is illustrated in Fig. 1.20.

The two precipitation datasets listed in Table 1.12 are used to develop kernel density estimates using resampling approach defined in Fig. 1.19. The value of N is equal to 100 and bootstrap samples generated for each value of N is 1400. The KD estimates for two datasets are shown in Fig. 1.21.

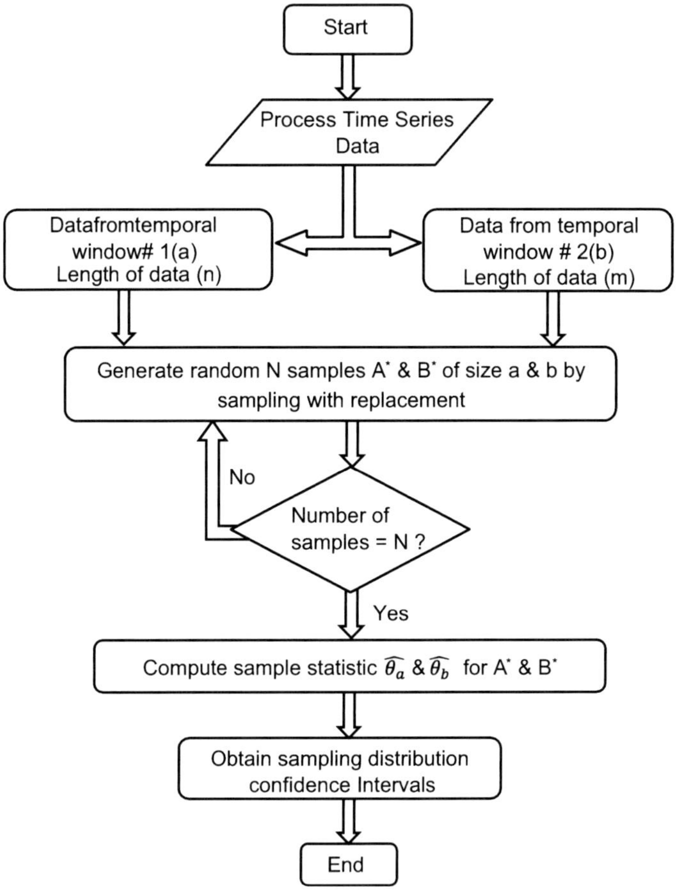

Figure 1.20 Bootstrap sampling approach for determination of confidence intervals. *(Modified from Teegavarapu, R.S.V., Goly, A., Obeysekera, J., 2013. Influences of Atlantic multidecadal oscillation phases on spatial and temporal variability of regional precipitation extremes. J. Hydrol. https://doi.org/10.1016/j.jhydrol.2013.05.003.)*

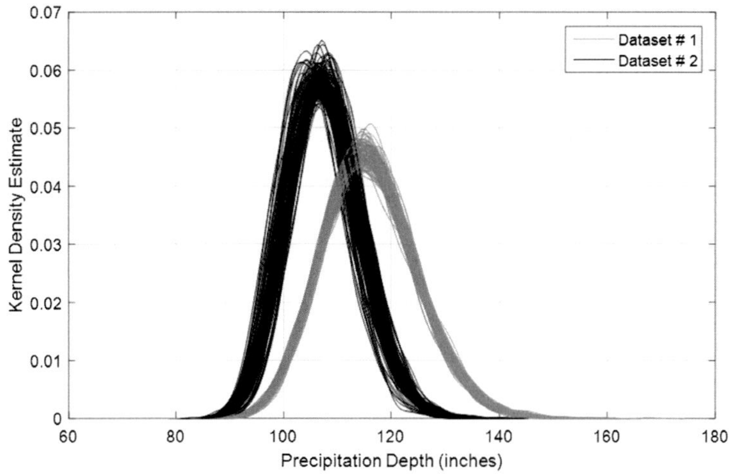

Figure 1.21 Kernel density estimates (KDEs) of mean precipitation depths from two datasets using random sampling without replacement and bootstrap sampling.

Bootstrap resampling based 95% level confidence intervals are calculated for the two datasets using the normal approximation, and the lower and upper limits for datasets # 1 and # 2 are 98.6, 133.6 and 94.9, 119.94 mm respectively. The number of bootstrap samples generated for this exercise is 1000.

20.3 Rank Sum Test or Mann-Whitney U Test

The Mann-Whitney U test can be used for evaluating differences in data (e.g., for any step change) from two different temporal windows. As the test is nonparametric, it is rank-based and does not require any assumptions about the distribution of the dataset. The test evaluates whether two independent data samples are different with respect to the median values with an assumption of the datasets are drawn from identical continuous distributions of the two populations except possibly for a location shift. The test is also referred to as Wilcoxon rank sum test.

The Mann-Whitney U statistic can be used to decide whether observations from two sample datasets of a hydrologic variable are from the same population. To apply the Mann-Whitney test, the raw data sequences with n elements should be divided into two sample groups with n_1 and n_2 elements, respectively. The raw dataset is then ranked from lowest to highest, including tied rank values where appropriate. The equations related to Mann-Whitney U statistic are as follows (Bobee and Ashkar, 1991; Corder and Foreman, 2009; Teegavarapu et al., 2013):

$$U_i = n_1 n_2 + \frac{n_i(n_i + 1)}{2} - \sum R_i \tag{1.45}$$

$$\overline{x}_u = \frac{n_1 n_2}{2} \tag{1.46}$$

$$S_u = \sqrt{\frac{n_1 n_2(n_1 + n_2 + 1)}{12}} \tag{1.47}$$

$$Z^* = \frac{U_i - \overline{x}_u}{S_u} \tag{1.48}$$

The variable S_u is the standard deviation and $\sum R_i$ is the sum of all the ranks from the sample of interest, U_i is the U statistic from the sample of interest and $\overline{x}_u$ is the mean. The index i refers to the either sample one or two and the smaller of the two U statistic values calculated is used for the test. The variable U_i can be examined for significance using a table of critical values provided by Corder and Foreman (2009). The variable Z^* is the z-score for a normal approximation of the data when the number of samples exceeds those available from the table of critical values. The null hypothesis, which is homogeneity, for the observations is rejected if the calculated Z^* statistic is greater than the selected critical value at 5% significance level obtained from the standard normal distribution table. One-tailed: (1) right tailed: median value of one dataset is higher than that of the other or (2) left tailed: median value of one dataset is lower than that of the other or (3) two-tailed: median values are not equal tests can be conducted. The two datasets from Table 1.1 are used for application of Rank Sum test. The Z^* value obtained is 0.5613 and the test results point to null hypothesis (median values of two data sets are equal) not being rejected with P-value of 0.575. The parametric two-sample unpaired t-test also resulted in the same result of null hypothesis being accepted.

21. ASSESSMENT OF CHANGES IN DISTRIBUTIONS OF DATA

21.1 Kolmogorov-Smirnov Two-Sample Test

Two-sample Kolmogorov-Smirnov (KS) test (Massey, 1951) can be used to compare the distributions of the observations from the two datasets. The null hypothesis (H_o) is that the two dataset values are from the same continuous distribution. The alternative hypothesis (H_a) is that these two

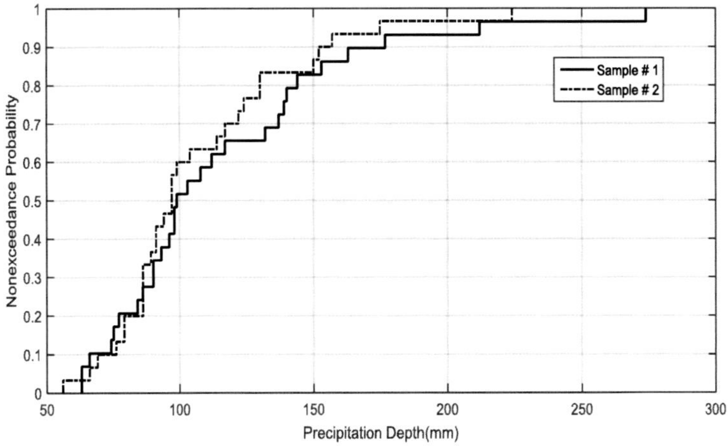

Figure 1.22 Empirical cumulative distribution function plots of two sample datasets.

datasets are from different continuous distributions. The hypothesis test can be carried out at a specific statistical significance level (e.g., 5%). Fig. 1.22 shows the two-empirical cumulative distribution function (CDF) plots of two datasets. Results of a two-sample KS test provide a KS statistic (D) value of 0.1782 and *P*-value of 0.694 and point to null hypothesis being accepted. The KS test performed using these two datasets at 5% significance level led to the result of null hypothesis is not being rejected suggesting that the two samples are drawn from the same continuous distribution. The test merely states that the two distributions are different—it does not indicate whether the change is an increase or decrease in the mean or due to a change in the variance or extremes (Kundzewicz and Robson, 2000) Another test that is similar to KS test for assessing difference in the distributions of two datasets is the Cramer-von-Mises test.

21.2 Ansari-Bradley Test

The Ansari-Bradley (AB) test can be used to evaluate the hypothesis that two independent samples from two datasets come from the same distribution (null hypothesis: H_o), against the alternative (alternative hypothesis: H_a) that they come from distributions that have the same median and shape but different dispersions (e.g., variances). The test can be conducted at a specific significance level (e.g., 5%). It is also regarded as a nonparametric equivalent of F-test for variances. The AB test conducted on the two datasets (from Table 1.12) resulted in a null hypothesis being not rejected with *P*-value of 0.5261. The test is carried out using statistics tool box of MATLAB.

22. INFLUENCE OF MISSING DATA ON TREND ANALYSIS

The occurrence of missing data in a specific part of the time-series and the length of missing data will influence the results of the trend analysis. When block maxima (e.g., annual maxima) are considered, the time intervals in which extreme values are known to have occurred are those intervals in which data is missing; the maximum values may be underestimated. Data filling can lead to changes in the probability distribution of data at a site and introduce significant biases when event-based analysis (such as daily or hourly extreme value analysis) is performed. In case of precipitation data, at temporal resolutions of a day or less, spatial interpolation alters the probability distribution of data, changes the autocorrelation structure and dry and wet-spell transitions (Teegavarapu, 2014). Many research efforts have focused on analyzing the trends in precipitation data, especially extremes, but biases introduced in these trends due to data infilling techniques are rarely investigated. A study by Teegavarapu (2014) used only a fixed length of missing data and analyzed the impact of infilling on precipitation characteristics and extremes.

In a recent study by Teegavarapu and Nayak (2017), influences of missing precipitation data on trend analysis are documented. Their study employed: (1) varying lengths of missing data for evaluation of statistically significant trends; (2) resampling methods to quantify the impacts of infilling on evaluation of extremes and (3) nonparametric hypothesis tests for assessment of differences in distributions of filled and unfilled precipitation datasets for different lengths of gaps. Their study indicated the presence of missing data and times-series with filled data may affect the results the time-series analysis, especially the trend assessments. Availability and quality of long-term gap-free precipitation data at temporal resolutions of day or less required to perform such analyses are often limited.

23. CHANGE DETECTION

Change points in the precipitation time-series can be identified using Pettitt's test (Pettitt, 1979). Step change in mean or median values can be evaluated using several tests (e.g., Distribution free CUSUM (cumulative sum control chart) (nonparametric test), cumulative deviation (parametric), Worsley likelihood ratio (parametric)). Parametric (e.g., two-sample t-test) and nonparametric (e.g., rank sum or Mann-Whitney) tests can also be used to evaluate statistically significant changes in mean and median values in two different data periods (two different temporal slices). Change detection analysis is a widely used technique to determine the location of the potential change in a data series (Mix et al., 2012). Change detection can be determined using two techniques: Cumulative Sum (CUSUM) plots and Change Point test. The following sections will discuss these two techniques in further detail.

23.1 Cumulative Sum

Several studies have used CUSUM as a sequential analysis technique used to observe change detection. The CUSUM concentrates on comparing two parts of a record and determining if they are significantly different. An example of CUSUM-based analysis is presented here. Fig. 1.23 shows the annual maximum streamflow data evaluated for variations. Also shown in the figure are the mean and standard deviation values obtained from a 30-year annual maximum flow (1981–2010). This average value is also referred to as climate normal (CN) and is used as the basis for determining the change or exceedance over a prespecified threshold. Any annual maximum streamflow's standard deviation that exceeds the threshold of five standard deviations, where the standard deviation threshold is based on the 30-year mean flow, constitutes a change in pattern. Fig. 1.23 shows an example of a CUSUM evaluation plot where change detection (or exceedance) is identified by the line with circular marker symbols.

Only the upper limit of the cumulative sum is obtained from Eq. (1.49).

$$\beta_i = \begin{cases} 0, & i = 1 \\ \max\left(0,\ \beta_{i-1} + \gamma_i - \mu_x - \frac{1}{2} n\sigma_x\right), & i > 1 \end{cases} \tag{1.49}$$

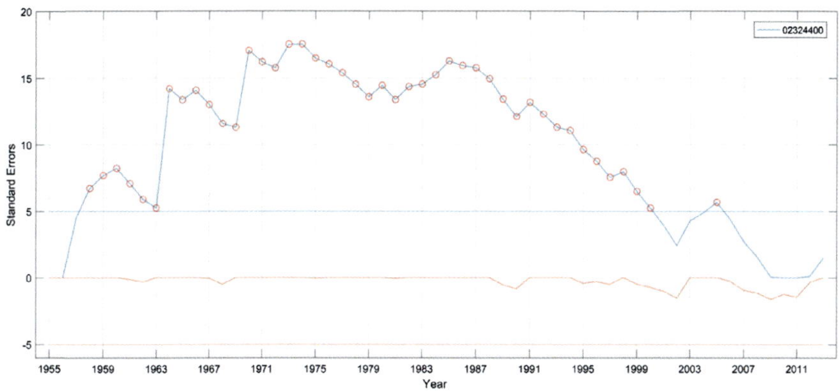

Figure 1.23 CUSUM evaluation plot of annual streamflow extremes at a site in southeastern United States.

where β_i = cumulative sum at year i, γ_i = data point at year i, μ_x = estimated average, σ_x = estimated standard deviation, n = number of standard deviations from the target mean that make a shift detectable.

23.2 Change Point Test

Change point identifies the point where a significant change occurs in a time-series. A simple illustration of existence of change point is provided in this section using annual maximum discharge values obtained from a United States Geological Survey (USGS) gauging station in the southeastern U.S. The standard deviation and mean of the annual maximum values, are then plotted and the change point (year) can be identified based on a significant change in the statistics being evaluated. Fig. 1.24 shows an example of a change point evaluation plot where a change point in mean of annual maxima is determined.

Kundzewicz and Robson (2000) provided details of distribution-free CUSUM and change point tests. Cavadias (1992) reported several tests or methodologies for detection of climate change and the multiple studies that have used them. These methodologies include likelihood ratio, Bayesian approach, Von Neumann ratio, Fuller's test of stochastic nonstationarity, regression, Kruskal-Wallis tests, student's t, autocorrelation, spectral analysis, orthogonal functions, CUSUM test, two-phase regression, Wald-Wolfowitz test, Pettitt test, and Mann-Kendall test. Changes in time-series of different hydroclimatic variables that include precipitation, air temperature, lake levels, and streamflows were evaluated in those studies.

24. VARIATIONS IN THE CHARACTERISTICS OF TIME-SERIES

Time-series evaluation for changes and trends is not only based on extremes derived from the fixed length of time-series but can also be based on a number of characteristics that define different aspects of the time-series. Descriptive indices derived based on the characteristics of hydrometeorological time-series that are already available or new ones can be developed and can be evaluated. These indices can be checked for long-term changes (i.e., trends), or variations in two or more temporal windows and for changes in spatial and temporal variability. The Expert Team on Climate Change Detection and Indices (ETCDI) defined a total of 27 indices for precipitation and temperature for evaluation of changes in extremes, and they are documented by WMO (2009). By defining these core set of indices and using the same definitions of extremes and analyzing the data in

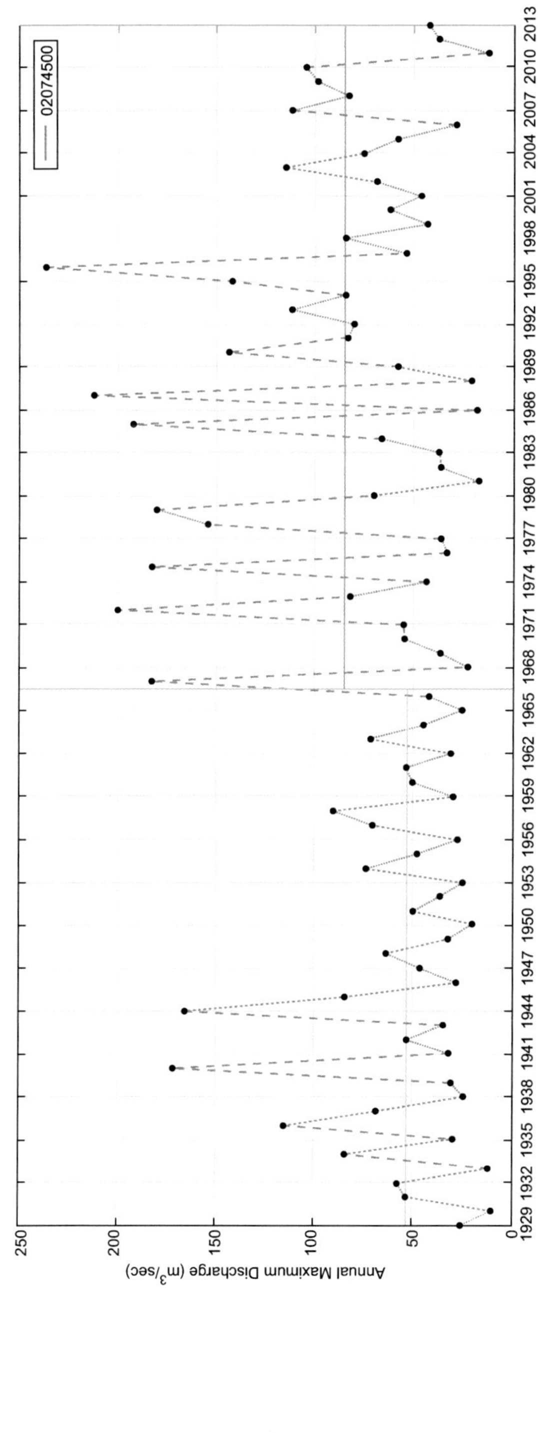

Figure 1.24 Change point analysis of annual streamflow extremes at a site in southeastern United States.

a standardized way, it is possible to compare results from different regions and to obtain coherent pictures of change around the world (WMO, 2009). Additional region-specific indices can be developed. According to WMO (2009), the core set of 27 indices describes different aspects of moderate temperature and precipitation extremes, including frequency, intensity, and duration. All these indices are widely used for monitoring changes in extremes, climate model evaluation, and assessments of future climate.

Calculations of several indices will involve counting of the number of days in a temporal window (e.g., season or a year) that exceed predefined specific threshold values, in particular, percentiles in the statistical distribution. Indices for water or environmental quality constituents primarily depend on the purpose for which the observations are used and they focus on: (1) evaluation of long-term trends; (2) changes in water or environmental quality due to execution of improvement projects (e.g., development and implementation of total maximum daily loads (TMDLs)) for streams due to point and nonpoint sources; and (3) assessment of water body conditions considering numeric and narrative water quality standards. Performance measures such as reliability, resiliency, and vulnerability (Teegavarapu and Simonovic, 2014) used for assessment of water resource systems can be used to evaluate hydrologic time also. Machiwal and Jha (2012) discuss the use and application of these measures for evaluating the characteristics of annual precipitation values. Variants of these measures can be developed to evaluate the hydroclimatological time-series in general.

25. CHARACTERISTICS AND INDICES FOR PRECIPITATION DATA

Descriptive indices for precipitation extremes were developed by ETCDI and are described by WMO (2009). These indices describe specific characteristics of extremes, including frequency, amplitude, and persistence (WMO, 2009). A total of 11 indices for precipitation are developed by ETCDI. Climate variability and change studies require time-series data of essential climatic variables to be gap-free (i.e., without missing data). Implications of using filled missing data existing in time-series on different indices need to be evaluated. Eleven indices for precipitation developed by ETCDI are used for evaluation in the current study, and they are RX1day, RX5day, SDII, R10mm, R20mm, Rnnmm, CDD, CWD, and PRCPTOT, R95pTOT, and R99pTOT. Definitions for these indices are adopted from WMO (2009) and as described by Teegavarapu (2012b) are reported here. RX1day is defined as the maximum 1-day precipitation (i.e., highest precipitation amount in 1-day period). RX5day refers to the maximum 5-day precipitation (i.e., highest precipitation amount in 5-day period). SDII is the simple daily intensity index (i.e., mean precipitation amount on a wet day greater than 1 mm). R10 mm is the number of heavy precipitation days (i.e., count of days where RR (daily precipitation amount) $\geq$ 10 mm). R20 mm is the number of heavy precipitation days (i.e., count of days where RR (daily precipitation amount) $\geq$ 20 mm). Rnnmm is the number of heavy precipitation days (i.e., count of days where daily precipitation amount $\geq$ user defined precipitation threshold [*nn*] expressed in mm). CDD is the consecutive dry days (i.e., maximum length of dry spell in which precipitation amount is less than 1 mm), CWD is the consecutive wet days index (i.e., maximum length of wet spell in which precipitation amount is greater than 1 mm) and finally PRCPTOT is the total precipitation in wet days (i.e., daily precipitation amount is greater than or equal to 1 mm). R95pTOT and R99pTOT refer to the values of total precipitation due to wet days (>95 percentile) and extremely wet days (>99 percentile). More detailed explanations of these indices along with their use in evaluation of the precipitation time-series with respect to climate variability are provided in the Chapter 2.

A number of characteristics other than indices previously described can be evaluated for precipitation data and they are: (1) interannual and intraannual variations; (2) seasonality index (SI); (3) spatial and temporal variability of extremes; (4) nature of extremes (based on type of storm events—e.g., convective, frontal, cyclonic); (5) transition states (transition probability states) as defined by a two state first-order Markov-chain are used to analyze dichotomous events (i.e., rain or no rain conditions) by defining four transition state probabilities; (6) persistence (serial autocorrelation); (7) intraevent temporal distribution of precipitation; (8) antecedent moisture conditions preceding extreme events; (9) temporal occurrences of extremes; (10) number of extremes over a specific threshold; (11) interevent time definition (IETD)–based events including assessment of depth, duration, and gap; (12) changes in average intensities associated with specific return period over time; (13) variations in the intensity-duration-frequency (IDF) or depth-duration-frequency (DDF) curves over time or in different select temporal windows; (14) standard precipitation index (SPI) and (15) transition probabilities estimated based on dry and wet conditions identified by SPI. Assessments using these indices and characteristics due to influences of climate change and variability for precipitation time-series are discussed in Chapter 2.

26. EXTREME INDICES FOR TEMPERATURE

The ETCDI defined 16 temperature indices for evaluation, and they are documented by WMO (2009). They are: (1) frost days; (2) summer days; (3) icing days; (4) growing season length; (5) monthly maximum value of daily maximum temperature; (6) monthly maximum value of daily minimum temperature; (7) monthly minimum value of daily maximum temperature; (8) monthly minimum value of daily maximum temperature; (9) monthly minimum value of daily minimum temperature; (10) cold nights; (11) cold day-times; (12) warm nights; (13) warm day-times; (14) warm spell duration index; (15) cold spell duration index; and (16) diurnal temperature range. The first four indices require a prespecified temperature value. Indices related to cold nights, cold day-times, warm days and warm nights require information from climatological normal period of 30 years. The warm and cold spell duration indices are based on information derived from maximum and minimum 5-day window centered around the calendar day in the base period (i.e., climate normal). At the time of the writing of this chapter, the normal or base period is 1981–2010. The diurnal temperature range is calculated based on the mean of the maximum and minimum daily temperature values.

27. STREAMFLOW INDICES AND CHARACTERISTICS

According to the United States Geological Survey (USGS), of all the threats posed by climatic variability and change, those associated with water resources are the most consequential for both society and the environment (Lins et al., 2010). In addition to increases in global average air and ocean temperatures, observations find increasing occurrences of extreme weather—including droughts (USGS, 2007). Understanding changes in the distribution, quantity, and quality of, and demand for water in response to climate variability and change is essential to planning for and adapting to future climatic conditions (Lins et al., 2010). A number of streamflow characteristics can be evaluated considering the lower- and higher-end extremes. To assess the influences of climate variability on these characteristics, watersheds that are with no anthropogenic disturbances need to be identified with long historical streamflow records. For example, the USGS in the United States compiled a list of sites that are part of the Hydro-Climatic Data Network (HCDN) and that are located in basins that are least anthropogenically disturbed. HCDN stations included in the network have to meet the following criteria (USGS, 2012): (1) sites are identified as being in current reference condition per the GAGES-II classification (i.e., minimally affected by direct human activities); (2) these sites have at least 20 years of complete and continuous discharge (or streamflow) data and (3) the basins have less than 5% impervious surface area as measured using the National Land Cover Data (NLCD) 2006. Lins et al. (2010) provide a list of questions that need to be answered in regard to streamflow characteristics, and they include:

- What are the patterns of changes in annual flood peak discharges as well as flood volumes?
- What is the variability in the timing of annual peak discharges?
- What are the changes in the number of largest floods (as defined by exceedances of a particularly high threshold)?
- Are there an unexpected number of new record peaks (either more or less than would be expected by chance)?
- Can any observed changes in peak flows be related unambiguously to changes in climatic factors?
- Is there a pattern of change in 1-day, 7-day, 14-day, or 28-day low flows?
- Do we see an unexpected number of new record low flows (more than expected or less than expected by chance)?
- Have multidecadal changes in mean discharge occurred and, if so, are these variations distinct from observed centennial changes? and
- Have changes occurred in the relation of annual or seasonal discharge to annual or seasonal precipitation?

A number of streamflow indices can be evaluated for short- and long-term changes and they include (1) annual maximum, mean, and median values; (2) peak over threshold (3) interquartile range (IQR); (4) Q-percentiles (Q1−Q99). Several variables classifying streamflow regimes as documented by Hisdal (2000) can also be evaluated and they are: (1) seasonal mean flow; (2) monthly mean flow; (3) timing of seasonal events: start of snowmelt; snowmelt snow-peak; maximum flow; freezing of rivers, ice break up and minimum flows; (4) stability of regimes.

Low flow analysis may focus on a number of issues, and they include: (1) understand the spatial and temporal variability of low flows in a region; (2) assess the spatial variability of streamflow extremes with the help of flow duration curves (FDCs) at a number of sites; (3) determine and compare low flow frequency statistics using N-day annual minimum mean flows for two temporal windows (e.g., coinciding with two phases of coupled oceanic and atmospheric oscillations) for climate variability studies;

(4) calculate the low flow indices such as 7Q10 and others. Streamflow deficit analysis can be used for identification, characterization, and management of multiyear hydrological droughts. A streamflow deficit occurs when the flow is below a specific threshold that defines a drought. The deficit starts when the flow goes below the threshold and ends as soon as the flow returns above the threshold. The deficit duration can be determined and evaluated for different temporal windows. The N-day annual minimum mean flows are fitted with a probability distribution (typically a Weibull distribution) and indices such as 7Q10 are derived. Once the values of these indices are derived, they can be evaluated for changes with respect to time or from one temporal window to another. Parametric and nonparametric hypothesis tests can be used to confirm the statistically significant changes from one period to another. Similarly, FDCs can be compared for changes in two periods. A flow duration curve (FDC) is the relationship between any given discharge (or flow) and the percentage of time that this discharge is equaled or exceeded (Smakhtin, 2001). FDCs can be evaluated for changes over time or for changes in two separate time intervals that coincide with temporal windows of warm and cool phases of coupled oceanic and atmospheric oscillations.

28. INDICES FOR ENVIRONMENTAL AND WATER QUALITY PARAMETERS

Environmental and water quality constituents can be evaluated for changes over time starting with trend analysis to split-data analysis that considers two temporal windows. In the former case, changes in the constituent concentrations with time are evaluated, whereas in the latter case statistically significant changes in the observations in two temporal windows are assessed. Assessment of water quality impairment of streams and water bodies in space and time is an essential task of watershed management. As many water quality standards are defined based on the type of water quality constituent measured, aggregation of observations over different temporal scales to ascertain average value (or central tendency of observed values) is often required. The geometric mean is a common measure used in the field of environmental quality monitoring, especially for those constituents which vary over several orders of magnitude (e.g., E-coli or fecal coliform concentrations). Geometric mean values in a specific period are compared to those measured in another period. Two-sample unpaired t-tests are used when observations in different time periods are not related (i.e., independent). Paired t-tests are used when the observations are related in cases when observations of water quality constituents are measured upstream and downstream of a project site (e.g., water treatment plant). However, parametric t-tests will require a number of assumptions to be fulfilled about the data. In such situations use of data transformation methods is sometimes beneficial. Environmental and water-quality data can be evaluated using the following:

- Long-term and short-term changes in characteristics of constituent observations.
- Spatial and temporal variability of water-quality observations.
- Trends in the observed data of water-quality constituents.
- Assessment of constituent concentrations conditioned on specific events (e.g., dry or wet; before and after remediation measures are taken up or implementation of TMDL strategies).
- Joint assessment of constituent concentrations and hydrometeorological variables (e.g., joint probability evaluation) that influence concentrations.
- Frequency and temporal occurrences of violations of environmental quality standards (qualitative or narrative and quantitative standards).
- Changes in joint variation of two or more environmental quality constituents.
- Conditional probability assessments.
- Changes in event mean concentrations (EMCs) of pollutants.
- Changes in observations from different temporal windows.
- Seasonal trend assessment.
- Confidence intervals on any statistic obtained based on bootstrap concept-based resampling of observations.
- Deterministic and probabilistic risk assessment due to environmental contaminants.

In many situations, variables other than the time such as temperature, streamflow, etc., influence water quality constituents. A parametric procedure using multiple linear regression model with time and covariables as predictors and the variable of interest as predictand can be developed. Regression diagnostics can be used to evaluate the coefficients of the predictors. If the t-test reveals that the coefficients for covariates are not statistically significant from zero, then variation in the variable can be assumed purely due to time.

A conceptually simple framework for the analysis of water quality data in space and time to assess the pathogen impairment of streams was documented by Teegavarapu et al. (2012). Observed fecal coliform data obtained through systematic and random sampling in the state of Kentucky were analyzed to assess stream quality. Their study dealt with: (1) point and regional assessment of water quality of streams; (2) identification of the spatial and temporal water quality trends and (3) evaluation of the impact of watershed quality improvement projects. The uniform guidance document developed by the United States Environmental Protection Agency (USEPA) EPA (2000) provides valuable advice on different types of statistical tests for evaluation of groundwater monitoring data at Resource Conservation and Recovery Act (RCRA) facilities. Although the guidance provided is specific to monitoring data at RCRA facilities, the documented methods are generic and essential for any environmental monitoring data analysis. EPA (2006a) and EPA (2006b) provide guidance on systematic planning using the data quality objectives process and directions on data assessment for practitioners.

Recent work of Hirsch et al. (2010) documented the need for new generation of trend analysis approach in the context of environmental monitoring that focuses on revealing the nature and magnitude of change, rather than strict hypothesis testing (Meals et al., 2011). Meals et al. (2011) list main attributes of such an approach and they include: (1) No assumptions are made about the constancy of flow-concentration relationship with respect to time; (2) No assumptions are made in regard to seasonal patterns that repeat exactly over the period of record, but allow the shape of seasonality to evolve over time; (3) Allows the shape of an estimated trend to be driven by the data and not constrained to follow a specific form such as linear or quadratic; trend patterns should be allowed to differ for different seasons, or flow conditions; (4) Provides consistent results describing trends in both concentration and load; (5) Provides not only estimates of trends in concentration and flux but also trend estimates where the variation in water quality due to variation in streamflow has been statistically removed; and (6) Includes diagnostic tools to assist in understanding the nature of the changes that have taken place over time, for example, to identify particular times of year or hydrologic conditions during which water quality changes are most pronounced. Hirsch et al. (2010) proposed and demonstrated an experimental trend analysis technique referred to as Weighted Regressions on Time, Discharge, and Season (WRTDS) that features the above discussed attributes.

29. SEA LEVEL VARIATIONS AND TRENDS

Changes and trends in observed sea levels can be evaluated using tidal gauge data that are available around the world. Since sea level data is like any other temporally measured data, both parametric and nonparametric approaches can be used for evaluation of changes in sea levels over time. The following questions can be asked about changes and trends in sea level measurements:

- What are the trends in annual maximum values?
- What are the trends in different characteristics of sea level data?
- How are sea level variations influenced by climate variability?
- Are there any intraannual changes in the sea levels?
- What are regional and global trends?
- Are there any abrupt changes in sea levels?

Trends in sea level data can be evaluated using nonparametric tests such as Spearman's Rho and Mann-Kendall tests. Parametric trend tests can also be carried out if the assumptions of the tests are fulfilled. Sea level variations influenced by different phases of coupled oceanic and atmospheric oscillations can be evaluated using either parametric (e.g., two sample unpaired t-test) or nonparametric (e.g., Mann-Whitney U or Ranksum test) tests. Observations are separated into two datasets using data from temporal windows that coincide with the two phases (warm or cool; El Niño or La Niña) of these oscillations. Change point tests can be used for assessment of abrupt changes in sea level observations. However, once the change point is identified, evaluation of time-series characteristics before and after the point needs to be carried out to confirm the changes.

30. CONCLUSIONS AND FUTURE RESEARCH DIRECTIONS

A comprehensive summary of approaches and techniques for analysis of variations and trends in hydroclimatic variables is provided in this chapter. Starting with basic exploratory data analysis approaches to complex statistical tests based on resampling methods are discussed in a detailed manner to help evaluate the time-series data of hydroclimatic variables, and assess and confirm changes and trends in these datasets. Parametric and nonparametric tests that will help evaluate statistically significant changes and trends in time-series are elaborated. Several illustrative examples are also provided to enhance the understanding of these techniques, limitations, and applications. Discussion about different evaluation indices for precipitation, temperature and environmental quality constituents, streamflows, and other hydrological variables is provided. Availability of long-term serially continuous and error-free data is critical for evaluation of trends and variations in hydroclimatic variables. Point data and gridded data are equally beneficial for assessment of trends and variations although the purposes they serve are different. Inferences about variations in characteristics and extremes of hydroclimatic variables can be drawn when both the datasets are used independently. Kundzewicz and Robson (2000) recommended the following directions to encourage and enhance the development of new techniques and methodologies for assessment of changes in discharge variable and they include: (1) Improved guidance as to which tests are best able to detect change under a realistic set of assumptions; (2) Development of multivariate methodologies that look at variables of interest alongside other hydrological variables and (3) Development and application of improved regional methodologies. These recommendations are still valid today, and new statistical methods are expected to be developed in future for assessment of time series of hydroclimatic variables in a changing climate.

REFERENCES

Abdou, A.E.A., 2014. Temperature trend on Makkah, Saudi Arabia. Atmos. Clim. Sci. 4 (3), 457–481.

Abghari, H., Tabari, H., Talaee, P.H., 2013. River flow trends in the west of Iran during the past 40 years: impact of precipitation variability. Global Planet. Change 101, 52–60.

Ahmad, I., Tang, D., Wang, T., Wang, M., Wagan, B., 2015. Precipitation trends over time using Mann-Kendall and Spearman's rho tests in Swat River Basin, Hindawi, Pakistan. Adv. Meteorol. 1–15. https://doi.org/10.1155/2015/431860.

Alexandersson, H., 1986. A homogeneity test applied to precipitation data. J. Climatol 6, 661–675.

Allison, P.D., 1998. Multiple Regression: A Primer. Pine Forge Press.

Amirabadizadeh, M., Huang, Y.F., Lee, T.S., 2015. Recent trends in temperature and precipitation in the Langat River Basin, Malaysia. Adv. Meteorol. 2015, 1–16, 579437.

Amo-Boateng, M., Zhijia, L., Yiqing, G., 2014. Inter-annual variation of streamflow, precipitation and evaporation in a small humid watershed (Chengcun Basin, China). Chin. J. Oceanol. Limnol. 32 (2), 455–468.

Anderson, R.L., 1942. Distribution of the serial correlation coefficient. Ann. Math. Stat.

Arora, M., Goel, N.K., Singh, P., 2005. Evaluation of temperature trends over India. Hydrol. Sci. 50 (1), 81–93.

ASCE, 2001. Hydrology Handbook, second ed. American Society of Civil Engineers, New York.

Ashkar, F., 1996. Extreme Floods in Hydrologic of Disasters. In: Singh, V.P. (Ed.), Water Science and Technology Library. Kluwer Academic Publishers, Netherlands.

Ashraf, M., Loftis, J.C., Hubbard, K.G., 1997. Application of geostatistics to evaluate partial weather station network. Agric. For. Meteorol. 84, 255–271.

Austria, P.F.M., Bandala, E.R., 2016. Maximum Temperatures and Heat Waves in Mexicali. Trends and Threshold Analysis, Mexico, pp. 21–28. Air, Soil and Water Research, 2016(9).

Bandyopadhyay, A., Bhadra, A., Raghuwanshi, N.S., Singh, R., 2009. Temporal trends in estimates of reference evapotranspiration over India. J. Hydrol. Eng. 14 (5), 508–515.

Bartels, R., 1982. The rank version of von Neumann's ratio test for randomness. J. Am. Stat. Assoc. 77 (377), 40–46.

Bartolini, G., Messeri, A., Grifoni, D., Mannini, D., Orlandini, S., 2014. Recent trends in seasonal and annual precipitation indices in Tuscany (Italy). Theor. Appl. Climatol. 118 (1), 147–157.

Basistha, A., Arya, D.S., Goel, N.K., 2009. Analysis of historical changes in rainfall in the Indian Himalayas. Int. J. Climatol. 29, 555–572.

Betancourt, J., 2009. Coping with non-stationarity in water and ecosystem management. In: 9th Annual SAHRA Meeting, September 23–24.

Birsan, M.-V., Molnar, P., Burlando, P., Pfaundler, M., 2005. Streamflow trends in Switzerland. J. Hydrol. 314 (1–4), 312–329.

Bobee, B., Ashkar, F., 1991. The Gamma Family and Derived Distributions Applied in Hydrology. Water Resources Publications, Littleton, Colorado.

Box, G.E.P., Jenkins, G.M., Reinsel, G.C., 1994. Time Series Analysis: Forecasting and Control, third ed. Prentice Hall, Englewood Cliffs, NJ.

Brimicombe, A., 2003. GIS, Environmental Modeling and Engineering. Taylor and Francis, London, UK.

Brunetti, M., Buffoni, L., Maugeri, M., Nanni, T., 2000. Trends of minimum and maximum daily temperatures in Italy from 1865 to 1996. Theor. Appl. Climatol. 66, 49–60.

Brunetti, M., Maugeri, M., Nanni, T., 2001. Changes in total precipitation, rainy days and extreme events in Northeastern Italy. Int. J. Climatol. 21, 861–871.

Buishand, T.A., 1982. Some methods for testing the homogeneity of rainfall records. J. Hydrol. 58, 11–27.

Caloiero, T., Coscarelli, R., Ferrari, E., Mancini, M., 2011. Trend detection of annual and seasonal rainfall in Calabria (Southern Italy). Int. J. Climatol. 31, 44–56.

Cannarozzo, M., Noto, L.V., Viola, F., 2006. Spatial distribution of rainfall trends in Sicily (1921–2000). Phys. Chem. Earth 31 (18), 1201–1211.

Cavadias, G.S., 1992. A Survey of Current Approaches to Modelling of Hydrological Time-Series with Respect to Climate Variability and Change. WCASP-23, WMOIFD-no. 534 (Geneva, Switzerland).

Chakrabarti, S., Cox, E., Frank, E., Güting, R., Han, J., Jiang, X., Kamber, M., Lightstone, S., Nadeau, T., Neapolitan, R.E., Pyle, D., Refaat, M., Schneider, M., Teorey, T., Witten, I., 2009. Data mining: know it all. Morgan Kaufmann, USA.

Chakraborty, S., Chaube, U.C., Pandey, R.P., K Mishra, S., 2013a. Analysis of reference evapotranspiration variability and trends in the Seonath river basin Chhattisgarh. Int. J. Adv. Eng. Sci. Technol. 2 (2), 144–152.

Chakraborty, S., Pandey, R.P., Chaube, U.C., Mishra, S.K., 2013b. Trend and variability analysis of rainfall series at Seonath River Basin, Chhattisgarh (India). Int. J. Appl. Sci. Eng. Res. 2 (4), 425–434.

Chang, K.-T., 2009. Introduction to Geographic Information Systems. McGraw Hill, New York.

Chebana, F., Ouarda, T.B.M.J., Duong, T.C., 2013. Testing for multivariate trends in hydrologic frequency analysis. J. Hydrol. 486, 519–530.

Chen, H., Guo, S., Xu, C.-Y., Singh, V.P., 2007. Historical temporal trends of hydro-climatic variables and runoff response to climate variability and their relevance in water resource management in the Hanjiang basin. J. Hydrol. 344, 171–184.

Chernick, M.R., 2007. Bootstrap Methods: A Guide for Practitioners and Researchers. Wiley-Interscience.

Cislaghi, M., Michele, C.D., Ghezzi, A., Rosso, R., 2005. Statistical assessment of trends and oscillations in rainfall dynamics: analysis of long daily Italian series. Atmos. Res. 77, 188–202.

Corder, W.G., Foreman, D.I., 2009. Nonparametric Statistics for Non-statisticians. Wiley, Hoboken.

Cristea, N.C., Lundquist, J.D., Loheide II, S.P., Lowry, C.S., Moore, C.E., 2014. Modelling how vegetation cover affects climate change impacts on streamflow timing and magnitude in the snowmelt- dominated upper Tuolumne Basin, Sierra Nevada. Hydrol. Process. 28 (12), 3896–3918.

Croitoru, A.-E., Piticar, A., Dragotă, C., Burada, D., 2013. 2013. Recent changes in reference evapotranspiration in Romania. Global Planet. Change 111, 127–132.

Cromwell, J.B., Labys, W.C., Terraza, M., 1994. Univariate Tests for Time Series Models, 07–99. Sage Publications, London.

Dahmen, E.R., Hall, M.J., 1990. Screening of Hydrological Data: Tests for Stationarity and Relative Consistency. Publication # 49. ILRI publication, Netherlands.

Daly, C., Neilson, R.P., Phillips, D.L., 1994. A statistical topographic model for mapping climatological precipitation over mountainous terrain. J. Appl. Meteorol. 33, 140–158.

Daly, C., Gibson, W.P., Taylor, G.H., Johnson, G.H., Pasteris, P., 2002. A knowledge-based approach to the statistical mapping of climate. Clim. Res. 22, 99–113.

Darshana, D., Pandey, A., Pandey, R.P., 2013. Analysing Trends in Reference Evapotranspiration and Weather Variables in the Tons River Basin in Central India. Stoch. Environ. Res. Risk. Assess. 27 (6), 1407–1421.

Davison, A.C., Hinkley, D.V., 1997. Bootstrap Methods and their Applications. Cambridge University Press.

Dery, S.J., Wood, E.F., 2005. Decreasing river discharge in northern Canada. Geophys. Res. Lett. 32, 1–4. L10401.

Detzel, D.H.M., Mine, M.R.M., 2014. Trends in hydrological series: methods and application. In: 11th International Conference on Hydroscience & Engineering. At Hamburg, Germany).

Dingman, S.L., 2002. Physical Hydrology. Prentice Hall, NJ.

Dinpashoh, Y., Jhajharia, D., Fakheri-Fard, A., Singh, V.P., Kahya, E., 2011. Trends in reference crop evapotranspiration over Iran. J. Hydrol. 399 (3–4), 422–433.

Dodge, Y., 2008. The Concise Encyclopedia of Statistics. Springer, Netherlands.

Duhan, D., Pandey, A., 2013. Statistical analysis of long term spatial and temporal trends of precipitation during 1901–2002 at Madhya Pradesh, India. Atmos. Res. 122, 136–149.

Durbin, J., Watson, G.S., 1950. Testing for serial correlation in least squares regression, I. Biometrika 37, 409–428.

Durbin, J., Watson, G.S., 1951. Testing for serial correlation in least squares regression, II. Biometrika 38, 159–177.

Efron, B., Gong, G., 1983. A leisurely look at the bootstrap, the jackknife, and cross validation. Am. Statistician 37, 36–48.

Efron, B., Tibshirani, R., 1993. An Introduction to the Bootstrap. Chapman and Hall, London.

Efron, B., 1979. Bootstrap methods: another look at the jackknife. Ann. Stat. 7, 1–26.

Eischeid, J.K., et al., 2000. Creating a serially complete, national daily time series of temperature and precipitation for the Western United States. J. Appl. Meteorol. 39, 1580–1591.

Ejder, T., Kale, S., Acar, S., Hisar, O., Mutlu, F., 2016. Effects of climate change on annual streamflow of Kocabas stream (Çanakkale, Turkey). J. Sci. Res. Rep. 11 (4), 1–11.

EPA, 2000. EPA/600/R-96/084). Guidance for Data Quality Assessment: Practical Methods for Data Analysis, EPA QA/G-9, QA00 Update.

EPA, 2006a. Guidance on Systematic Planning Using the Data Quality Objectives Process. EPA QA/G-4. United States Environmental Protection Agency, Office of Environmental Information, Washington, DC, USA.

EPA, 2006b. Data Quality Assessment: Statistical Methods for Practitioners. EPA QA/G-9S. United States Environmental Protection Agency, Office of Environmental Information, Washington, DC, USA.

Espadafor, M., Lorite, I.J., Gavilán, P., Berengena, J., 2011. An analysis of the tendency of reference evapotranspiration estimates and other climate variables during the last 45 years in Southern Spain. Agric. Water Manag. 98, 1045–1061.

Fathian, F., Morid, S., Kahya, E., 2015. Identification of trends in hydrological and climatic variables in Urmia Lake basin. Iran. Theor. Appl. Climatol. 19, 443–464.

Fathian, F., Aliyari, H., Kahya, E., Dehghan, Z., 2016. Temporal trends in precipitation using spatial techniques in GIS over Urmia Lake Basin, Iran. Int. J. Hydrol. Sci. Technol. 6 (1), 62–81.

Gadgil, A., Dhorde, A., 2005. Temperature trends in twentieth century at Pune, India. Atmos. Environ. 39, 6550–6556.

Gajbhiye, S., Meshram, C., Mirabbasi, R., Sharma, S.K., 2016. Trend analysis of rainfall time series for Sindh river basin in India. Theor. Appl. Climatol. 125, 593–608.

Gallego, M.C., Garcia, J.A., Vaquero, J.M., Mateos, V.L., 2006. Changes in frequency and intensity of daily precipitation over the Iberian Peninsula. J. Geophys. Res. 111, 1–15. D24105.

Garbrecht, J., Liew, M.V., Brown, G.O., 2004. Trends in precipitation, streamflow, and evapotranspiration in the great plains of the United States. J. Hydrol. Eng. 9 (5), 360–367.

Gautam, M.R., Acharya, K., 2012. Streamflow trends in Nepal. Hydrol. Sci. J. 57 (2), 344–357.

Gautam, M.R., Acharya, K., Tuladhar, M.K., 2010. Upward trend of streamflow and precipitation in a small, non-snow-fed, mountainous watershed in Nepal. J. Hydrol. 387, 304–311.

Gauthier, T.D., 2001. Detecting trends using Spearman's Rank Correlation Coefficient. Environ. Forensics 2, 359–362.

Gemmer, M., Becker, S., Jiang, T., 2004. Observed monthly precipitation trends in China 1951–2002. Theor. Appl. Climatol. 77 (1), 39–45.

Gibbons, J.D., 1997. Nonparametric Methods for Quantitative Analysis. American Sciences Press, Inc., Ohio-USA.

Gilbert, R.O., 1987. Statistical Methods for Environmental Pollution Monitoring. Van Nostrand Reinhold Company, Inc, New York.

Gocic, M., Trajkovic, S., 2013. Trend analysis of water quality parameters for the Nisava river. Archit. Civ. Eng. 11 (3), 199–210.

Gonzalez-Hidalgo, J.C., Lopez-Bustins, J.-A., Martin-Vide, J., Luis, M.D., 2009. Monthly precipitation trends on the Mediterranean fringe of the Iberian Peninsula during the second-half of the twentieth century (1951–2000). Int. J. Climatol. 29 (10), 1414–1429. P. Štepánek.

Goovaerts, P., 2000. Geostatistical approaches for incorporating elevation into the spatial interpolation of rainfall. J. Hydrol. 228, 113–129.

Graham, J.W., 2012. Missing Data: Analysis and Design. Springer, New York.

Grayson, R., Gunter, B., 2001. Spatial Patterns in Catchment Hydrology: Observations and Modeling. Cambridge University Press.

Grubb, H., Robson, A., 2000. Exploratory/Visual Analysis. In: Kundezwewicz, Z.W., Robson, A. (Eds.), Detecting Trend and Other Changes in Hydrological Data. UNESCO, WMO, Geneva, Switzerland. WMO/TD-No. 1013.

Guo, L., Li, L., 2015. Variation of the proportion of precipitation occurring as snow in the Tian Shan Mountains, China. Int. J. Climatol. 35, 1379–1393.

Guo, L., Xia, Z., 2014. Temperature and precipitation long-term trends and variations in the Ili-Balkhash Basin. Theor. Appl. Climatol. 115 (1), 219–229.

Hair, J.F., Black, W.C., Babin, B.J., Anderson, R.E., 2010. Multivariate Data Analysis. Prentice Hall.

Hajani, E., Rahman, A., Haddad, K., 2014. Trend analysis for extreme rainfall events in new South Wales, Australia. Int. J. Environ. Chem. Ecol. Geol. Geophys. Eng. 8 (12), 834–839.

Hamed, K.H., Rao, A.R., 1998. A modified Mann-Kendall trend test for autocorrelated data. J. Hydrol. 204, 182–196.

Hamed, K.H., 2008. Trend detection in hydrologic data: the Mann–Kendall trend test under the scaling hypothesis. J. Hydrol. 349, 350–363.

Hamed, K.H., 2016. The distribution of Spearman's rho trend statistic for persistent hydrologic data. Hydrol. Sci. J. 61 (1), 214–223.

Han, J., Kamber, M., 2006. Data Mining: Concepts and Techniques. Morgan Kaufmann, Waltham.

Hanel, M., Pavlásková, A., Kyselý, J., 2016. Trends in characteristics of sub-daily heavy precipitation and rainfall erosivity in the Czech Republic. Int. J. Climatol. 36, 1833–1845.

Hirsch, R.M., Slack, J.R., Smith, R.A., 1982. Techniques of trend analysis for monthly water quality data. Water Resour. Res. 18 (1), 107–121.

Hirsch, R.M., Alexander, R.B., Smith, R.A., 1991. Selection of methods for the detection and estimation of trends in water quality. Water Resour. Res. 27 (5), 803–813.

Hisdal, H., 2000. Tests for changes in flow regimes. In: Kundezwewicz, Z. W., Robson, A. (Eds.), Detecting Trend and Other Changes in Hydrological Data. UNESCO, WMO, Geneva, Switzerland. WMO/TD-No. 1013.

Hirsch, R.M., Moyer, D.L., Archfield, S.A., 2010. Weighted regressions on time, discharge, and season (WRTDS), with an application to Chesapeake Bay river inputs. J. Am. Water Resour. Assoc. 46 (5), 857–880.

Hodgkins, G.A., Dudley, R.W., Huntington, T.G., 2003. Changes in the timing of high river flows in New England over the 20th Century. J. Hydrol. 278, 244–252.

Hollander, M., Wolfe, D.A., 1973. Nonparametric Statistical Methods. John Wiley and Sons, Inc., Hoboken, NJ.

Hu, Q., Wang, Y., 2009. Impact assessment of climate change and human activities on annual highest water level of Taihu Lake. Water Sci. Eng. 2 (1), 1–15.

Hu, Y., Maskey, S., Uhlenbrook, S., 2012. Trends in temperature and rainfall extremes in the Yellow River source region, China. Climatic Change 110, 403–429.

Hutchinson, M.F., 1995. Interpolating mean rainfall using thin-plate smoothing splines. Int. J. Geogr. Inf. Sys. 9, 385–403.

Irizarry-Ortiz, M.M., Obeysekera, J., Park, J., Trimble, P., Barnes, J., Park-Said, W., Gadzinski, E., 2012. Historical trends in Florida temperature and precipitation. Hydrol. Process. 27 (16), 2225–2246.

Jhajharia, D., Singh, V.P., 2011. Trends in temperature, diurnal temperature range and sunshine duration in Northeast India. Int. J. Climatol. 31, 1353–1367.

Jhajharia, D., Shrivastava, S.K., Sarkar, D., Sarkar, S., 2009. Temporal characteristics of pan evaporation trends under the humid conditions of northeast India. Agric. For. Meteorol. 149 (5), 763–770.

Jhajharia, D., Dinpashoh, Y., Kahya, E., Singh, V.P., Fakheri-Fard, A., 2012. Trends in reference evapotranspiration in the humid region of northeast India. Hydrol. Process. 26, 421–435.

Jhajharia, D., Dinpashoh, Y., Kahya, E., Choudhary, R.R., Singh, V.P., 2014. Trends in temperature over Godavari river basin in southern Peninsular India. Int. J. Climatol. 34 (5), 1369–1384.

Jiang, T., Su, B., Hartmann, H., 2007. Temporal and spatial trends of precipitation and river flow in the Yangtze River Basin, 1961–2000. Geomorphology 85 (3–4), 143–154.

Jones, J.R., Schwartz, J.S., Ellis, K.N., Hathaway, J.M., Jawdy, C.M., 2015. Temporal variability of precipitation in the upper Tennessee valley. J. Hydrol. Reg. Stud. 3, 125–138.

Kahya, E., Kalayci, S., 2004. Trend analysis of streamflow in Turkey. J. Hydrol. 289, 128–144.

Kalra, A., Ahmad, S., 2011. Evaluating changes and estimating seasonal precipitation for the Colorado River Basin using a stochastic nonparametric disaggregation technique. Water Resour. Res. 47, 1–26. W05555.

Kampata, J.M., Parida, B.P., Moalafhi, D.B., 2008. Trend analysis of rainfall in the headstreams of the Zambezi river basin in Zambia. Phys. Chem. Earth 33, 621–625.

Karpouzos, D.K., Kavalieratou, S., Babajimopoulos, C., 2010. Trend analysis of precipitation data in Pieria region (Greece). Eur. Water 30, 31–40.

Kaushal Sujay, S., Gene Likens, E., Jaworski Norbert, A., Pace Michael, L., Sides Ashley, M., Seekell, David, Belt Kenneth, T., Secor David, H., Wingate Rebecca, L., 2010. Rising stream and river temperatures in the United States. Front. Ecol. Environ. 8 (9), 461–466.

Kendall, M.G., 1975. Rank Correlation Methods. Charles Griffin and Co Ltd, London, U.K.

Kisi, O., 2015. An innovative method for trend analysis of monthly pan evaporations. J. Hydrol. 527, 1123–1129.

Krajewski, W.F., 1987. Co-kriging of radar and rain gauge data. J. Geophys. Res. 92 (D8), 9571–9580.

Kumar, S., Merwade, V., Kam, J., Thurner, K., 2009. Streamflow trends in Indiana: effects of long term persistence, precipitation and subsurface drains. J. Hydrol. 374, 171–183.

Kundu, S., Khare, D., Mondal, A., Mishra, P.K., 2015. Analysis of spatial and temporal variation in rainfall trend of Madhya Pradesh, India (1901–2011). Environ. Earth Sci. 73 (12), 8197–8216.

Kundzewicz, Z.W., Robson, A., 2000. Detecting Trend and Other Changes in Hydrological Data. World Climate Program – Water, WMO/UNESCO, WCDMP-45, WMO/TD 1013. Geneva, 157 pp.

Larson, L.W., Peck, E.L., 1974. Accuracy of precipitation measurements for hydrologic forecasting. Water Resour. Res. 156, 1687–1696.

Lettenmaier, D.P., 1988. Multivariate nonparametric tests for trend in water quality. Water Resour. Bull. 24 (3), 505–512.

Li, Z., Sun, D., 2015. Trend and abrupt analysis of rainfall change during last 50 years in the WeiHe Basin. Earth Sci. 4 (6), 228–234.

Li, Z.L., Xu, Z.X., Li, J.Y., Li, Z.J., 2008. Shift trend and step changes for runoff time series in the Shiyang River basin, northwest China. Hydrol. Process. 22, 4639–4646.

Li, Y., Horton, R., Ren, T., Chen, C., 2010. Agricultural water management. Agric. Water Manag. 97, 300–308.

Libiseller, C., Grimvall, A., 1999. Performance of partial Mann-Kendall tests for trend detection in the presence of covariates. Environmetrics 13, 71–84.

Lins, H.F., Slack, J.R., 1999. Streamflow trends in the United States. Geophys. Res. Lett. 26 (2), 227–230.

Lins, H.F., Hirsch, R.M., Kiang, J., 2010. Water – the Nation's Fundamental Climate Issue: A White Paper on the U.S. Geological Survey Role and Capabilities; Circular 1347. U.S. Geological Survey, pp. 1–14.

Little, R.J.A., Rubin, D.B., 2002. Statistical Analysis with Missing Data. Wiley, New York.
Ljung, G.M., Box, G.E.P., 1978. On a measure of a lack of fit in time series models. Biometrika 65, 297–303.
Llyod, C.D., 2005. Assessing the effect of integrating elevation data into the estimation of monthly precipitation in Great Britain. J. Hydrol. 308, 128–150.
Loader, C., 1999. Local Regression and Likelihood. Springer Verlag.
Longobardi, A., Villani, P., 2010. Trend analysis of annual and seasonal rainfall time series in the Mediterranean area. Int. J. Climatol. 30, 1538–1546.
Lu, G.Y., Wong, D.W., 2008. An adaptive inverse-distance weighting spatial interpolation technique. Comput. Geosci. 34 (9), 1044–1055.
Machiwal, D., Jha, M.K., 2012. Hydrologic Time Series Analysis: Theory and Practice. Springer, New York, USA.
Mann, H.B., 1945. Non-parametric tests against trend. Econometrica 13, 245–259.
Marengo, J.A., Tomasell, J., Uvo, C.R., 1998. Trends in streamflow and rainfall in tropical South America: Amazonia, eastern Brazil, and northwestern Peru. J. Geophys. 103 (D2), 1775–1783.
Marin, L., Birsan, M.-V., Bojariu, R., Dumitrescu, A., Micu, D.M., Manea, A., 2014. An Overview of Annual Climatic Changes in Romania: Trends in Air Temperature, Precipitation, Sunshine Hours, Cloud Cover, Relative Humidity and Wind Speed during the 1961–2013 Period.
Martino, G.D., Fontana, N., Marini, G., Singh, V.P., 2013. Variability and trend in seasonal precipitation in the continental United States. J. Hydrol. Eng. 18 (6), 630–640.
Massey, F.J., 1951. The Kolmogorov-Smirnov test for goodness-of-fit. J. Am. Stat. Assoc. 46 (253), 68–78.
Masson, D., Frei, C., 2016. Short Communication Long-term variations and trends of mesoscale precipitation in the Alps: recalculation and update for 1901–2008. Int. J. Climatol. 36 (1), 492–500.
McBean, A.E., Rovers, F.A., 1998. Statistical Procedures for Analysis of Environmental Monitoring Data and Risk Assessment. Prentice Hall, Upper Saddle River.
McCabe, G.J., Wolock, D.M., 1997. Climate change and the detection of trends in annual runoff. Clim. Res. 8, 129–134.
McCuen, R., 2003. Modeling Hydrologic Change: Statistical Methods. CRC Press, Boca Raton, USA.
McLeod, I., Hipel, K.W., Bodo, B.A., 1991. Trend analysis methodology for water quality time series. Environmetrics 2 (2), 169–200.
Meals, D.W., Spooner, J., Dressing, S.A., Harcum, J.B., 2011. Statistical Analysis for Monotonic Trends, Tech Notes 6, November 2011. Developed for U.S. Environmental Protection Agency by Tetra Tech, Inc., Fairfax, VA, 23 p.
Memarian, H., Balasundram, S.K., Talib, J.B., Sood, A.M., Abbaspour, K.C., 2012. Trend analysis of water discharge and sediment load during the past three decades of development in the Langat basin, Malaysia. Hydrol. Sci. J. 57 (6), 1207–1222.
Méndez-Lázaro, P.A., Nieves-Santiango, A., Miranda-Bermúdez, J., 2014. Trends in total rainfall, heavy rain events, and number of dry days in San Juan, Puerto Rico, 1955–2009. Ecol. Soc. 19 (2), 50.
Miao, L., Jun, X., Dejuan, M., 2012. Long-term trend analysis of seasonal precipitation for Beijing, China. J. Resour. Ecol. 3 (1), 064–072.
Miller, J.F., 1972. Physiographically Adjusted Precipitation Frequency Maps: Distribution of Precipitation in Mountainous Areas. WMO-No. 326 (11), 264–277.
Milly, P.C.D., Betancourt, J., Falkenmark, M., Hirsch, R.M., Kundzewicz, Z.W., Lettenmaier, D.P., Stouffer, R.J., 2008. Stationarity is dead: whither water management. Science 319, 573–574.

Min, L., Yanjun, S., Yan, Z., Changming, L., 2010. Trend in pan evaporation and its attribution over the past 50 years in China. J. Geogr. Sci. 20 (4), 557–568.
Mix, K., Vicente, L.L., Walter, R., 2012. Environmental drivers of streamflow changes in the Upper Rio Grande. Water Res. Manag. 26 (1), 253–272.
Modarres, R., Silva, V.D.P.R.D., 2007. Rainfall trends in arid and semi-arid regions of Iran. J. Arid Environ. 70, 344–355.
Mohsin, T., Gough, W.A., 2010. Trend analysis of long-term temperature time series in the Greater Toronto Area (GTA). Theor. Appl. Climatol. 101 (3), 311–327.
Mondal, A., Khare, D., Kundu, S., 2015. Spatial and temporal analysis of rainfall and temperature trend of India. Theor. Appl. Climatol. 122, 143–158.
Myatt, G.J., Johnson, W.P., 2009. Making Sense of Data II: A Practical Guide to Data Visualization, Advanced Data Mining Methods, and Applications. Wiley, New Jersey.
Nafarzadegan, A.R., Hossein, A., Singh, V.P., Kherad, M., 2013. Parametric and non-parametric trend of reference evapotranspiration and its key influencing climatic variables. Ecopersia 1 (2), 123–144.
Naghettini, M., 2017. Fundamentals of Statistical Hydrology. Springer, Switzerland.
Naoum, S., Tsanis, I.K., 2003. Temporal and spatial variation of annual rainfall on the island of Crete, Greece. Hydrol. Process. 17 (10), 1899–1922.
NIST, 2018. http://www.itl.nist.gov/div898/handbook/eda/eda.htm.
Nott, J., 2006. Extreme Events: A Physical Reconstruction and Risk Assessment. Cambridge University Press, Cambridge.
O'Sullivan, D., Unwin, D.J., 2010. Geographical Information Analysis. John Wiley & Sons, Inc., NJ.
Ogolo, E.O., Adeyemi, B., 2009. Variations and trends of some meteorological parameters at Ibadan, Nigeria. Pac. J. Sci. Technol. 10 (2), 981–987.
Onoz, B., Bayazit, M., 2003. The power of statistical tests for trend detection. Turk. J. Eng. Environ. Sci. 27, 247–251.
Ott, W.R., 1995. Environmental Statistics and Data Analysis. CRC Press, Boca Raton.
Ouarda, T.B.M.J., Charrona, C., Kumar, K.N., Marpu, P.R., Ghedira, H., Molini, A., Khayal, I., 2014. Evolution of the rainfall regime in the United Arab Emirates. J. Hydrol. 514 (6), 258–270.
Partal, T., Kahya, E., 2006. Trend analysis in Turkish precipitation data. Hydrol. Process. 20 (9), 2011–2026.
Petrow, T., Merz, B., 2009. Trends in flood magnitude, frequency and seasonality in Germany in the period 1951–2002. J. Hydrol. 371, 129–141.
Pettitt, A.N., 1979. A non-parametric approach to the change-point problem. Appl. Stat. 28 (2), 126–135.
Piccarreta, M., Capolongo, D., Boenzi, F., 2004. Trend analysis of precipitation and drought in Basilicata from 1923 to 2000 within a southern Italy context. Int. J. Climatol. 24, 907–922.
Pranuthi, G., Dubey, S.K., Tripathi, S.K., Chandniha, S.K., 2014. Trend and change point detection of precipitation in urbanizing districts of Uttarakhand in India. Indian J. Sci. Technol. 7 (10), 1573–1582.
Rahmat, S.N., Jayasuriya, N., Adnan, M.S., Bhuiyan, M., 2016. Analysis of spatio-temporal trends using standardised precipitation index (SPI). ARPN J. Eng. Appl. Sci. 11 (4), 2387–2392.
Ramos-Calzado, P., et al., 2008. A novel approach to precipitation series completion in climatological datasets: application to Andalusia. Int. J. Climatol. 28, 1524–1535.
Regonda, S.K., Rajagopalan, B., Clark, M., Zagona, E., 2006. A Multi-model ensemble forecast framework: application to spring seasonal flows in the Gunnison River Basin. Water Resour. Res. 42 (9), W09404.

Reza, Y.M., Javad, K.D., Mohammad, M., Ashish, S., 2011. Trend detection of the rainfall and air temperature data in the Zayandehrud basin. J. Appl. Sci. 11 (12), 2125–2134.

Rodda, H.J.E., Little, M.A., 2015. Understanding Mathematical and Statistical Techniques in Hydrology: An Examples-based Approach. Wiley Blackwell.

Rubin, D.B., 1987. Multiple Imputation for Nonresponse in Surveys. John Wiley, New York.

Rubin, D.B., 1996. Multiple imputation after 18+ years. J. Am. Stat. Assoc. 91, 473–489.

Sabzevari, A.A., Zarenistanak, M., Tabari, H., Moghimi, S., 2015. Evaluation of precipitation and river discharge variations over southwestern Iran during recent decades. J. Earth Syst. Sci. 124 (2), 335–352.

Salas, J.D., 1993. Analysis and modeling of hydrological time series. In: Maidment, D.R. (Ed.), Handbook of Hydrology. McGraw-Hill, New York.

Santos, M., Fragoso, M., 2013. Precipitation variability in Northern Portugal: data homogeneity assessment and trends in extreme precipitation indices. Atmos. Res. 131, 34–45.

Saptomo, S.K., Setiawan, B.I., Yuge, K., 2009. Climate change effects on paddy field thermal environment and evapotranspiration. Paddy Water Environ. 7, 341–347.

Satterthwaite, F.E., 1946. An approximate distribution of estimates of variance components. Biometrics Bull. 2, 110–114.

Sawaske, S.R., Freyberg, D.L., 2014. An analysis of trends in baseflow recession and lowflows in rain-dominated coastal streams of the pacific coast. J. Hydrol. 519, 599–610.

Sayemuzzaman, M., Jha, M.K., 2014. Seasonal and annual precipitation time series trend analysis in North Carolina, United States. Atmos. Res. 137, 183–194.

Sayemuzzaman, M., Jha, M.K., Mekonnen, A., Schimmel, K.A., 2014. Subseasonal climate variability for North Carolina, United States. Atmos. Res. 145–146, 69–79.

Sayemuzzaman, M., Mekonnen, A., Jha, M.K., 2015a. Diurnal temperature range trend over North Carolina and the associated mechanisms. Atmos. Res. 160, 99–108.

Sayemuzzaman, M., Jha, M.K., Mekonnen, A., 2015b. Spatio-temporal long-term (1950–2009) temperature trend analysis in North Carolina, United States. Theor. Appl. Climatol. 120, 159–171.

Seleshi, Y., Zanke, U., 2004. Recent changes in rainfall and rainy days in Ethiopia. Int. J. Climatol. 24 (8), 973–983.

Sen, P.K., 1968. Estimates of the regression coefficient based on Kendall's tau. J. Am. Stat. Assoc. 63, 1379–1389.

Şen, Z., 2014. Trend identification simulation and application. J. Hydrol. Eng. 19 (3), 635–642.

Seo, D.-J., Smith, J.A., 1993. Rainfall estimation using rain gauges and radar: a Bayesian approach. J. Stoch. Hydrol. Hydraul. 5 (1), 1–14.

Seo, D.-J., Krajewski, W.F., Bowles, D.S., 1990a. Stochastic interpolation of rainfall data from rain gauges and radar using cokriging - 1. Design of experiments. Water Resour. Res. 26 (3), 469–477.

Seo, D.-J., Krajewski, W.F., Bowles, D.S., 1990b. Stochastic interpolation of rainfall data from rain gauges and radar using cokriging: 2. Results. Water Resour. Res. 26 (5), 915–924.

Seo, D.-J., 1996. Nonlinear estimation of spatial distribution of rainfall - an indicator cokriging approach. Stoch. Hydrol. Hydraul. 10, 127–150.

Seo, D.J., 1998. Real-time estimation of rainfall fields using radar rainfall and rain gage data. J. Hydro. 208, 37–52.

Serrano, V., Mateos, L., Garcia, J.A., 1999. Trend analysis of monthly precipitation over the Iberian Peninsula for the period 1921–1995. Phys. Chem. Earth 24 (1–2), 85–90.

Shadmani, M., Marofi, S., Roknian, M., 2012. Trend analysis in reference evapotranspiration using Mann-Kendall and Spearman's rho tests in arid regions of Iran. Water Resour. Manag. 26, 211–224.

Shan, N., Shi, Z., Yang, X., Gao, J., Cai, D., 2015. Spatiotemporal trends of reference evapotranspiration and its driving factors in the Beijing–Tianjin sand source control project region, China. Agric. For. Meteorol. 200, 322–333.
Shepard, D., 1968. A Two-dimensional interpolation function for irregularly spaced data. In: Proceedings of the Twenty-third National Conference of the Association for Computing Machinery, pp. 517–524.
Shimazaki, H., Shinomoto, S., 2007. A method for selecting the bin size of a time histogram. Neural Comput. 19 (6), 1503–1527.
Shimazaki, H., Shinomoto, S., 2010. Kernel bandwidth optimization in spike rate estimation. J. Comput. Neurosci. 29 (1–2), 171–182.
Simanton, J.R., Osborn, H.B., 1980. Reciprocal-distance estimate of point rainfall. J. Hydraul. Eng. Div. 106HY7.
Singh, V.P., Chowdhury, K., 1986. Comparing some methods of estimating mean aerial rainfall. Water Resour. Bull. 222, 275–282.
Smakhtin, V.U., 2001. Low flow hydrology: a review. J. Hydrol. 147–186, 1991.
Smith, J.A., 1993. In: Maidment, D.R. (Ed.), Precipitation, Chapter 3, Handbook of Hydrology. McGraw Hill, New York.
Smith, L.C., 2000. Trends in Russian arctic river-ice formation and breakup, 1917 to 1994. Phys. Geogr. 21 (1), 46–56.
Some'e, B.S., Ezani, A., Tabari, H., 2012. Spatiotemporal trends and change point of precipitation in Iran. Atmos. Res. 113, 1–12. Statistics, 13,1-13.
Su, B.D., Jiang, T., Jin, W.B., 2006. Recent trends in observed temperature and precipitation extremes in the Yangtze River basin, China. Theor. Appl. Climatol. 83, 139–151.
Tabari, H., Aghajanloo, M.-B., 2013. Temporal pattern of aridity index in Iran with considering precipitation and evapotranspiration trends. Int. J. Climatol. 33 (2), 396–409.
Tabari, H., Talaee, P.H., 2011a. Temporal variability of precipitation over Iran: 1966–2005. J. Hydrol. 396 (3–4), 313–320.
Tabari, H., Talaee, P.H., 2011b. Recent trends of mean maximum and minimum air temperatures in the western half of Iran. Meteorol. Atmos. Phys. 111 (3), 121–131.
Tabari, H., Talaee, P.H., 2011c. Analysis of trends in temperature data in arid and semi-arid regions of Iran. Global Planet. Change 79, 1–10.
Tabari, H., Nikbakht, J., Talaee, P.H., 2012. Identification of trend in reference evapotranspiration series with serial dependence in Iran. Water Resour. Manag. 26, 2219–2232.
Tabios III, G.Q., Salas, J.D., 1985. A comparative analysis of techniques for spatial interpolation of precipitation. Water Resour. Bull. 21, 365–380.
Talaee, P.H., 2014. Iranian rainfall series analysis by means of nonparametric tests. Theor. Appl. Climatol. 116, 597–607.
Tan, P.-N, Steinbach, M., Kumar, V., 2006. Introduction of Data Mining. Pearson, USA.
Tanarhte, M., Hadjinicolaou, P., Lelieveld, J., 2012. Intercomparison of temperature and precipitation data sets based on observations in the Mediterranean and the Middle East. J. Geophys. Res. 117, 1–24. D12102.
Taxak, A.K., Murumkar, A.R., Arya, D.S., 2014. Long term spatial and temporal rainfall trends and homogeneity analysis in Wainganga basin, Central India. Weather Clim. Extremes 4, 50–61.
Teegavarapu, R.S.V., Chandramouli, V., 2005. Improved weighting methods, deterministic and stochastic data-driven models for estimation of missing precipitation records. J. Hydrol. 312, 191–206.

Teegavarapu, R.S.V., Peters, D., Pathak, C., 2008. Evaluation of functional forms of rain gauge—radar (NEXRAD) data relationships. In: Babcock, R.W., Walton, R. (Eds.), Proceedings of World Environmental and Water Resources Congress. Hawaii. ASCE, pp. 1—9. https://doi.org/10.1061/40976(316)413.

Teegavarapu, R.S.V., Tufail, M., Ormsbee, L., 2009. Optimal functional forms for estimation of missing precipitation records. J. Hydro. 374, 106—115.

Teegavarapu, R.S.V., Nayak, A., 2017. Evaluation of long-term trends in extreme precipitation: implications of infilled historical data and temporal-window based analysis. J. Hydrol. 550, 614—634.

Teegavarapu, R.S.V., Simonovic, S.P., 2014. Simulation of multiple hydropower reservoir operations using system dynamics approach. Water Resour. Manag. 28, 1937—1958.

Teegavarapu, R.S.V., Meskele, T., Pathak, C., 2012. Geo-spatial grid-based transformations of precipitation estimates using spatial interpolation methods. Comput. Geosci. 40, 28—39. https://doi.org/10.1016/j.cageo.2011.07.004.

Teegavarapu, R.S.V., Goly, A., Obeysekera, J., 2013. Influences of Atlantic multidecadal oscillation phases on spatial and temporal variability of regional precipitation extremes. J. Hydrol. https://doi.org/10.1016/j.jhydrol.2013.05.003.

Teegavarapu, R.S.V., 2007. Use of universal function approximation in variance-dependent interpolation technique: an application in hydrology. J. Hydrol. 332, 16—29.

Teegavarapu, R.S.V., 2008. Innovative spatial interpolation methods for estimation of missing precipitation records: concepts and applications. In: Prague, Bruthans, J., Kovar, K., Hrkal, Z. (Eds.), Proceedings of HydroPredict 2008, pp. 79—82.

Teegavarapu, R.S.V., 2009. Estimation of missing precipitation records integrating surface interpolation techniques and spatio-temporal association rules. J. Hydroinf. 11 (2), 133—146.

Teegavarapu, R.S.V., 2012b. Floods in a Changing Climate: Extreme Precipitation. Cambridge University Press.

Teegavarapu, R.S.V., 2013. Missing precipitation data estimation using optimal proximity metric-based imputation, nearest neighbor classification and cluster-based interpolation methods. Hydrol. Sci. J. https://doi.org/10.1080/02626667.2013.862334.

Teegavarapu, R.S.V., 2014. Statistical corrections of spatially interpolated precipitation estimates. Hydrol. Process. 28 (11), 3789—3808. https://doi.org/10.1002/hyp.9906.

Thompson, S.A., 1999. Hydrology for Water Management. Balkema Publishers, Netherlands.

Tomczak, M., 1998. Spatial Interpolation and its uncertainty using automated anisotropic inverse distance weighting (IDW) — cross-validation/jackknife approach. J. Geogr. Inf. Decis. Anal. 2, 18—30.

Topaloğlu, F., Özfidaner, M., Aydin, F., 2012. Regional trends in Turkish pan evaporation. J. Food Agric. Environ. 10 (3&4), 960—962.

Tsanis, I., 2004. Orographic precipitation modeling with multiple linear regression. J. Hydro. Eng. 9 (2), 79—102.

Tung, Y.K., 1983. Point rainfall estimation for a mountainous region, Journal of Hydraulic Engineering. Am. Soc. Civ. Eng. 109 (10), 1386—1393.

USGS, 2007. Facing tomorrow's challenges—U.S. Geological Survey science in the decade 2007—2017. U.S. Geological Survey Circular 1309.

USGS, 2012. USGS Hydro-climatic Data Network 2009 (HCDN-2009); Fact Sheet 2012-3047. U.S. Geological Survey.

Vasiliev, I.R., 1996. Visualization of spatial dependence: an elementary view of spatial autocorrelation. In: Arlinghaus, S. (Ed.), Practical Handbook of Spatial Statistics. CRC Press, Boca Raton, FL.

Vergni, L., Todisco, F., 2011. Spatio-temporal variability of precipitation, temperature and agricultural drought indices in Central Italy. Agric. For. Meteorol. 151 (3), 301—313.

Vieux, B.E., 2001. Distributed Hydrologic Modeling Using GIS. Water Science and Technology Library, Kluwer Academic Publishers.
Villarini, G., Serinaldi, F., Smith, J.A., Krajewski, W.F., 2009. On the stationarity of annual flood peaks in the continental United States during the 20th century. Water Resour. Res. 45, 1−17, 08417.
Von Neumann, J., 1941. Distribution of the ratio of the mean square successive difference to the variance. Ann. Math. Stat. 12, 367−395.
Von Storch, H., Zwiers, F.W., 1999. Statistical Analysis in Climate Research. Cambridge University Press, New York.
Wagesho, N., Goel, N.K., Jain, M.K., 2013. Temporal and spatial variability of annual and seasonal rainfall over Ethiopia. Hydrol. Sci. J. 58 (2), 354−373.
Wahlin, K., Grimvall, A., 2010. Roadmap for assessing regional trends in groundwater quality. Environ. Monit. Assess. 165, 217−231.
Wald, A., Wolfowitz, J., 1943. An exact test for randomness in the non-parametric case based on serial correlation. Ann. Math. Statist. 14 (4), 378−388.
Wang, W., Chen, Y., Becker, S., Liu, B., 2015. Linear trend detection in serially dependent hydrometeorological data based on a variance correction Spearman Rho method. Water 7, 7045−7065.
Wang, F., 2006. Quantitative Methods and Applications in GIS. CRC Press, Boca Raton, FL.
Wei, T.C., McGuinness, J.L., 1973. Reciprocal Distance Squared Method: A Computer Technique for Estimating Area Precipitation. Technical Report ARS-Nc-8. U.S. Agricultural Research Service, North Central Region, Ohio, USA.
Wilks, D.S., 2011. Statistical Methods in the Atmospheric Sciences. Elsevier Publication.
WMO, 1988. Analyzing Long Time Series of Hydrologic Data with Respect to Climate Variability, WCAP-3, WMO/TD-no. 224. World Meteorological Organization, Geneva, Switzerland.
WMO (World Meteorological Organization), 2009. Guidelines on Analysis of Extremes in a Changing Climate in Support of Informed Decisions for Adaptation, Climate Data and Monitoring, WCDMP-no. 72 (Geneva).
Woodridge, J., 2003. Introductory Econometrics: A Modern Approach. South-Western College Publication.
Xia, Y., Fabian, P., Stohl, A., Winterhalter, M., 1999. Forest Climatology: estimation of missing values for Bavaria, Germany. Agric. For. Meteorol. 96, 131−144.
Xia, Y., et al., 2001. Forest climatology: estimation and use of daily climatological data for Bavaria, Germany. Agric. For. Meteorol. 106, 87−103.
Xu, C.-Y., Singh, V.P., 1998. A review on monthly water balance models for water resources investigations. Water Resour. Manag. 12, 31−50.
Xu, Z.X., Takeuchi, K., Ishidaira, H., 2003. Monotonic trend and step changes in Japanese precipitation. J. Hydrol. 279, 144−150.
Xu, C.-Y., Gong, L., Jiang, T., Chen, D., Singh, V.P., 2006. Analysis of spatial distribution and temporal trend of reference evapotranspiration and pan evaporation in Changjiang (Yangtze River) catchment. J. Hydrol. 327 (1−2), 81−93.
Xu, C.-Y., Zhang, Q., Tahir, M.E.H.E., Zhang, Z., 2010. Statistical properties of the temperature, relative humidity, and net solar radiation in the Blue Nile-eastern Sudan region. Theor. Appl. Climatol. 101, 397−409.
Yeh, C.-F., Wang, J., Yeh, H.-F., Lee, C.-H., 2015. Spatial and temporal streamflow trends in northern Taiwan. Water 7 (2), 634−651.
Yilmaz, G., Perera, B.J.C., 2015. Spatiotemporal trend analysis of extreme rainfall events in Victoria, Australia. Water Resour. Manag. 29 (12), 4465−4480.
Young, F.W., Valero-Mora, P.M., Friendly, M., 2006. Visual Statistics: Seeing Data with Dynamic Interactive Graphics. Wiley.

Yue, S., Hashino, M., 2003. Long term trends of annual and monthly precipitation in Japan. J. Am. Water Resour. Assoc. 1752–1688.

Yue, S., Pilon, P., 2003. Canadian streamflow trend detection: impacts of serial and cross-correlation. Hydrol. Sci. 48 (1), 51–63.

Yue, S., Pilon, P., 2004. A comparison of the power of the t test, Mann-Kendall and bootstrap tests for trend detection. Hydrol. Sci. 49 (1), 21–37.

Yue, S., Wang, C.Y., 2002. Regional streamflow trend detection with consideration of both temporal and spatial correlation. Int. J. Climatol. 22, 933–946.

Yue, S., Wang, C., 2004. The Mann-Kendall test modified by effective sample size to detect trend in serially correlated hydrological series. Water Resour. Manag. 18, 201–218.

Yue, S., Pilon, P., Cavadias, G., 2002. Power of the Mann–Kendall and Spearman's rho tests for detecting monotonic trends in hydrological series. J. Hydrol. 259 (1–4), 254–271.

Yurekli, K., 2015. Impact of climate variability on precipitation in the upper Euphrates–Tigris rivers basin of Southeast Turkey. Atmos. Res. 154, 25–38.

Zhang, X., Harvey, K.D., Hogg, W.D., Yuzyk, T.R., 2001. Trends in Canadian streamflow. Water Resour. Res. 37 (4), 987–998. https://doi.org/10.1029/2000WR900357.

Zhang, Q., Liu, C., Xu, C., Xu, Y., Jiang, T., 2006. 2006. Observed trends of annual maximum water level and streamflow during past 130 years in the Yangtze River basin, China. J. Hydrol. 324, 255–265.

Zhang, Y., Shao, Q., Xia, J., Bunn, S.E., Zuo, Q., 2011. Changes of flow regimes and precipitation in Huai River Basin in the last half century. Hydrol. Process. 25, 246–257.

Zhao, W., Yu, X., Ma, H., Zhu, Q., Zhang, Y., Qin, W., Ai, N., Wang, Y., 2015. Analysis of precipitation characteristics during 1957–2012 in the semi-arid loess plateau, China. PLoS One 10 (11), 1–13.

Zheng, C., Wang, Q., 2014. Spatiotemporal variation s of reference evapotranspiration in recent five decades in the arid land of Northwestern China. Hydrol. Process. 28, 6124–6134.

Zimmerman, D.C., et al., 1999. An experimental comparison of ordinary and universal kriging and inverse distance weighting. Math. Geol. 31, 375–390.

FURTHER READING

Duhan, D., Pandey, A., Pandey, R.P., 2013. Analysing trends in reference evapotranspiration and weather variables in the Tons River Basin in Central India. Stoch. Environ. Res. Risk Assess. 27 (6), 1407–1421.

Helsel, D.R., Hirsh, R.M., 1991. Statistical Methods in Water Resources. Techniques of Water-Resources Investigations of the USGS Book 4. Chapter A3.

Hodgson, M.E., 1989. Searching methods for rapid grid interpolation. Prof. Geogr. 411, 51–61.

Likens, G.E., Jaworski, N.A., Pace, M.L., Sides, A.M., Seekell, D., Belt, K.T., Secor, D.H., Wingate, R.L., 2010. Rising stream and river temperatures in the United States. Front. Ecol. Environ. 8 (9), 461–466.

Rahmat, S.N., Jayasuriya, N., Bhuiyan, M., 2012. Trend analysis of drought using standardised precipitation index (SPI) in Victoria, Australia. In: 34th Hydrology and Water Resources Symposium, pp. 441–448.

Teegavarapu, R.S.V., 2012a. Spatial interpolation using non-linear mathematical programming models for estimation of missing precipitation records. Hydrol. Sci. J. 57 (3), 1–24.

Teegavarapu, R.S.V., Aly, A., Pathak, C.S., Ahlquist, J., Fuelberg, H., Hood, J., 2017. Infilling missing precipitation records using variants of spatial interpolation and data-driven methods: use of optimal weighting parameters and nearest neighbor-based corrections. Int. J. Climatol. https://doi.org/10.1002/joc.5209.

Teegavarapu, R.S.V., 2016. Spatial and temporal estimation and analysis of precipitation. In: Handbook of Applied Hydrology. McGraw Hill.

Theil, H., 1950. A rank-invariant method of linear and polynomial regression analysis, 1,2, and 3: Ned. In: Akad. Wentsch Proc., vol. 53, pp. 386–392, 521-525, and 1397-1412.

Velez-Pereira, A.M., Linares, C.D., Delgado, R., Belmonte, J., 2016. Temporal trends of the airborne fungal spores in Catalonia (NE Spain), 1995–2013. Aerobiologia 32, 23–37.

Zhi, L., Dongyong, S., 2015. Trend and abrupt analysis of rainfall change during last 50 years in the WeiHe basin. Earth Sci. 4 (6), 228–234.

CHAPTER 2

Changes and Trends in Precipitation Extremes and Characteristics: Links to Climate Variability and Change

Ramesh S.V. Teegavarapu
Department of Civil, Environmental and Geomatics Engineering, Florida Atlantic University, Boca Raton, FL, United States

Contents

Trends and Changes in Hydroclimatic Variables
ISBN 978-0-12-810985-4
https://doi.org/10.1016/B978-0-12-810985-4.00002-5

1. INTRODUCTION

Understanding and evaluating the influences of coupled oceanic and atmospheric oscillations and climate change on variations and trends in precipitation characteristics and extremes are the focus of this chapter. Precipitation regime changes are known to be influenced by climate change and also several coupled oceanic and atmospheric oscillations around the globe. Major oscillations include Atlantic Multidecadal Oscillation (AMO), Pacific Decadal Oscillation (PDO), North Atlantic Oscillation (NAO), El Niño Southern Oscillation (ENSO). Exhaustive evaluations of individual and combined influences of these oscillations, through descriptive indices-based assessment of precipitation extremes and changes in rainfall characteristics, identification of spatially varying influences on dry and wet spell transition states, antecedent precipitation prior to extreme events, intra-event temporal distribution of precipitation, and changes in temporal occurrences of extremes are essential tasks to study climate variability influences on precipitation. Evaluation of these oscillations and their influences at different spatial scales and temporal resolutions considering multiple duration–specific precipitation extremes is also critical for flood control and water resources management. Assessment of influences of oscillations on precipitation with regional hydroclimatology, physiography, and topography defining or confining in some cases the spatial extent of these influences is important to establish possible physical basis for spatial and temporal changes in precipitation patterns.

2. CHANGES AND TRENDS

Spatial and temporal changes and trends in precipitation extremes and characteristics due to climate variability and change can be evaluated using a number of indices. These indices reflect different properties of precipitation time-series and they include: (1) interannual and intraannual variations;

(2) seasonality; (3) spatial and temporal variability of extremes; (4) nature of extremes (based on events related to type of storm: convective, frontal, etc.); (5) transition states as defined by rain or no rain dichotomous events; (6) persistence as defined and quantified by serial autocorrelation; (7) intraevent temporal distribution of precipitation; (8) antecedent moisture conditions preceding extreme events; (9) temporal occurrences of extremes; (10) number of extremes over a specific threshold; (11) interevent time definition (IETD)—based events and (12) individual and coupled influences of internal modes of climate variability manifested by coupled oceanic and atmospheric oscillations. One of the major questions that need to be answered related to precipitation extremes and characteristics is: How the interannual, decadal, quasidecadal and multidecadal climate variability affect the occurrence of precipitation extremes of a specific magnitude and frequency? Many of these indices for changes in precipitation due to climate variability were evaluated in multiple recent studies (e.g., Teegavarapu et al., 2013; Goly and Teegavarapu, 2014).

Evaluation of precipitation characteristics and extremes will involve a number of steps (Teegavarapu, 2016) ranging from data collection to development of inferences about the influences of climate change or variability using one or more parametric or nonparametric statistical hypothesis tests. Two types of data are generally available for analysis, and they are (1) point data (i.e., rain gauge measured precipitation data) and (2) gridded data (i.e., radar or satellite estimated precipitation data). Gridded data are often available as a spatially interpolated product at a specific spatial resolution. The following steps are primarily recommended for analysis of point precipitation data:

- Collect and evaluate the precipitation data and assess missing data lengths, nonhomogeneity issues, and erroneous or anomalous data using station or site metadata.
- Fill missing data with the help of appropriate spatial or temporal interpolation methods using available data from single or multisensor precipitation estimates.
- Check the homogeneity of data after any infilling and note change points (if any) in the time-series using visual diagnostics and appropriate statistical tests.
- Apply corrections to precipitation estimates (i.e., infilled data) and reevaluate the homogeneity of the time-series if needed.
- Identify temporal windows for climate variability analysis or two or more nonoverlapping periods.

- Develop new or use available indices that can help characterize the changes in the variables (or time-series of variables).
- Select statistical methods for analysis of these indices: examples of these include statistical inference tests (parametric and nonparametric).
- Identify, select, and execute parametric and nonparametric tests.
- Report statistically significant results from inference and trend analysis tests.
- Assess the influences of climate variability and change on precipitation in space and time.
- Understand and document potential implications associated with the influences (noted in the previous step) of climate variability and change on hydrologic design and water resources management.

Gridded precipitation data based on spatially interpolated estimations from point ground-based observations from single-sensor (i.e., rain gauge) or multisensor estimates (e.g., radar- and satellite-based combined estimates) can be used for evaluation of spatial precipitation characteristics. It is important to note that using such data may result in underestimation of higher-end extremes and overestimation of lower-end extremes. The data needs to be corrected or homogenized before it is used in any study. If bias is noted, local and global bias corrections based on single or multiple rain gauge(s) can be adopted. Quantile-matching (QM)-based corrections can also be employed for bias corrections. Goly and Teegavarapu (2014) indicated that both point observations and spatially complete gridded precipitation datasets are valuable for analysis. However, inferences based on gridded precipitation data should be interpreted with an abundance of caution.

3. CLIMATE VARIABILITY IMPACTS ON PRECIPITATION CHANGES: INFLUENCES OF COUPLED OCEANIC AND ATMOSPHERIC OSCILLATIONS

Characterizing of extreme precipitation events considering the magnitude, duration, and location is crucial for the design and long-term operation of water resources and hydrological systems (Teegavarapu et al., 2013). Several studies (Cronin, 2010; Enfield et al., 2001; Lyon and Barnston, 2005; Tootle et al., 2005) have confirmed the link between teleconnections described as associations between climatic variations (anomalies) at distance locations (Glantz, 2001; Trimble et al., 2005) and extreme precipitation events. Some of the major teleconnections driven by coupled oceanic-atmospheric processes (or oscillations) affecting the precipitation patterns around the globe are: Atlantic Multidecadal Oscillation (AMO) (Enfield et al., 2001; Curtis, 2008), El Niño-Southern Oscillation (ENSO) (Ropelewski and Halpert, 1987), Pacific Decadal Oscillation (PDO) (Schneider and Cornuelle, 2005), and North Atlantic Oscillation (NAO) (Cronin, 2010). A comprehensive review of these oscillations is provided by Teegavarapu (2013b).

Variability of extreme precipitation events in several parts of the world was investigated by several researchers. These studies have focused on a specific region of the world with or without teleconnections that are used to explain the variability. Long-term historical data available in several research studies were used to assess these trends. Studies relevant to variability assessments in the United States were done by Karl et al. (1995) and Groisman et al. (1999). Goswami et al. (2006) and Manton (2001) focused on trends in heavy precipitation in Indian subcontinent and studies relevant to Southeast Asia and South Pacific. Trends in extreme rainfall and dry events in Australia were investigated by Suppaih and Hennessy (1998). Similar studies are focusing on Europe (Haylock and Goodess, 2004), Japan (Iwashima and Yamamoto, 1993), and Sweden (Hellstrom and Malmgren, 2004). Recently the spatiotemporal variability of dependence among precipitation extremes was investigated over entire South America for the period 1940–2004 by Khan et al. (2007).

Several research studies have reported changes in extreme precipitation and flood events due to possible reasons associated with global warming. Changnon and Kunkel (1995) examined trends in the heavy multiday precipitation events associated with the floods for the period from 1921 to 1985 and found upward trends in both floods and heavy precipitation events over portions of the central United States. Karl and Knight (1998) noted about 8% increase in precipitation across the contiguous United States since 1910 which was primarily due to increases in the intensity of heavy to extreme precipitation events. Kunkel et al. (1999) found statistically significant upward trends in 1-year return period, 7-day duration events of about 3% per decade and in 5-year, 7- day events of about 4% per decade for the period from 1931 to 1996. Groisman et al. (2001) reported a 50% increase in the frequency of extreme precipitation days (with precipitation exceeding 101.6 mm) in the upper midwest United States during the 20th century.

Large-scale multidecadal Sea Surface Temperature (SST) anomalies have had significant impacts on regional and global climate. Linkages of SSTs with the variability in the regional hydrologic phenomenon, such as temperature, precipitation, streamflows, and floods and droughts, have been well documented. McCabe et al. (2004) reported that the frequencies of drought in the conterminous United States are highly correlated with the PDO and AMO. They concluded that these two components explain about 52% variance in droughts over the conterminous United States. Rogers and Coleman (2003) showed that the high (low) streamflow in the upper Mississippi River basin are consistent with cold (warm) phases of the AMO winter signal. Strong relationship between SSTs and variability in precipitation over the United States and other regions of the world has been discussed in several research studies (Enfield et al., 2001; Goodrich and Ellis, 2008; Lachniet et al., 2004). Enfield et al. (2001) demonstrated that the cool (warm) AMO phases are directly related to the above (below) normal rainfall over most of the United States. They also showed that the inflows to the Mississippi River basin, in the eastern United States, varies by about 10% and the inflow to the Lake Okeechobee, in south Florida, varies by about 40% in response to the changes in AMO phases. Strong correlation of El Niño (La Niña) southern oscillation phases with the increase (decrease) in precipitation and snowfall over the United States and Canada has been documented in several research findings (e.g., Groisman and Easterling, 1994; Groisman et al., 1994; Latif and Barnett, 1994). Teegavarapu et al. (2013) and Goly and Teegavarapu (2014) have evaluated the impacts of AMO and ENSO on changes in precipitation extremes and several characteristics in southeastern region of the U.S.

In addition to precipitation, multiple studies have demonstrated a strong relationship between streamflow variability and the atmospheric-oceanic circulations (Rogers and Coleman, 2003; Kahya and Dracup, 1993; McCabe, 1996). A strong influence of ENSO on streamflow patterns over the western United States has been documented in numerous research findings (e.g., Clark et al., 2001; Kahya and Dracup, 1993; Tootle et al., 2005; Dettinger and Diaz, 2000). Tootle et al. (2005) found a strong influence of ENSO on streamflow variability over the western United States, while a weaker influence due to other atmospheric-oceanic circulations such as PDO, NAO, and AMO was also reported. McCabe et al. (2004) reported the influence of AMO and PDO on the droughts over the United States. Analyses of precipitation data are vital for various hydrologic design and management applications. However, variability due to SSTs should be considered while performing analyses of long-term data. Long-term analyses of precipitation data are also required to study the changes due to global warming. Increase in frequency of heavy-precipitation events over the eastern United States has been predicted under the warmer climate scenarios. Therefore, long-term trend analyses of extreme events, along with SSTs-related variability should be investigated. Availability and quality of long-term precipitation data required to perform such analyses are often limited. Most of the long-term precipitation data contain gaps due to instrument malfunctioning, the inability of the operator to collect data, discontinuity, etc. Continuous time-series data is a preliminary requirement before analyses of long-term data can be performed.

4. CLIMATE CHANGE: INFLUENCES ON PRECIPITATION EXTREMES AND CHARACTERISTICS

Precipitation events of a specific magnitude that lead to catastrophic floods are evaluated by many research studies aimed at understanding the influence of climate change and variability of these events. The Intergovernmental Panel on Climate Change (IPCC) fourth (IPCC, 2007) and most recent fifth (IPCC, 2018) assessments indicate the increases in the frequency of heavy precipitation events (frequency or proportion of total precipitation from heavy falls) over most areas are indicated likely based on the trends in the later part of 20th century, especially after 1960. The human contribution to the trends in these events is described as most likely in the past IPCC assessments and several studies. Evidence to changes in the global hydrologic cycle was suggested by Huntington (2006) as an ongoing intensification of the water cycle and increased precipitation in the past few decades (Wentz et al., 2007). Lohmann (2008) indicates anthropogenic influences have shown to have both increasing and decreasing effect on convective precipitation in many regions of the world.

Climate-induced changes in precipitation intensity and rain-snow partitioning are two critical issues affecting future water resources management (Nicholls et al., 2008). The IPCC in its fourth assessment has concluded that increases in the amount of precipitation are very likely in the high latitudes, while decreases are likely in most subtropical land regions (IPCC, 2007). Climate change models suggest an increase in global average annual precipitation during the 21st century, although changes in precipitation may vary from one region to another (USEPA, 2008). Min et al. (2011) showed that human-induced increases in greenhouse gases have contributed to the observed intensification of heavy precipitation events. They indicate changes in extreme precipitation projected by models, and thus the impacts of future changes in extreme precipitation are underestimated as models seem to underestimate the observed increase in heavy precipitation with warming.

5. MULTIMODEL, MULTIPLE SCENARIO-BASED PROJECTIONS FOR FUTURE

Assessment of changes in precipitation extremes and characteristics will require the use of outputs from General Circulation Models (GCMs) that are downscaled to obtain an estimate of an essential climate variable of interest (i.e., precipitation) at a specific spatial scale. Statistical and dynamic downscaling models have been developed in the past couple of decades to assess the hydrologic impacts of climate change by projecting the station-scale hydrological variables from large-scale atmospheric variables simulated by GCMs (Goly et al., 2014; Teegavarapu, 2013b; Teegavarapu and Goly, 2018). Climate impact studies on a regional scale typically use the simulation results from GCMs to assess past and future precipitation trends. GCMs account for various internal and external atmospheric forcings, including various scenarios for socioeconomic, greenhouse gases and population changes.

Challenges of downscaling precipitation using different statistical downscaling models and optimal selection of predictor variables for these models were discussed by Goly et al. (2014) and Teegavarapu and Goly (2018). Goly et al. (2014) used simulations from a third-generation single GCM, the Canadian Centre for Climate Modelling and Analysis (CCCma) Coupled Global Climate Model, version 3 (CGCM3), with an A1B scenario for their analysis of downscaled precipitation in Florida, USA. The Special Report on Emissions Scenarios (SRES) scenario A1B is selected for the estimation of monthly rainfall at different rain gauge stations. The CGCM3 represents a 720-ppm stabilization experiment. Although GCM runs are available at time scales shorter than 1 month, there is little confidence in those GCM outputs (Goly et al., 2014). Teegavarapu (2013b) discussed the issues with downscaled precipitation from multiple GCMs using Bias-Correction Spatial Disaggregation (BCSD) approach (BCSD, 2018). Results from these GCMs were available to the public through Coupled Model Intercomparison Project (CMIP), an online portal (BCSD, 2018). The most recent phase of this project, phase 5, referred to as CMIP5 provides downscaled monthly and daily precipitation products for the U.S. Currently, 134 downscaled GCM simulations are available that are obtained by Bias Corrected Constructed Analog (BCCA) methodology. The current CMIP, the CMIP5, provides simulation from 21 GCMs, four emission scenarios referred to as representative concentration pathways (RCPs): four RCPs (viz., RCP2.6, RCP4.5, RCP6.0, RCP8.5), and 12 ensemble runs.

Details of these four RCPs are available from Moss et al. (2010). Bias correction of downscaled projections is a critical step and will involve the use of different quantile matching-based methodologies (Teegavarapu, 2013b). Intergovernmental Panel on Climate Change (IPCC) (IPCC, 2018) has provided guidance on the formulation of climate change scenarios.

Five criteria that should be met by climate scenarios if they are to be useful for impact researchers and policy makers are suggested by IPCC and are reported below verbatim from IPCC (2018):

- Criterion 1: Consistency with global projections. They should be consistent with a broad range of global warming projections based on increased concentrations of greenhouse gases. This range is variously cited as 1.4–5.8°C by 2100, or 1.5–4.5°C for a doubling of atmospheric CO_2 concentration (otherwise known as the "equilibrium climate sensitivity").
- Criterion 2: Physical plausibility. They should be physically plausible; that is, they should not violate the basic laws of physics. Hence, changes in one region should be physically consistent with those in another region and globally. In addition, the combination of changes in different variables (which are often correlated with each other) should be physically consistent.
- Criterion 3: Applicability in impact assessments. They should describe changes in a sufficient number of variables on a spatial and temporal scale that allows for impact assessment. For example, impact models may require input data on variables such as precipitation, solar radiation, temperature, humidity, and windspeed at spatial scales ranging from global to site and at temporal scales ranging from annual means to daily or hourly values.
- Criterion 4: Representative. They should be representative of the potential range of future regional climate change. Only in this way can a realistic range of possible impacts be estimated.
- Criterion 5: Accessibility. They should be straightforward to obtain, interpret, and apply for impact assessment. Many impact assessment projects include a separate scenario development component which specifically aims to address this last point.

Once the bias corrected downscaled projections from GCMs are available at a spatial resolution of interest, the precipitation time-series can be analyzed to:

- Evaluate trends in precipitation extremes and changes in distributional characteristics over time.

- Assess changes in precipitation characteristics over time using constant or moving temporal windows.
- Estimate extreme events in future with a prespecified return period.
- Establish bounds on the precipitation extremes to help in hydrologic design under nonstationary climate.
- Assess the risk of possible occurrences of rare extremes.

The following sections will provide details of different indices and other metrics that can be used for changes in precipitation time-series due to climate variability and change.

6. EVALUATION OF TRENDS

Trends in precipitation extremes and characteristics defined by different extreme precipitation indices can be evaluated using parametric and nonparametric methods discussed in Chapter 1. The significance of temporal trends present in precipitation extremes can be evaluated using the Mann-Kendall (MK) trend test (Hirsch and Slack, 1984; Lettenmaier et al., 1994; Yue et al., 2002) described in Chapter 1. The MK test has been used in several past studies (Dufek and Ambrizzi, 2008; Pal and Al-Tabbaa, 2009; Keggenhoff et al., 2014) for evaluation of trends in extreme precipitation indices. A recent study by Teegavarapu and Nayak (2017) has focused on evaluation of influence of missing precipitation data on long-term trends and extreme precipitation indices. They have analyzed over 50 sites in Florida for evaluation of trends in long-term precipitation data for biases due infilled precipitation data on the detection of long-term change using parametric and nonparametric statistical techniques. Their study indicated that analysis of precipitation time-series with missing data gaps in filled and unfilled data will lead to different conclusions about precipitation extremes and characteristics in different temporal windows. Underestimation of precipitation higher-end extremes was noted when filled data was used. Caution should be exercised when data from temporal windows which coincide with phases of oceanic and atmospheric oscillations are used for making inferences about changes to precipitation extremes in those phases and in general about climate variability influences on those extremes.

7. ISSUES RELATED TO MISSING PRECIPITATION DATA

Availability and quality of long-term gap-free precipitation at different temporal resolutions are required to evaluate changes in extremes and characteristics. Long-term precipitation datasets in many instances contain gaps due to a variety of other reasons. Statistical analyses of long-term precipitation data require filling of these data gaps with appropriate values using spatial or temporal interpolation to obtain a serially complete dataset. Data filling techniques are normally selected based on location and desired accuracy of estimates required for the hydrologic analyses. Content for this section is adopted from a recent survey of methods for spatial and temporal estimation and analysis of precipitation by Teegavarapu (2016).

The mechanisms by which missing data occur are defined as: missing at random (MAR), missing completely at random (MCAR), and missing not at random (MNAR) (Little and Rubin, 2002). In many missing precipitation data estimation studies, gaps can be attributed to being as data missing completely at random (MCAR) thus enabling estimation of missing precipitation data using temporal and spatial interpolation methods. The former relies on the persistence defined by strong serial autocorrelation, and the latter depends on reliable chronologically serial observations at rain gauges other than the one at which missing data exists. Estimation using spatial interpolation requires chronological pairing (CP) of data at all the available rain gauges. Strong persistence or autocorrelation in rainfall observations benefits temporal interpolation. This refers to infilling of data using temporal interpolation and two procedures suggested by Buuren (2012) as last observation carry forward (LOCF) and baseline observation carry forward (BOCF) approaches can be used. BOCF method is also referred to as single imputation (i.e., replacing a missing value with one value as opposed to multiple imputation where multiple values are suggested for filling).

Conceptually simple or naïve missing data estimation methods use rainfall information in space and time from all rain gauges to estimate missing data at the base station (i.e., site or gauge at which missing data exists). Gauge mean estimator (GME), single best estimator (SBE), and climatological mean estimator (CME) methods use an average value of observations from the nearby rain gauges, observations from one rain gauge, and mean of historical observations available for that specific temporal slice at the rain gauge which is known to be having the missing data, respectively. The optimal number of rain gauges in GME and a most representative rain gauge in SBE can be obtained by optimization or by use of any proximity criterion that is Euclidean or a predefined distance metric or using the correlation between the base station and any other station. The accuracy of these simple estimation methods is debatable when the rainfall variability is high in the region of interest. These methods serve as benchmark methods against whom more advanced spatial interpolation methods are compared and evaluated.

A distance-dependent approach such as Inverse Distance Weighting Method (IDWM) (Simanton and Osborn, 1980; Wei and McGuinness, 1973) is widely used in many missing precipitation estimation studies. In the United States, especially in the operational hydrology literature, IDWM is often referred to as the National Weather Service (NWS) method. The IDWM is routinely used for estimating missing rainfall data (ASCE, 1996) and several variants of this method are used in different studies. Hodgson (1989) modified IDWM to include a learned search approach that reduces the number of distance calculations. To incorporate topographical aspects, Shepard (1968) proposed a modified IDWM that is referred to as a barrier method. The quadrant method (McCuen, 2004) uses the nearest gauge to the base gauge in each quadrant for estimation of missing precipitation data in weighting scheme similar to IDWM. Nearest gauge in each quadrant is defined by Euclidean distance. Normal ratio method (Teegavarapu and Chandramouli, 2005; McCuen, 2004; Tung, 1983) is another conceptually simple method for estimation of missing data that relies on the assumption related to average annual precipitation values at sites.

Kriging or variance-dependent stochastic interpolation and its several variants have been used to estimate missing precipitation data (Ashraf et al., 1997; Teegavarapu and Chandramouli, 2005), as well as to interpolate precipitation from point measurements (Dingman, 2011; Vieux, 2001) and for aerial mapping of precipitation (Karnieli, 1990). Regression and time-series analysis methods (Salas, 1993), global interpolation methods that use trend surface analysis and regression (Chang, 2006), regression models using elevation information (Lloyd, 2009), locally weighted polynomials (Regonda et al., 2006), thin-splines (Xia et al., 1999), and local and global trend surface models (Teegavarapu, 2012) have been used in the past for estimation of missing precipitation data. Other techniques such as polynomial interpolation by least-squares and Lagrange approach, inverse distance, multiquadric interpolation, optimal interpolation, and kriging were also used for estimation of missing data by Guillermo et al. (1985).

Teegavarapu and Chandramouli (2005) developed and investigated several variants of inverse distance method and data-driven methods for estimation of missing precipitation records at a site. These methods are coefficient of correlation weighing method (CCWM), artificial neural network estimation method (ANNEM), modified Thiessen polygon method, nearest neighbor weighting method (NNWM) and its' revised version, and kriging estimation method (KEM). All the methods evaluated were data intensive and sensitive, with kriging estimation method being computationally most intensive of all the methods. Use of conceptually acceptable surrogate measures for distances and improvised weighting factors improved the estimates from these variants. CCWM has proven to be a better weighting method compared to other interpolation methods in several studies (e.g., Kim et al., 2008; Westerberg et al., 2009) The kriging method is considered a reliable interpolation technique (O'Sullivan and Unwin, 2010), but is plagued by several limitations. These include selecting the appropriate semivariogram to characterize spatial variability, assignment of arbitrary values to the sill and nugget parameters and distance intervals, observation value-insensitive variance estimates, and the computational burden to interpolate the surfaces.

Universal functional approximators such as artificial neural networks (ANN) were used for fitting a semivariogram model using the raw data in ordinary kriging to estimate missing precipitation data by Teegavarapu (2007). The use of ANN eliminates the need for the predefined authorized semivariogram models to capture the spatial variation of data and the trial-and-error process involved in estimation of semivariogram parameters. Use of modular neural networks for estimation of missing precipitation is reported by Kajornrit et al. (2012). Kuligowski and Barros (1998) report improved estimates of missing data using ANN compared to those from distance-weighted and linear regression models at a number of rain gauge sites. A similar conclusion about improvements in estimates was reported by Teegavarapu and Chandramouli (2005). Optimal functional forms for estimation of missing precipitation records using genetic algorithms were developed by Teegavarapu et al. (2009). Association Rule Mining (ARM) - based spatial interpolation (Teegavarapu, 2009) is used to improve the precipitation estimates provided by traditional and improved deterministic and stochastic spatial interpolation techniques. The ARM methodology besides offering insights into the spatiotemporal precipitation data patterns and the associations among observations also helped in addressing one major ubiquitous limitation of all spatial interpolation techniques in accurately estimating missing precipitation records. Considerable improvements in the estimates were achieved when the ARM was used in conjunction with the interpolation techniques. Copula-based methods for infilling are found to be superior compared to a number of interpolation methods (nearest neighbor, linear regression, fuzzy rules, ordinary kriging, multiple linear regression [MLR], MLR using the expectation maximization [EM] algorithm) in a study by Bardossy and Pegram (2014). Landot et al. (2008) used a copula-based method for infilling long-term gaps in precipitation data in Florida with the help of support vector machine (SVM). However, the copula-based method reported by Landot et al. (2008) was found to be inferior to CCWM and linear optimization method in a study by Teegavarapu et al. (2018). Presti et al. (2010) investigated simple substitution, parametric regression, ranked regression, and rank-invariant method of linear and polynomial regression analysis (Theil method). They concluded that Theil method and simple substitution method provided better results among all those compared.

Optimization methods applicable to a specific range of precipitation values and cluster-based approaches in space and time were developed by Teegavarapu (2012). Models using mixed integer nonlinear programming formulations (MINLP) were developed for estimation of missing precipitation with the use of binary variables for selection of rain gauges that participate in the spatial interpolation process and also in the process of selection of the optimum cluster of rain gauges for estimation of missing data. Several variants of these mathematical programming formulations are proposed by Teegavarapu (2012) to improve the estimates of missing precipitation data. The variants include (1) inverse distance weighting method with an optimal number of gauges; (2) optimal quadrant method; (3) inverse distance weighting method with optimal parameters and number of gauges; and (4) linear weighting method with optimal clusters of rain gauges. These variants investigated in their study provided improved estimates of precipitation data compared to those obtained from different traditional distance-based weighting and kriging methods. A nonlinear mathematical programming model using binary variables was used by Teegavarapu and Pathak (2008) and Teegavarapu et al. (2010) to infill missing precipitation records using radar-based rainfall estimates. Clusters of radar-based precipitation estimates from grid surrounding the grid with the gauge with missing values are optimally selected for the infilling process.

The similarity of two observation datasets based on real or categorical attribute values can be evaluated using proximity metrics based on numerical taxonomy concepts (Ling, 2010). These metrics were used to assign weights in the interpolation schemes to estimate missing data by Teegavarapu (2014) for the estimation of missing precipitation data. A recent study by Pappas et al. (2014) provided a conceptually simple method using nearest neighbors (observations in time at one site) for estimation of missing data. Corrections of spatial interpolation–based precipitation estimates considering rain or no-rain conditions at the nearest rain gauge selected by any proximity metric or correlation can be used. Corrections suggested by Teegavarapu (2009) using data mining–based association rules were proved to be useful. However, overestimation of precipitation magnitudes in many instances was noted. The support vector machine (SVM)-based approach was used to select rain gauges for correction by Teegavarapu et al. (2018). The single best estimator (SBE) method and its optimal variant are developed by Teegavarapu et al. (2011) and Teegavarapu (2014), respectively, for post corrections of missing data estimates at a gauge. Preserving site-specific statistics along with an accurate estimation of missing data is essential when data gaps at a gauge are filled. Postcorrection methods using quantile-quantile (Q-Q), equi-ratio quantile matching (ERQM), and single best estimation methods were suggested by Teegavarapu (2013b). These methods preserved site statistics and also provided improved estimates of several WMO-based extreme precipitation indices that are closer to observed series than those obtained without these corrections. Teegavarapu (2013b) and El-Sharif and Teegavarapu (2012) proposed and evaluated several multiobjective optimization formulations for preserving site-specific statistics and correlations among different sites and spatial variability of precipitation data evaluated using a variogram.

Several methods ranging from conceptually simple to computationally intensive are available that can be used for estimation of missing precipitation data. Selection and use of one or more methods for estimation is an exhaustive task and will depend on several constraints and is also dictated by numerous objectives related to any specific study. The focus, in general, should be on the selection of those methods that are conceptually sound, field tested, and robust. The geographical region of study, ease of method development by modelers, transferability, and utility of a part or complete computational code from a similar model available elsewhere, practical applicability, data availability, and time constraints are a few important factors that are considered in the selection process. Also, the objectives and goals set forth by study and agency needs will primarily direct the identification and finally the selection process. Computationally efficient algorithms are required for implementation of improved estimation methods based on auxiliary data other than precipitation datasets. Surrogate measures of correlation among observations are required for applying weighting techniques. Use of correlation coefficients (Teegavarapu and Chandramouli, 2005), statistical distance (Ahrens, 2006), and proximity metrics (Teegavarapu, 2014) are few recommendations for the development of conceptually simple weighting methods. Universal functional approximation methods such as artificial neural networks (ANN) are viable techniques for estimation of missing precipitation. However, overfitting, lack of generalization, unrealistic numeric results, replicability, and repeatability are some of the main concerns of these techniques. Reliable data-driven, artificial intelligence (AI) or statistical techniques are required for prediction of "rain" or "no-rain" conditions. These techniques precede the application of spatial interpolation techniques for estimating missing precipitation records.

Rain-radar relationships need to be characterized via functional forms to effectively utilize them in estimating missing precipitation. Optimum clustering of gauges for efficient and robust estimation of missing precipitation records can be identified. Mathematical programming formulations with binary variables can be used for this purpose for determination of an optimal number of sites (i.e., neighborhood size) and clusters. Spatial and temporal variability of rainfall processes and local and regional climatology are essential factors that need to be considered when estimating missing data at a site. In few data-sparse situations, naïve methods described in this chapter are the only way out. In general, filling missing precipitation data more than 5% of the entire dataset leads to underestimation of higher-end extremes and an increase in bias in the estimation of precipitation extremes (Teegavarapu et al., 2011, Teegavarapu and Nayak, 2017). Multiple imputation (infilling) methods (Buuren, 2012) based on bootstrap sampling and optimal interpolation (Teegavarapu, 2013b) are useful in quantifying the uncertainties associated with estimates provided by spatial interpolation methods. Software focused on estimation of missing data is rarely generic. Variance deflation, over- and underestimation of lower- and higher-end extremes, respectively, and alteration of statistical distributions are inevitable consequences of any spatial interpolation method (Goly and Teegavarapu, 2014; Teegavarapu, 2013b; Teegavarapu, 2012) often used for generation of gridded datasets. Point and spatially complete gridded precipitation datasets are equally valuable for analysis of precipitation extremes and characteristics. Justifiable inferences about variability in precipitation characteristics and extremes can be drawn when both the datasets are used independently (Goly and Teegavarapu, 2014).

Application of any spatial interpolation method for infilling missing precipitation data will require time-consistent precipitation measurements at all the rain gauges. If one or more gauges are missing data, they need to be filled first before any spatial interpolation method can be used for estimation. The filling can be carried out using a conceptually simple estimation method, temporal interpolation, or data from sources other than rain gauge if available. If one or more rain gauges have a substantial amount of missing data, then parameters of the interpolation method need to be reestimated using the available rain gauges with serially complete precipitation records. If the amount of data missing is negligible and missing data can be explained as MCAR, and the time-series can be used for analysis, then listwise and pairwise deletion of data methods common in social sciences arena (Enders, 2010) can be used. However, if missing data are not estimated or gaps in data are not filled, meaningful conclusions about the data cannot be made for any hydroclimatic studies.

7.1 Changes to Moments and Distributions of Data

Data filling can lead to changes in the probability distribution of data at a site and introduce significant biases when event-based analysis (such as daily or hourly extreme precipitation analysis) is performed. At temporal resolutions of a day or less, spatial interpolation methods alter the probability distribution of data, changes the autocorrelation structure, and dry and wet spell transitions (Teegavarapu, 2014). Nonexceedance probability curves (or cumulative distribution function plots) of filled and unfilled data at four rain gauges with different lengths of the data field by an optimal spatial interpolation are shown in Fig. 2.1. It is evident from the plots that as the lengths of data filled increase, the deviations from filled and unfilled data become large. Many research efforts have focused on analyzing the trends in precipitation data, especially extremes, but biases introduced in these trends due to data infilling techniques are rarely investigated. A study by Teegavarapu (2014) used a fixed length of missing data and analyzed the impact of infilling on precipitation characteristics and extremes. In a recent study, Teegavarapu and Nayak (2017) employed: (1) varying lengths of missing data for evaluation of statistically significant trends; (2) resampling methods to quantify the impacts of infilling on evaluation of extremes; and (3) nonparametric hypothesis tests for assessment for differences in distributions of filled and unfilled precipitation datasets for different lengths of gaps.

7.2 Stationarity Assessment

According to WMO (2009) stationarity means that, excluding random fluctuations, the data-series is invariant with respect to time. Precipitation

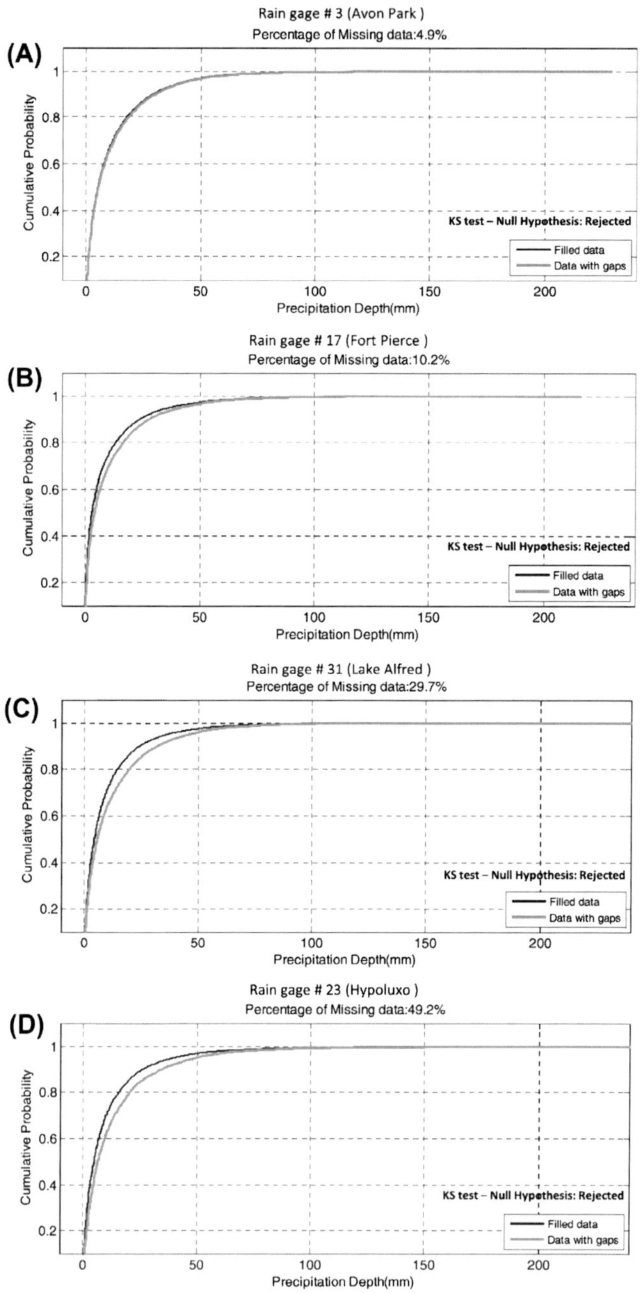

Figure 2.1 A–D Cumulative distribution function plots of observed and filled precipitation data-series at four gauges in Florida. *(From Teegavarapu, R.S.V., Nayak, A., 2017. Evaluation of long-term trends in extreme precipitation: implications of infilled historical data and temporal-window based analysis. J. Hydrol. 550, 616–634.)*

extremes are evaluated as block maxima or a single extreme value per year. Annual precipitation extremes time-series should be checked for stationarity when used for frequency analysis and derivation of design rainfall intensities. This is an important element of the statistical analysis conducted after the initial phase of the data collection. Trend analyses should be conducted for all the available annual extreme data for all durations and tests for statistically significant trends should be based on MK or other tests. According to Ashkar (1996), two important forms of nonstationarity are introduced in a time-series (e.g., streamflow time-series) and they are (1) jumps and (2) trends. Another form of nonstationarity is the existence of cycles that are associated with long-term climatic oscillations. Statistical tests for detecting stationarity include the Mann–Whitney U test for jumps and Wald-Wolfowitz test for trend (Bobee and Ashkar, 1991).

Assessment of stationarity or nonstationarity of any hydrometeorological variable time-series is also an essential task to confirm the nonexistence or existence of climate change or anthropogenic influences, respectively. The first step in the assessment of stationarity of any hydrometeorological time-series is to visually evaluate the time-series plot with components such as level (average value), trend, seasonality, and noise (any random variation). Presence of trend and changes in the mean and variance of the data in different time intervals suggests a sign of nonstationarity. Seasonal patterns at fixed and known frequency may also affect the stationarity of time-series. Also, the presence of seasonality does necessarily make a time-series nonstationary. Stationarity is defined by Kundzewicz and Robson (2000) as a property of a random variable whose statistical moments do not change with time (strict stationarity—first and second moments are constants with time, weak stationarity—first moments are constant with time). Also, the autocovariance does not depend on the time when the time-series is stationary. A time-series is stationary if its characteristics are time invariant.

Nonstationarity in a time-series can occur due to changes or jumps in mean and nonconstant variance (i.e., referred to as heteroskedasticity) over time. The time-series can be stationary in the mean if there is no change in the mean value over time and stationary in the variance if there is constant variance over time. Gujarati (2004) states that "a stochastic process is said to be stationary if its mean and variance are constant over time and the value of the covariance between the two time periods depends only on the distance or gap or lag between the two time periods and not the actual time at which the covariance is computed."

A preliminary visual evaluation of stationarity of any time-series can be carried out using the following plots:

1. Run sequence plot
2. Autocorrelation plot (or Autocorrelation function [ACF] plot)
3. Partial autocorrelation function plot (PACF)

A run sequence or time-series plot will reveal any possible upward or downward trends, changes in the mean (i.e., jumps in the mean), and the variance and seasonal variations. ACF provides serial autocorrelation values at different lags and shows the time-dependence of time-series observations. The ACF for a series with a trend will show a slow decrease in the positive autocorrelations as the lags increase. For time-series with seasonality, the autocorrelation values peak at seasonal time lags or at lags linked with seasonal variations. A large spike in the autocorrelation value in a PACF plot may suggest that the series is nonstationary. A time-series with the presence of stable seasonality can be regarded as cyclostationary (Gardner et al., 2006).

Trend analysis using Spearman's Rho (SR) or variants of MK tests can be used to confirm the existence or nonexistence of trends in time-series data. A time-series with statistically significant either upward or downward trend is nonstationary, and is stationary when there is a lack of any statistically significant trend in a time-series. Also, a time-series is strictly stationary if all moments of its probability distributions are invariant over time (Gujarati, 2011). To confirm stationarity or nonstationarity, time-series can be broken into two or more nonoverlapping subseries of equal or unequal lengths so that the characteristics of each of these subseries can be evaluated. Summary statistics from these subseries can be compared visually and with statistical tests for changes in statistics from one time period to another. Parametric and nonparametric tests that can compare a statistic between two groups (i.e., two subseries) or across the groups can be used. Parametric statistical hypothesis tests such as student's *t*-test and two sample *t*-test described in Chapter 1 can be used when several distribution-specific requirements are fulfilled. Mann—Whitney U test, a nonparametric test, can be used in instances if parametric tests cannot be applied.

The author of this chapter has evaluated the use of several nonparametric hypothesis tests to assess the stationarity of time-series when time-series data is broken into subseries of equal or unequal lengths. The tests are used to confirm whether the statistical and distributional characteristics of the subseries are different or not different. The tests that can be used for evaluation of stationarity are (1) Kruskal—Wallis (KW) and Ansari-Bradley

(AB); and (2) two-sample Kolmogorov–Smirnov (KS) (Massey, 1951). The KW test is also referred to as Kruskal–Wallis H test as the statistic calculated for this test is referred by this alphabet (i.e., H). The null and alternative hypotheses for KW test are that the data in each subseries come from the same distribution, and data from not all subseries come from the same distribution. The null and alternative hypotheses for AB test are that the data from two subseries come from the same distribution and data come from distributions with same median and shape but different dispersions. The AB test is considered as a nonparametric alternative to the two-sample F-test used for assessment of equality of variances. The null and alternative hypotheses for KS test are that the data from two different subseries are from the same continuous distribution and the two subseries are from different distributions. KS test can be conducted for two subseries selected from a number of subseries derived from the main time-series. The number of KS tests needed is equal to ${}_{NS}C_2$, where NS is the number of subseries. The KS test can also be conducted with two subseries randomly selected from the complete time-series with different sizes (i.e., size defined by the number of elements in the subseries). Another alternative to two-sample KS test that can be used is the Cramer Von-Mises test.

A two-sample Chi-square test can also be used in place of two-sample KS test noting that the number of bins used for the test will affect the results. The null hypothesis for this test is that the two samples come from a common distribution and the alternative hypothesis is that they do not come from a common distribution. The number of bins used for both the samples in the test should be the same. Also, Levene test (Levene, 1960) as a diagnostic tool to evaluate equality of variances across multiple subseries can be conducted. The test is less stringent on the requirement of normality for the sample datasets compared to Bartlett test. In case of Levene test the within group variability is based on the mean of the sample data of the group. The Brown and Forsythe test (1974) is an extension of Levene test which utilizes median or trimmed mean in addition to the mean used in Levene test. A simple visual diagnostic test for equality of variances across different groups is a plot of mean–standard deviation scatter plot.

Several tests are available to assess the stationarity of time-series which do not require partitioning of the series into subseries and they include: (1) Dickey Fuller or Augmented Dickey Fuller (ADF) test (Dickey and Fuller, 1979); (2) Kwiatkowski Phillips Schmidt Shin (KPSS) (Kwiatkowski et al., 1992) test; and (3) Phillips Perron (PP) test (Phillips and Perron, 1988) test. The null hypothesis of ADF test checks for the existence of unit root. The

KPSS test can be used to detect trend stationarity. The tests can help assess the null hypothesis that a time-series is trend stationary against the alternative that it is a nonstationary unit-root process. The PP test evaluates the null hypothesis that a series is trend stationary and the alternative hypothesis that series is a nonstationary unit-root process.

8. EVALUATION OF CHANGES IN PRECIPITATION EXTREMES

Extreme values from a precipitation time-series at single or multiple sites can be evaluated by defining the region or period of interest. If the changes in extremes over time need to be evaluated, the extremes can be obtained from a block of time (e.g., season, a year, or a decade or a period that coincided with a specific phase of natural cycles of climate) or from a number of observation sites in a region. In some studies, values above a specific threshold referred to peak over the threshold (POT) values are obtained and analyzed. Precipitation extremes are also evaluated considering the magnitudes, as well as the durations (i.e., time intervals of extreme storm events) during which extremes occur.

9. DESCRIPTIVE INDICES FOR PRECIPITATION EXTREMES

Descriptive indices for precipitation extremes were developed by Expert Team on Climate Change Detection and Indices (ETCDI) and are described in a report by World Meteorological Organization (WMO) (WMO, 2009). These indices describe specific characteristics of extremes, including frequency, amplitude, and persistence (WMO, 2009). A total of 11 indices for precipitation developed by ETCDI are provided in Table 2.1. The indices can be used for evaluation of trends and also for assessment of changes in two separate time periods. The notation and descriptions used for these indices are adopted from Teegavarapu (2013b). According to WMO technical regulations, standard normals are averages of climatological data computed for the following consecutive periods of 30 years: January 1, 1901 to December 31, 1930; January 1, 1931 to December 31, 1960; and so forth. A 30-year period needs to be specified for climate normal for calculation of a few indices provided in Table 2.1. These climatological baseline periods are nonoverlapping and the most recent WMO normal period is 1961–90 (Lu, 2010).

Table 2.1 Descriptive Indices for Precipitation Extremes

Index	Description
RX1day	Maximum one-day precipitation
RX5day	Maximum five-day precipitation
SDII	Simple daily intensity index
R10mm	Count of precipitation days with PR greater than 10 mm
R20mm	Count of precipitation days with PR greater than 20 mm
Rnnmm	Count of days where PR greater than a threshold value
CDD	Consecutive dry days (PR $<$ 1 mm)
CWD	Consecutive wet days (PR $\geq$ 1 mm)
R95pTOT	Total precipitation due to wet days ($>$95th percentile)
R99pTOT	Total precipitation due to extremely wet days ($>$99 percentile)
PRCPTOT	Total precipitation in wet days ($>$1 mm)

PR. Daily Rainfall.

The following definitions for the indices are adopted from WMO (2009).

$RX1day_j$ is the maximum 1-day precipitation (i.e., highest precipitation amount in 1-day period). If PR_{ij} is the daily precipitation amount on day i in period j. The maximum 1-day value for period j is given by Eq. (2.1).

$$RX1day_j = \max\{PR_{ij}\} \quad \forall j \tag{2.1}$$

$RX5day_j$ is the maximum 5-day precipitation (i.e., highest precipitation amount in 5-day period). If PR_{kj} is the precipitation amount for the 5-day interval k in period j, where k is defined by the last day. The maximum of 5-day values for period j is given by Eq. (2.2).

$$RX5day_j = \max\left\{\sum_{n=1}^{k} PR_{nj}\right\} \quad \forall j \tag{2.2}$$

SDII is the simple daily intensity index (i.e., mean precipitation amount on a wet day. Let PR_{ij} be the daily precipitation amount on wet day w (PR $\geq$ 1 mm) in period j. If w represents the number of wet days in j, then the simple precipitation intensity index is given by Eq. (2.3).

$$SDII_j = \frac{1}{W}\left\{\sum_{n=1}^{w} PR_{wj}\right\} \quad \forall j \tag{2.3}$$

R10mm is the number of heavy precipitation days (i.e., count of days where PR (daily precipitation amount) $\geq$ 10 mm). Let PR_{ij} be the daily precipitation amount on day i in period j. The index R10mm is given by Eq. (2.4).

$$R10mm_j = \sum_{i=1}^{nT} RC_i \quad \{RC_i = 1, \quad \text{if} \;\; PR_{ij} > 10 \text{ mm}; \quad \text{else } RC_i = 0\} \quad \forall j \tag{2.4}$$

R20mm is the number of heavy precipitation days (i.e., count of days where PR (daily precipitation amount) $\geq$ 20 mm). Let PR_{ij} be the daily precipitation amount on day i in period j. The index R20mm is given by Eq. (2.5).

$$R20mm_j = \sum_{i=1}^{nT} RC_i \quad \{RC_i = 1, \;\; \text{if} \;\; PR_{ij} > 20 \text{ mm}; \quad \text{else } RC_i = 0\} \quad \forall j \tag{2.5}$$

Rnnmm is the number of heavy precipitation days (i.e., count of days where PR (daily precipitation amount) $\geq$ user defined precipitation threshold (mm)). Let PR_{ij} be the daily precipitation amount on day i in period j. The index Rnnmm is given by Eq. (2.6).

$$Rnnmm_j = \sum_{i=1}^{nT} RC_i \quad \{RC_i = 1, \quad \text{if} \;\; PR_{ij} > \text{nn mm}; \quad \text{else } RC_i = 0\} \quad \forall j \tag{2.6}$$

CDD index is the consecutive dry days (i.e., the maximum length of a dry spell (PR < 1 mm)). Let PR_{ij} be the daily precipitation amount on day i in period j. The count of the largest number of consecutive days where $PR_{ij} < 1$ mm is given by Eq. (2.7).

$$CDD_j = \sum_{i=1}^{nT} RCC_i \quad \{RCC_i = 1, \;\; \text{if} \;\; PR_{ij} < 1 \text{ mm}; \quad \text{else } RCC_i = 0\} \quad \forall j \tag{2.7}$$

CWD index is the consecutive wet days (i.e., the maximum length of the wet spell (PR > *1 mm*)). Let PR_{ij} be the daily precipitation amount on day i in period j. The count of the largest number of consecutive days where $PR_{ij} > 1$ mm is given by Eq. (2.8).

$$CWD_j = \sum_{i=1}^{nT} RCC_i \quad \{RCC_i = 1, \text{ if } PR_{ij} > 1 \text{ mm}; \text{ else } RCC_i = 0\} \tag{2.8}$$

The indices R95pTOT and R99pTOT are calculated based on 95th and 99th percentiles of rainfall depths based on the normal (base period 1961–90) are given by Eqs. (2.9) and (2.10).

$$R95pTOT = \sum_{j=1}^{nT} PR_j \quad \{PR_j > RRn95\} \quad \forall j \tag{2.9}$$

$$R99pTOT = \sum_{j=1}^{nT} PR_j \quad \{PR_j > RRn99\} \quad \forall j \tag{2.10}$$

PRCPTOT is the total precipitation in wet days (i.e., PR (daily precipitation amount) ≥ 1 mm). Let PR_{ij} be the daily precipitation amount on day i in period j. The index PRCPTOT is given by Eq. (2.11).

$$PRCPTOT = \sum_{i=1}^{nT} PR_{ij} \quad \{\text{if } PR_{ij} > 1 \text{ mm}\} \quad \forall j \tag{2.11}$$

Chronologically continuous (i.e., without gaps) daily precipitation data is essential for accurate estimation of some of the extreme precipitation indices described in Table 2.1. For example, CDD, CWD, PRCPTOT, RX1day and SDII will require gap-free data. Some of the indices can be calculated for a year, season, or a specific number of months in a year. For example, the Rx1day index can be derived and can be evaluated for a precipitation time-series filled with gaps. Missing precipitation data, if exists, can be filled using spatial or temporal interpolation provided the infilling does not change the statistics of the time-series. The extreme precipitation indices discussed in this section were used recently in two studies (Teegavarapu et al., 2013; Goly and Teegavarapu, 2014) to evaluate statistically significant changes from one phase to another of two major coupled atmospheric and oceanic oscillations (viz., AMO and ENSO) influencing precipitation characteristics in the southeastern United States. All indices excepting two (viz., R95pTOT and R995pTOT) were used for evaluation of bias corrected estimates of daily precipitation data using spatial interpolation in a recent study (Teegavarapu, 2014). Fig. 2.2 shows a comparison of indices for observed, spatially interpolated (i.e., estimated), and bias corrected precipitation datasets. Bias corrections of the estimated precipitation data are carried using single best estimate (SBE), Quantile Matching (QM), and Equi-Ratio Quantile Matching (ERQM) methods.

In a recent study for evaluation of trends based on filled and unfilled daily precipitation time-series, Teegavarapu and Nayak (2017) used two indices, RX1day and Rnnmm. Two threshold precipitation values for Rnnmm index were considered, and they are 25 and 50 mm. Absolute differences in the median values of Rx1day calculated based on unfilled and filled data with respect to the percentage of missing data are shown in Fig. 2.3. The differences generally increase with the increase in the amount of missing data. The threshold values can be region-specific as defined by any regional hydrometeorological agency. For example, in case of India, different precipitation thresholds are qualitatively described based on the magnitude of rain. Different categories defined by Indian Meteorological Department (IMD) are: no rain (0 mm), very light rain (0.1−2.4 mm), light rain (2.5−7.5 mm), moderate rain (7.6−35.5 mm), rather heavy (35.6−64.4 mm), heavy rain (64.5−124.4 mm), very heavy rain (124.5−244.4 mm), and extremely heavy rain (244.5 mm).

10. EVALUATION OF CHANGES IN PRECIPITATION CHARACTERISTICS: STATISTICAL INFERENCE TESTS

Parametric and nonparametric tests elaborated in Chapter 1 can be used to ascertain statistically significant changes in precipitation extremes and characteristics in two different temporal windows (i.e., time periods). The temporal windows can be selected based on information about changes in the time-series data, or they could coincide with different phases of coupled oceanic and atmospheric oscillations cycles (e.g., AMO warm phase or cool phase). The parametric unpaired two-sample *t*-test can be used to evaluate the differences between the extreme precipitation indices and characteristics during different temporal windows at prespecified significance level as long as conditions related to this test are satisfied. The test requires normality of sample datasets. The normality of the datasets can be confirmed using visual checks based on normal probability plots (Benson, 1962; Mage, 1982) and several statistical hypothesis tests such as Lilliefors (1967), Jarque and Bera (1987), and Chi-square goodness-of-fit (Corder and Foreman, 2009). Visual tests alone are not recommended for confirmation of normality.

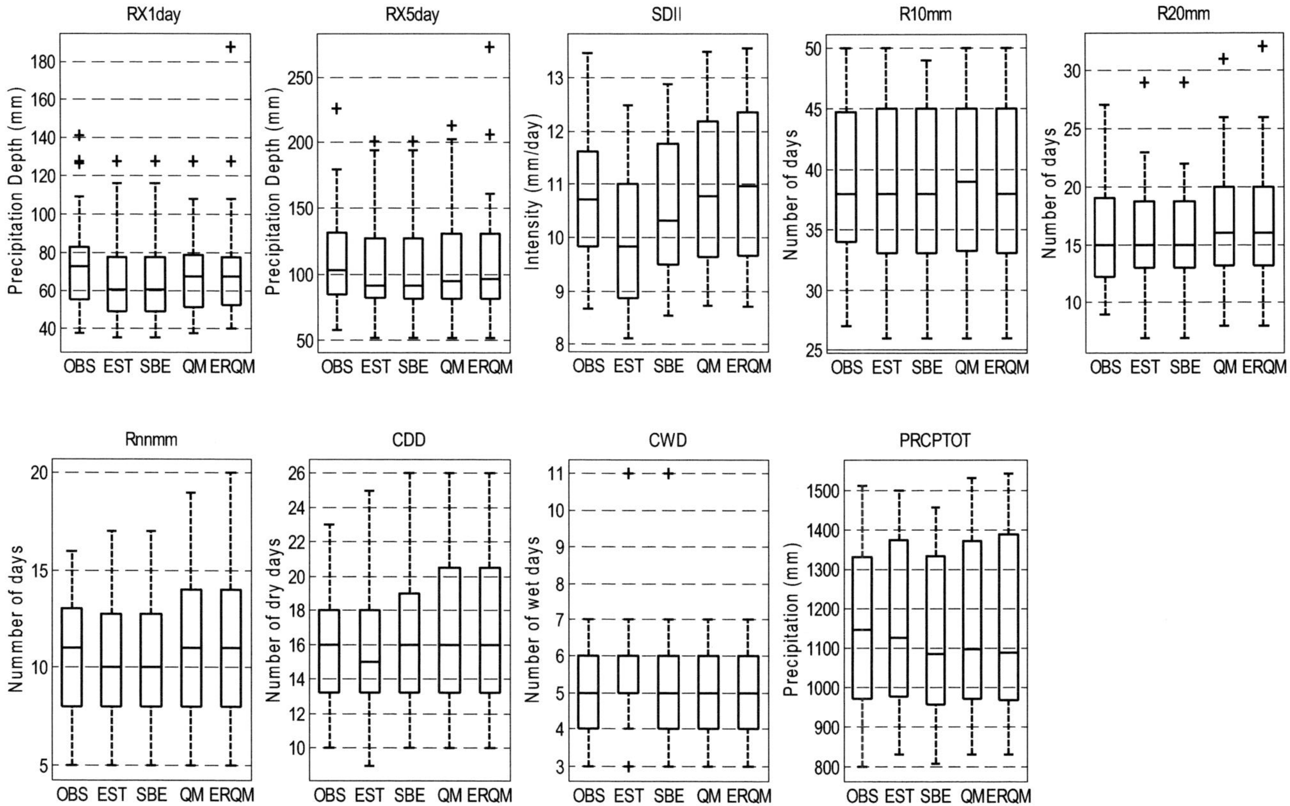

Figure 2.2 Variations of different precipitation extreme indices derived using derived and bias-corrected data. *(From Teegavarapu, R.S.V., 2014. Missing precipitation data estimation using optimal proximity metric-based imputation, nearest neighbor classification and cluster-based interpolation methods. Hydrol. Sci. J. 59 (11), 2009–2026.)*

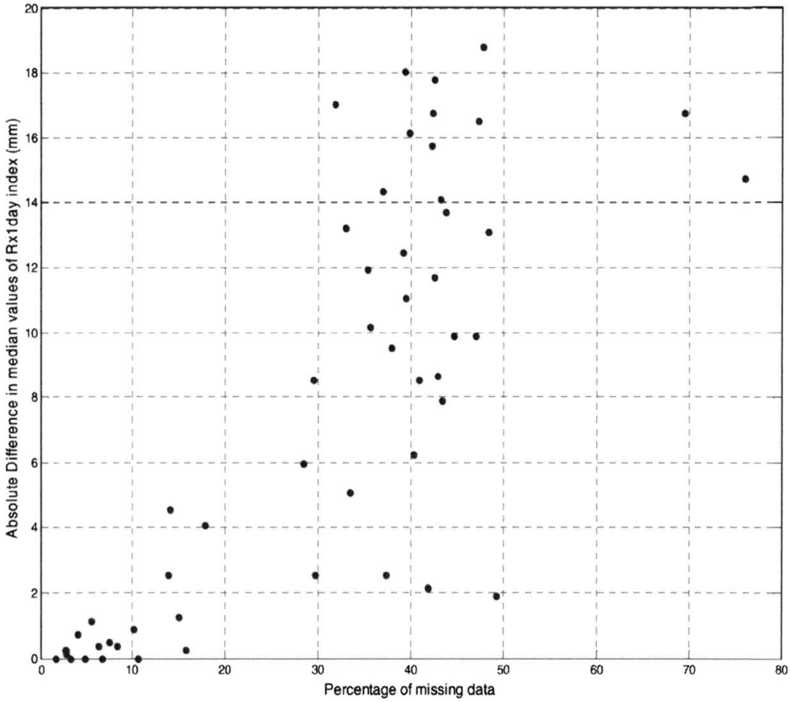

Figure 2.3 Variations in the median Rx1day index values calculated based on unfilled and filled data with respect to percentage of missing data. *(From Teegavarapu, R.S.V., Nayak, A., 2017. Evaluation of long-term trends in extreme precipitation: implications of infilled historical data and temporal-window based analysis. J. Hydrol. 550, 616–634.)*

Lilliefors test as a two-sided goodness-of-fit test for normality is applicable for situations where a fully specified null distribution is not known, as opposed to the Kolmogorov–Smirnov (KS) test (Massey, 1951) that requires that the null distribution be completely specified. Jarque-Bera test helps to check the validity of the null hypothesis that the data comes from a normal distribution with an unknown mean and variance and Chi-square test checks the null hypothesis that the data follows a normal distribution using parameters (mean and variance) estimated from the sample. Chi-square test is sensitive to the number of bins used for grouping the sample data.

In the case when the datasets (samples) do not conform to normality, Box–Cox (Box and Cox, 1964) transformation can be used as a last resort after all other possible data transformations, including those recommended by Helsel and Hirsch (2002) are checked and are found to be not viable. Logarithmic, square, square-root, and several other transformations should be initially evaluated before adopting any complex transformations such as Box−Cox. When Box−Cox transformation is used, the only one parameter of such power transform is obtained by optimization of a log-likelihood function. Before the two-sample unpaired *t*-test can be used, a two-sample F-test (or Fisher's test) (Snedecor and Cochran, 1989) needs to be carried out to evaluate the equality or inequality of sample variances. The F-test evaluates the null hypothesis that the two sample datasets come from normal distributions with the same variance. If the F-test confirms the null hypothesis, the two-sample parametric *t*-test can be used to test the null hypothesis that the two sample datasets come from normal distributions with equal means but unknown variances. Satterthwaite's modified *t*-test (Satterthwaite, 1946) is used when variances of the two samples are noted to be unequal based on F-test. Descriptions of two-sample parametric *t*-test and its modified version along with details of the calculation of *t*-statistics are discussed in Chapter 1. Goly and Teegavarapu (2014) and Teegavarapu et al. (2013) in two studies have used two-sample unpaired *t*-tests to confirm statistically significant changes in precipitation extreme and indices in two phases of AMO and ENSO. In both the studies, Box−Cox transformations were used for sample datasets that do not conform to normality.

11. WET AND DRY TRANSITION STATES

A two state first-order Markov-chain can be used to analyze dichotomous events (i.e., rain or no rain events) by estimating four transition probabilities referred by variables P_{11}, P_{10}, P_{01}, and P_{00}. The variables P_{11} and P_{10} refer to transition probabilities of occurrences of positive and no precipitation in a given time interval given the occurrence of positive precipitation values in the previous interval, respectively. Similarly, the variables P_{01} and P_{00} refer to probabilities of occurrences of positive precipitation value and no precipitation in a time interval given no precipitation in the previous interval, respectively. Events with precipitation depth greater than (or equal to) zero or predefined threshold are categorized as wet (dry) days in computing the transition probabilities. Transition probabilities $P_{00,}$ and P_{01}; P_{10} and P_{11} each sum up to 1 (one). Thus, only estimates for P_{01} and P_{10} are needed to analyze the four transitions. Definitions of these probabilities are provided in Fig. 2.4.

Rain or no-rain states for determination of transition probabilities

		Time interval (i+1)	
Time	$\theta_i > 0$	$\theta_{i+1} > 0$ [Wet-Wet]	$\theta_{i+1} = 0$ [Wet-Dry]
Interval (i)	$\theta_i = 0$	$\theta_{i+1} > 0$ [Dry-Wet]	$\theta_{i+1} = 0$ [Dry-Dry]

$$P_{11} = P_r(\theta_{i+1} > 0 \mid \theta_i > 0)$$

$$P_{10} = P_r(\theta_{i+1} = 0 \mid \theta_i > 0)$$

$$P_{01} = P_r(\theta_{i+1} > 0 \mid \theta_i = 0)$$

$$P_{00} = P_r(\theta_{i+1} = 0 \mid \theta_i = 0)$$

P_r: Probability, θ_i : Precipitation in time interval i;
θ_{i+1} : Precipitation in time interval $i+1$.

Figure 2.4 Calculation of transition probabilities for rain or no-rain states.

12. SPATIAL AND TEMPORAL OCCURRENCES OF EXTREMES

Point (i.e., rain gauge) and gridded precipitation data (i.e., spatially interpolated data available at a specific resolution) can be used for evaluation of spatial and temporal occurrences of extremes. In the current context, temporal occurrence refers to the realization of an extreme event during a specific time (viz., month, season, etc.). Changes in temporal occurrences of extremes within a year are critical for many water and disaster management agencies to control and manage or mitigate floods resulting from extreme precipitation events. Gridded precipitation data will provide the spatial variation of the changes in the extremes. Teegavarapu et al. (2013) and Goly and Teegavarapu (2014) have evaluated changes in the spatial and temporal occurrences in precipitation extremes including dry-wet transitions in different phases of two oscillations, namely AMO and ENSO. Temporal shift in the occurrences of precipitation extremes from wet season in AMO warm phase to dry season in cool phase was noted. Increase in the magnitudes of these extremes was observed in AMO warm phase compared with cool phase. Dry season precipitation in Florida, USA, was augmented by El Niño and during combined AMO and El Niño phases. Higher precipitation extremes are seen in different rainfall homogeneous areas in South Florida in warm phase compared to cool phase of AMO especially at durations longer than 24 h. Spatial occurrences can be confined to specific physiographic or climatic regions. Conditions prevailing before time during which extremes occur are also important and need to be evaluated. For example, Goly and Teegavarapu (2014) evaluated 10-day rainfall accumulations preceding storm events linked to annual extremes and noted that wetter conditions preceded during warm phases compared to cool phases of AMO. Exceedance probabilities for 10-day rainfall amounts were higher in case of warm phase compared to cool phase.

13. EVALUATION OF DROUGHTS: USE OF STANDARD PRECIPITATION INDEX (SPI)

Standard Precipitation Index (SPI) (McKee et al., 1993, 1995) that needs only information about precipitation can be used for evaluation of meteorological droughts. Monthly observed precipitation data is typically used for estimation of SPI. SPI calculation involves fitting a probability distribution (generally a Gamma distribution) to 1, 3, 6, 12-month precipitation totals and then use standard normal distribution to obtain SPI values. SPI is widely used by drought planners for quantifying precipitation deficit (WMO, 2012). The probability density function of a two parameter Gamma distribution is given by Eq. (2.12).

$$f(x) = \frac{1}{\beta^{\alpha}\Gamma(\alpha)}(x)^{\alpha-1}e^{-(x)/\beta} \tag{2.12}$$

The parameters of gamma distribution can be estimated by maximum likelihood estimation (MLE) method. Precipitation data may contain zero values, and in such instances gamma function is undefined (McKee et al., 1993, 1995). However, alternative procedures suggested by Thom (1966) can be used to estimate cumulative probability values based on incomplete gamma distribution functions (Edwards and McKee, 1997).

$$\widehat{\alpha} = \frac{1}{4A}\left(1 + \sqrt{1 + \frac{4A}{3}}\right) \tag{2.13}$$

$$\widehat{\beta} = \frac{\overline{x}}{\widehat{\alpha}} \tag{2.14}$$

where

$$A = \ln(\overline{x}) - \frac{\sum \ln(x)}{n} \tag{2.15}$$

n = number of precipitation observations.

Based on the estimated parameters $(\widehat{\alpha}, \widehat{\beta})$, the cumulative probability of an observed precipitation value can be obtained by

$$F(x) = \frac{1}{\widehat{\beta}^{\hat{\alpha}}\Gamma(\widehat{\alpha})}\int_0^x x^{\hat{\alpha}-1}e^{-x/\hat{\beta}}dx \tag{2.16}$$

Letting $t = x/\widehat{\beta}$, Eq. (2.16) becomes the incomplete gamma function:

$$F(x) = \frac{1}{\Gamma(\widehat{\alpha})}\int_0^x t^{\hat{\alpha}-1}e^{-t}dt \tag{2.17}$$

Since the gamma function is undefined for $x = 0$ and a precipitation distribution may sometimes contain zero values, the cumulative probability then can be expressed as:

$$D(x) = p + (1 - p)F(x) \tag{2.18}$$

where p is the probability of a zero value. If m is the number of zeros in a precipitation time-series, the value of p can be estimated by m/n (Thom, 1966). The Z value can be computationally estimated using a rational approximation provided by Abramowitz and Stegun (1972) that converts cumulative probability to the standard normal random variable (Z):

$$Z = -\left(t - \frac{c_0 + c_1t + c_2t^2}{1 + d_1t + d_2t^2 + d_3t^3}\right) \quad 0 < D(x) \leq 0.5 \tag{2.19}$$

$$Z = \left(t - \frac{c_0 + c_1t + c_2t^2}{1 + d_1t + d_2t^2 + d_3t^3}\right) \quad 0.5 < D(x) \leq 1.0 \tag{2.20}$$

$$t = \sqrt{-2\ln(D(x))} \quad 0 < D(x) \leq 0.5 \tag{2.21}$$

$$t = \sqrt{-2\ln(1 - D(x))} \quad 0.5 < D(x) \leq 1.0 \tag{2.22}$$

The values of c_0, c_1, d_1, d_2, d_3 are 2.515517, 0.802853, 0.010328, 1.432788, 0.189269, and 0.001308, respectively. The maximum absolute error possible (Abramowitz and Stegun, 1972) due to rational approximation is $4.5 \cdot 10^{-4}$.

The empirical cumulative probabilities are obtained by sorting the precipitation data in the increasing order of magnitude (Panofsky and Brier, 1958). The smooth curve shown in Fig. 2.5B is the cumulative distribution function of the fitted two-parameter gamma distribution, and the curve

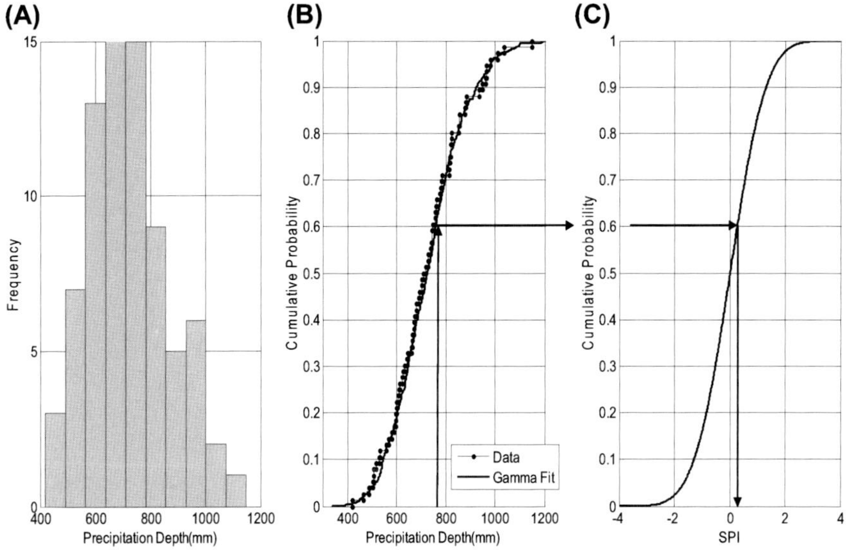

Figure 2.5 Illustration of 2-month SPI calculation using fitted gamma distribution. (A) Histogram of precipitation data; (B) Cumulative distribution function plot and data; (C) standard normal plot for SPI calculation.

with markers is empirical cumulative distribution function (ECDF) of the precipitation data. The cumulative probability distribution of a standard normal random variable is shown in Fig. 2.5C. SPI can be obtained using the 2-month precipitation value shown in Fig. 2.5B. Guttman (1994) suggests that 50–60 years (or more) are preferred for calculation of SPI. Distributions other than gamma can also be used to characterize the precipitation data. Guttman (1999) evaluated several candidate distributions which included: 2-parameter gamma; 3-parameter Pearson Type III, 3-parameter generalized extreme value (GEV), the 4-parameter Kappa, and the 5 parameter Wakeby. Based on the study conducted by Guttman, it was concluded that the Pearson Type III distribution is the "best" universal model and that the reliability of the SPI is sample size dependent and SPIs with time scales longer than 24 months may be unreliable considering data limitations. SPI values can be used to define wet and dry conditions based on the classification system proposed by McKee et al. (1993) as extremely wet (2 or higher); very wet (1.5–1.99); moderately wet (1.0–1.49); near normal (−0.99 to 0.99); moderately dry (−1.0 to −1.49); severely dry (−1.5 to −1.99); and extremely dry (−2 and less).

14. EVALUATION OF INTRAANNUAL VARIATIONS: USE OF SEASONALITY INDEX

The Seasonality Index (SI) (Walsh and Lawler, 1981) can be used to determine the intraannual monthly distribution of precipitation. SI provides an assessment of the precipitation regime in the region of interest. The index, SI, is the sum of the absolute deviations of the monthly rainfall from the mean monthly rainfall, divided by the total annual precipitation of the given year. The SI is given by Eq. (2.23).

$$SI_i = \frac{1}{R_i} \sum_{n=1}^{n=12} \left| x_{in} - \frac{R_i}{12} \right| \tag{2.23}$$

The variable R_i is the total annual precipitation in a particular year, and x_{in} is the actual monthly rainfall in month n. Precipitation regimes and associated seasonality index values are documented by Walsh and Lawler (1981) are provided in Table 2.2.

15. CHANGES IN PRECIPITATION RATES: EVALUATIONS OF INTENSITY DURATION FREQUENCY CURVES

Precipitation depths for different durations and return periods are generally characterized by Depth-Duration-Frequency (DDF) or Intensity-Duration-Frequency (IDF) curves. Accurate DDF curved—based estimates of precipitation values are essential and critical for hydrologic design (Teegavarapu, 2013b,c; Teegavarapu et al., 2013, 2005; Asquith, 1998). Rainfall frequency analysis is conducted using extreme precipitation data for different durations (typically 1, 2, 6, 12, 24, 48, 72, 96, and 120 h) for different return periods ranging from 2 to 100 years. Intensities for larger return periods are sometime

Table 2.2 Classification of Seasonality Index Values (Walsh and Lawler, 1981)

Seasonality Index	Precipitation Regime
<0.19	Precipitation spread throughout the year
0.20—0.39	Precipitation spread throughout the year, but with a definite wetter season
0.40—0.59	Rather seasonal with a short drier season
0.60—0.79	Seasonal
0.80—0.99	Marked seasonal with a long dry season
1.00—1.19	Most precipitation in <3 months
>1.20	Extreme seasonality, with almost all precipitation in 1—2 months

required for use in hydrologic design, although the reliability of these estimates is questionable when data is limited. Hydrologic design, in general, assumes stationarity of time-series of precipitation. However, changes in precipitation extremes due to climate variability can be evaluated by separating data from two or more phases of coupled oceanic and atmospheric oscillations. Establishment of precipitation DDF curves involved three steps. First, a probability density function (PDF) or cumulative distribution function (CDF) is fitted to each group comprised of the data for any given duration. The maximum precipitation depth for each time interval is related with the corresponding return period from the CDF. The maximum precipitation depth can be determined using a theoretical distribution function that is used to characterize the distribution of precipitation extremes.

A recent survey of rainfall frequency estimation methods by Svenson and Jones (2010) indicated that GEV is the most frequently used distribution for rainfall extremes. In a recent study by Teegavarapu et al. (2013), GEV distribution was used to characterize the distribution of the precipitation extremes for all durations (Teegavarapu and Goly, 2011). GEV, lognormal, 3-parameter log-normal, Pearson, and log-Pearson distributions were also evaluated in their study for characterizing extreme precipitation data in this study. Two indices, mean absolute deviation (MAD) and mean square deviation (MSD) (Jain and Singh, 1987), were used to measure the relative goodness-of-fits of distributions to the data. GEV with a flexible three-parameter model expressed by a probability density function (PDF) given by Eq. (2.24) was found to be most appropriate for characterizing 24-h duration precipitation extremes in the region (i.e., Florida).

$$f(x) = \begin{cases} \frac{1}{\sigma}\exp\left(- (1 + kz)^{-\frac{1}{k}}\right)(1 + kz)^{-1-\frac{1}{k}} & k \neq 0 \\ \frac{1}{\sigma}\exp(- z - \exp(-z)) & k = 0 \end{cases} \tag{2.24}$$

The variables k, σ, μ are the shape, scale, and location parameters, respectively, and $z = (x - \mu/\sigma)$. The parameters of the distribution can be estimated using maximum log-likelihood estimation method.

16. WET AND DRY SPELLS

Precipitation magnitude above a specific threshold value can be used for assessment of dry and wet spells. In many studies, a zero value of precipitation on a given day or in a time interval is used for the dry condition in that particular interval. The threshold value can also be chosen based on guidelines available from a regional meteorological agency. Once the value of the threshold is defined, consecutive wet and dry days of specific lengths can be referred to as wet and dry spell lengths respectively. The length of each wet and dry spell in a block of time (i.e., year) can be used to calculate the mean length of wet and dry spells individually. The lengths of spells considering a threshold level are representative of regional rainfall patterns over time and can be evaluated for changes. The number of wet or dry spells that is equal to or longer than a prefixed threshold value can be evaluated in different temporal windows. Wet and dry spells can provide further information about the precipitation characteristics, flooding and drought conditions in a watershed.

17. EXAMPLES OF ASSESSMENTS FROM DIFFERENT STUDIES

Evaluation of extreme precipitation events considering the magnitude, duration, and geographical location is crucial for the design of hydrologic systems and long-term operation of water resource systems. There are several studies confirming the link between internal modes of climate variability and extreme precipitation events at different spatial and temporal scales. Some of the major internal modes of climate variability affecting the precipitation around the globe are coupled oceanic and atmospheric oscillations such as AMO, ENSO, PDO, and NAO. Oscillations such as ENSO and AMO cycles or phases influence rainfall patterns and droughts in several parts of the continental United States and other regions of the world. The hydroclimatology of the southeastern region, especially the state of Florida in the United States, is influenced by all these oscillations and therefore provides an ideal region to study changes in precipitation. The state of Florida is surrounded by multiple water bodies with Gulf of Mexico on the west, Atlantic Ocean on the east, and straits of Florida on the south. Approximately 18% of the state is covered by water bodies, mainly large lakes and wetlands. Climate and soil moisture regimes and physiographic features are different in continental (Northern part of Florida) and peninsular (Southern part of Florida) regions of the state. The state of Florida frequently experiences tropical storms and hurricanes impacting the longest densely populated coastline in the United States. The following sections will present assessments related to precipitation from different studies conducted by the research group of the author of this chapter and others.

17.1 Changes to Extreme Precipitation Intensities

Larger precipitation totals (especially for long temporal durations greater than 24 h) have been attributed to different phases of these oscillations in the state of Florida (Teegavarapu et al., 2013). Generalized Extreme Value (GEV), log-normal, 3-parameter log-normal, Pearson, and log-Pearson distributions were evaluated for characterizing extreme precipitation data in their study. GEV with a flexible three-parameter model was found to be most appropriate for characterizing 24-h duration precipitation extremes in the region. Depth duration curves for different sites in Florida, USA, for two phases of AMO were evaluated by Teegavarapu et al. (2013). Their study suggested the following key points in the development of design storms or intensity duration frequency (IDF) curves using extreme precipitation datasets considering the influences of coupled oceanic and atmospheric oscillations:

- Conventional wisdom of using entire dataset of available precipitation extremes for determination of design storms is now debatable considering the influences of different phases of oscillations on a whole or parts of a region spatially and temporally. Multiple oscillations influencing regional hydrology simultaneously in different temporal windows may increase or decrease the frequency and magnitudes of extreme precipitation events and influence intraannual temporal occurrences of extremes.
- Emphasis should be placed on those temporal windows in which the combined influences of two or more oscillations lead to rare extremes and data selection for design should be representative of these extremes. The idea of using site-specific extremes for hydrologic design confined to one specific region is not new. However, consideration of region-specific influences of climate variability at different spatial and temporal scales is advocated for hydrologic design.

17.2 Drought Analysis Using SPI

Assessment of droughts using SPI values in the state of Florida, the USA, using 11 sites is presented in this section. Chronological time-series of SPI values for different temporal scales are evaluated for all 11 rain gage stations in the state of Florida. The threshold limit of above (below) +1 (−1) is considered for wet condition (drought condition) to evaluate the influences of AMO phases on precipitation. Results based on SPI values for 3-, 6- and 12-month timescales point to a higher number of wetter conditions that

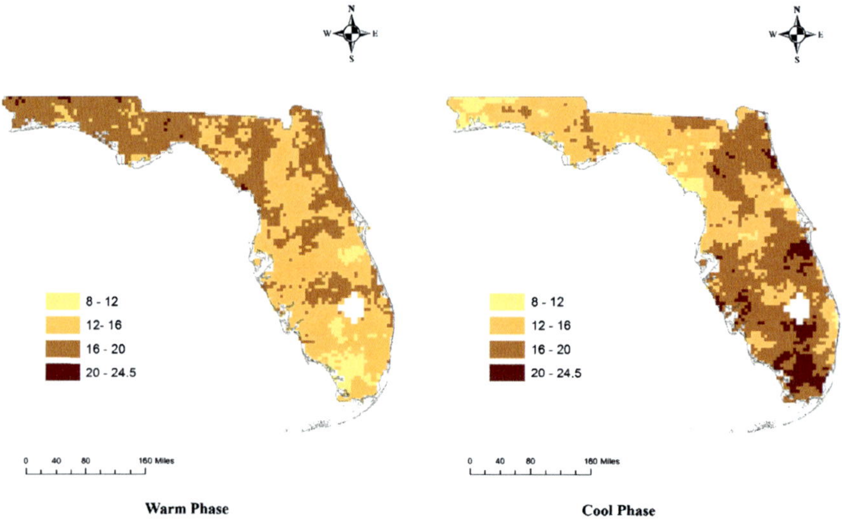

Figure 2.6 Variation in percentage occurrences of drought based on 3-month SPI values in two phases of AMO. *(From Goly, A., Teegavarapu, R.S.V., 2014. Individual and coupled influences of AMO and ENSO on regional precipitation characteristics and extremes. Water Resour. Res. 50 (6), 4686–4079.)*

occur in peninsular (continental) region during warm (cool) phase. Similarly, a higher number of drought conditions occur in peninsular (continental) region during cool (warm) phase, but with the exception of one rain gauge location in southeast Florida. Gridded data are used to show the spatial variability of percentage number of drought occurrences calculated based on 3-month SPI values for two phases of AMO in Fig. 2.6. The peninsular region in general experiences lower drought occurrences in AMO warm phase than compared to those in a cool phase, whereas the continental region experiences somewhat opposite conditions. These dry and wet cycles follow a similar trend as observed for all other precipitation indices in these two regions.

Combined influences of AMO and ENSO are assessed using 3-month SPI values estimated between years 1950–2010. The AMO warm and cool phases are represented by the *red* and *blue lines* respectively. Additionally, El Niño (*red circles*) and La Niña (*blue circles*) episodes based on a threshold of ±0.5° C for the Oceanic Niño Index (ONI) (3-month running mean of extended reconstructed sea surface temperature (ERSST).v3b SST anomalies in the Niño 3.4 region ([5° N–5° S, 120°–170° W]), based on centered 30-year base period from 1981 to 2010 are presented on top of the SPI time series. Two stations were arbitrarily chosen, one from the

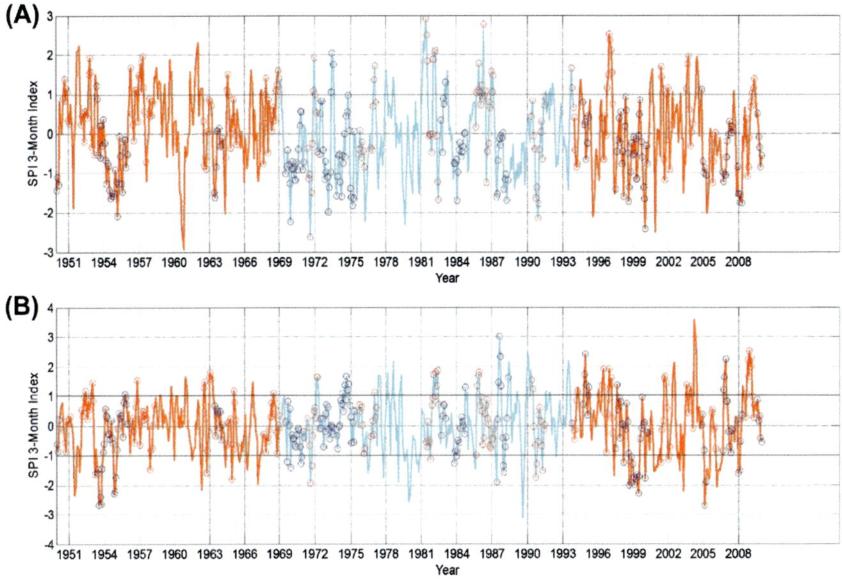

Figure 2.7 Variation of 3-month SPI values at (A) gauge 1 and (B) gauge 7. Time series lines (*red and blue lines* represent warm and cool phases, respectively). *Red (blue) colored circles* represent El Niño and La Niña conditions, respectively. *(From Goly, A., Teegavarapu, R.S.V., 2014. Individual and coupled influences of AMO and ENSO on regional precipitation characteristics and extremes. Water Resour. Res. 50 (6), 4686–4079.)*

peninsular region and the other from the continental region of Florida. Based on 3-month SPI values, it has been noticed that ENSO has profound significance in the peninsular region (Fig. 2.7A), El Niño is contributing toward wetter conditions and La Niña toward more dry conditions in both warm and cool phases. On the other hand, whereas in the continental region (Fig. 2.7B), the significance of ENSO is less intense. During warm phase more or less La Niña affects the droughts but during cool phase, La Niña is equally responsible for wetter conditions (SPI value above 1) as El Niño. There is no significant difference in El Niño, and La Niña influences during cool phase considering the 3-month SPI value and ONI. It is apparent and especially noticed during the time periods 1954–57, 1963–66, 1972–75, and 1987–90. Moreover, it is important to note that AMO with no effect of ENSO is largely responsible for both the extremely drought and wetter conditions. In fact, based on two figures (Fig. 2.7A and B) one can notice that extreme droughts and wetter conditions defined by SPI values are purely based on influences of two different phases of AMO with no effect of ENSO.

17.3 Dry and Wet Spell Transition Probabilities

Three-state (i.e., dry, normal and wet) first-order Markov chain transition probabilities were evaluated using SPI values at different time scales at all the rain gauge stations in Florida by Goly and Teegavarapu (2014). They evaluated SPI values in two phases of a major coupled atmospheric and oceanic oscillation, AMO that influences the hydoclimatology of Florida. They concluded, in general, the transition probabilities: dry-to-wet (P_{DW}); wet-to-wet (P_{WW}) were found to be higher at stations located in the continental region of Florida in AMO cool phase than those in the same region in the warm phase. Exactly opposite trends are evident when similar probabilities are analyzed using data from rain gauges in the peninsular region of Florida.

17.4 Temporal Distribution of Precipitation

Temporal distribution of the rainfall within the extreme storm events can be obtained by estimating the ratio of accumulated rainfall (P) over a specific subperiod of the storm and the total rainfall (P_{tot}). The $\frac{P}{P_{tot}}$ ratio values for storm events in two different periods can be compared. Goly and Teegavarapu (2014) compared these ratios derived from different storm events in Florida, USA, for two periods that coincide with two phases of ENSO using cumulative distribution functions (CDFs) or nonexceedance probability curves. Fig. 2.8 shows the nonexceedance probability plots of P/P_{tot} values for maximum 5-day event for dry season during El Niño and La Niña. Many of the long-duration extreme events in the study region are of 3-day and 5-day durations. These long-duration (3-day or 5-day) events are of interest for hydrologic design rainfall as these are common in many regions of the world due to slow moving tropical storms and hurricane (cyclone or typhoon) landfalls. The temporal distributions as shown in Fig. 2.8 exhibit higher exceedance probabilities during El Niño for days 1—4, suggesting monotonic peaking in the ratios and late peaking during El Niño.

17.5 Spatial Variation of Extremes

Variation of precipitation extremes at multiple sites can be assessed for mapping the changes in the region with respect to climate variability or change. Statistically significant changes in seven extreme precipitation indices, two transition probabilities and variance in two phases of AMO (i.e., cool and warm phases) are shown in Fig. 2.9. Statistically significant

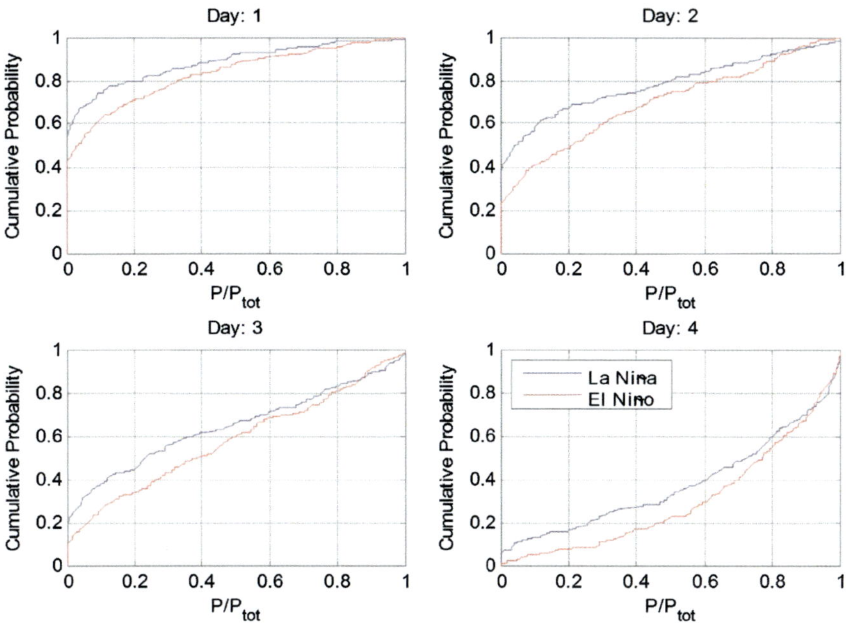

Figure 2.8 Nonexceedance probability plots of P/P_{tot} values for maximum 5-day event for dry season during El Niño and La Niña. *(From Goly, A., Teegavarapu, R.S.V., 2014. Individual and coupled influences of AMO and ENSO on regional precipitation characteristics and extremes. Water Resour. Res. 50 (6), 4686–4079.)*

results of RX1day, RX5day, R10mm, CDD, CWD, PRCPTOT, P01, P10, and variances based on parametric t tests between the two phases of AMO for wet season are shown in Fig. 2.9. Extreme indices RX1day, RX5day, and PRCPTOT display strong and significant positive differences throughout Florida, suggesting that the AMO warm phase is responsible for the higher extremes and totals. Threshold-based indices R10mm, R20mm, and Rnnmm present significant positive differences throughout Florida except at two gauges in the northwest region which present significant negative differences. Two distinct regions: east (negative differences) and northwest to central west (positive differences) were identified for CWD. For CDD, most of the gauges show positive differences except three gauges on the west which display otherwise. P01 shows a trend similar to that of annual precipitation but with varying magnitude in the differences. Negative differences are evident for wet-to-dry transition state (P10) at most of the gauges except in the southeast. Spatial variations of distributions of monthly occurrences of precipitation extremes in two phases of AMO are shown in Fig. 2.10. The temporal occurrence (i.e., the month in which

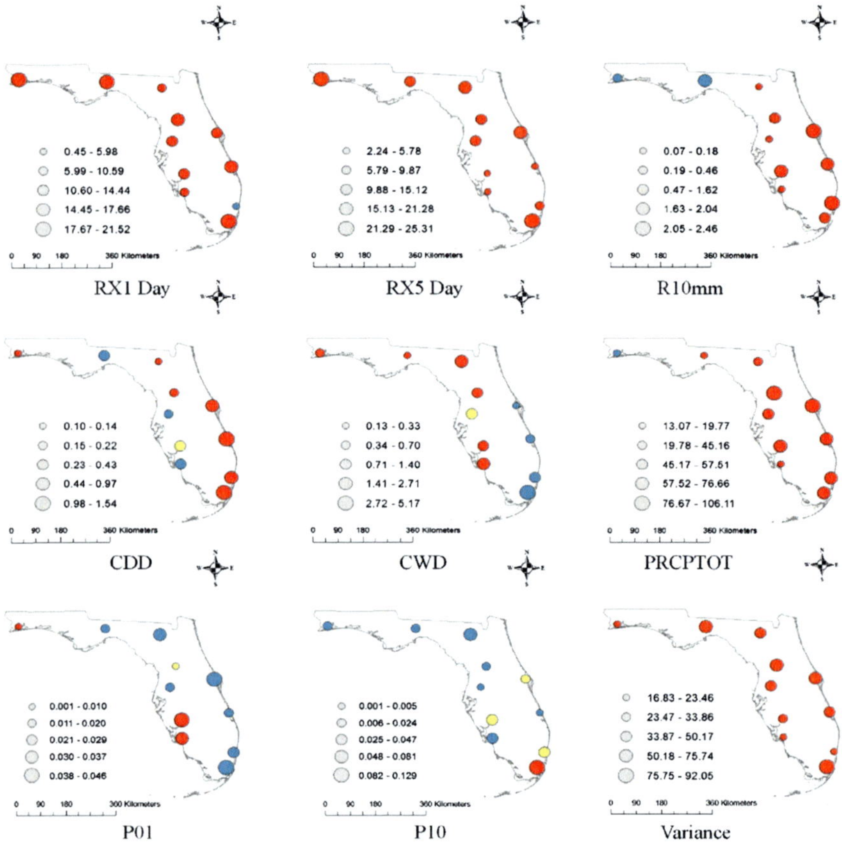

Figure 2.9 Significant (5%) difference in precipitation indices, dry and wet spell transitions, and variances for AMO warm—AMO cool episodes during wet season. Positive (negative) significance (warm (cool)) is represented by *red (blue) circles. Yellow circles* represent insignificance. *(From Goly, A., Teegavarapu, R.S.V., 2014. Individual and coupled influences of AMO and ENSO on regional precipitation characteristics and extremes. Water Resour. Res. 50 (6), 4686—4079.)*

extreme 1-day precipitation occurs) in each year is obtained at each rain gage location and the distributions of these occurrences are shown in Fig. 2.10. The median of occurrences combining all thegages is around March-April for cool phase and June-July for warm phase, suggesting that most of 1-dayextremes occur in the early part of year for cool phase and later part of the year for the warm phase. This temporal shift in the occurrences of extremes along with spatial variability will require changes in operational strategies of large network of canals and drainage structures to be modified to handle runoff volume resulting from these extremes.

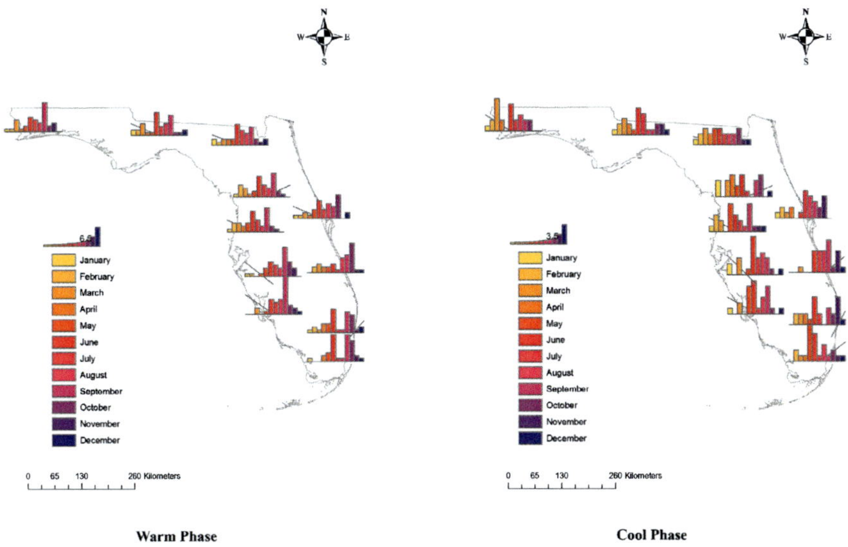

Figure 2.10 Spatial variation of distributions of temporal occurrences of RX1day extreme precipitation events during AMO warm and cool phases. *(From Goly, A., Teegavarapu, R.S.V., 2014. Individual and coupled influences of AMO and ENSO on regional precipitation characteristics and extremes. Water Resour. Res. 50 (6), 4686–4079.)*

Most of the extremes in the south and east regions of Florida occur in the months of June and September during warm phase and this might be due to higher number of hurricane landfalls. Approximately 20% (30%) of the annual extremes are observed in the dry season during AMO warm (cool) phase.

18. PHYSICAL BASIS OF CLIMATE VARIABILITY INFLUENCES: FLORIDA CASE STUDY EXAMPLE

Statistical inference tests can help confirm the trends or changes in precipitation time-series evident from results of several studies discussed in this chapter. However, there is also a need for establishing a physical basis for influences due to climate change and variability on precipitation. Hydroclimatology, spatially varying soil temperature regimes, physiography, land use, and land cover and other factors are expected to affect the climate variability and change influences on precipitation and other essential climate variables in a region. For example, a strong influence of regional climatology restricting the spatial influence of one of the oscillations (i.e., AMO) on precipitation in the state of Florida, USA was evident from the analysis reported in work by Teegvarapu et al. (2013). Soil temperature regimes

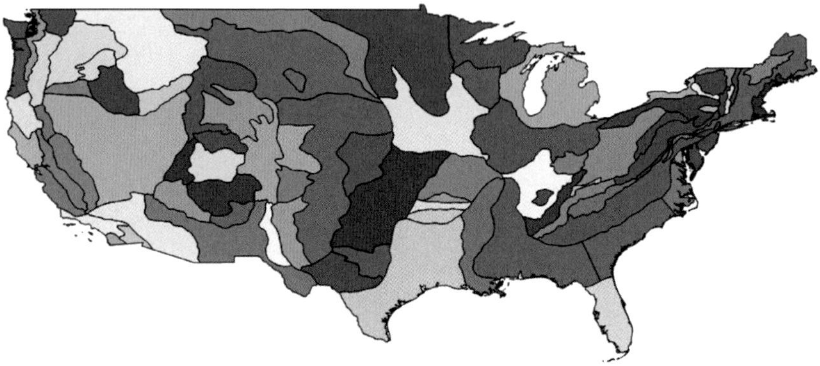

Figure 2.11 Physiographic regions of the United States. *(Data source: United States Geological Survery.)*

(thermic and hyperthermic) and physiographic features of the region separating the state of Florida in continental and peninsular regions were attributed to be possible factors for restriction of the spatial extent of influences of AMO on precipitation. Fig. 2.11 shows the physiographic regions in the continental United States. The line separating the lower (Southern) and upper (Northern) physiographic region of Florida also approximately coincides with the soil temperature regime separation line.

The varying nature of spatial influences in the continental and peninsular regions are also dependent on the temporal window (i.e., annual, wet and dry season) used for analysis. Higher correlations (based on Pearson correlation coefficient, Kendall's τ, and Spearman's ρ) between monthly gridded precipitation values and an average of Sea Surface Temperatures (SSTs) over the Atlantic region were evident in the peninsular region in warm phase than those in cool phase in continental region. Differences in precipitation characteristics due to two AMO phases are stronger during wet season than the dry season. Florida's wet season precipitation is mostly attributed to localized convective storms, tropical storms, and frequent hurricane landfalls.

Winsberg (2003) analyzed the ENSO's effect on Florida seasonal weather events and indicted that ENSO amplifies weather events in the spring. During winter, the pressure systems move jet streams, and their accompanying storm tracks closer to Florida than in other seasons and because of that, the amplification in the other seasons is not nearly as great as in the winter months (December–February). O'Brien et al. (1999) indicated that during El Niño, jet streams that guide fronts and storms more

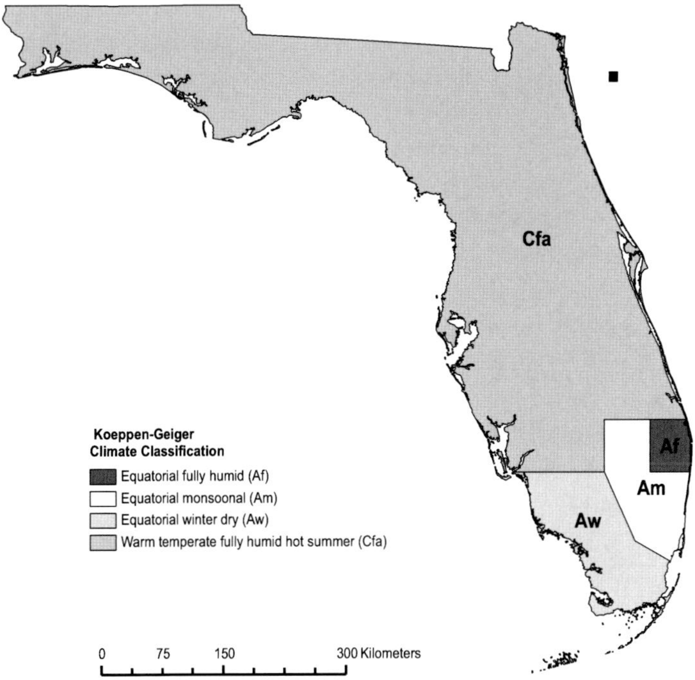

Figure 2.12 Climate zones in the state of Florida, USA, based on to Köppen-Geiger climate classification scheme.

frequently pass close to or over Florida. Due to this above average rainfall and reduced frequency of hurricanes, below normal temperatures are observed. During La Niña episode jet streams and their accompanying fronts and storms pass farther to the north and affect Florida's weather less than in neutral years, increasing the chances of hurricanes, above average temperatures and generally dry conditions.

According to Köppen-Geiger climate classification scheme (Kottek et al., 2006), Florida's climate is defined by four classes as shown in Fig. 2.12: equatorial fully humid (Af) (southeastern region), equatorial monsoonal (Am) (southeastern region), equatorial winter dry (Aw) (southwest), and warm temperate fully humid hot summer (Cfa) (most part of the state). The three classes (Af, Am, and Aw) are confined to three areas in the south with minor influences of these climatic zones on the variability of precipitation. However, two rain gauges: stations 2 and 8 located in Am and Aw regions, respectively, show a reversal of trends or no significance for few indices that are analyzed in this study. Hurricane landfall frequency and spatial density of storm paths in the region will affect the precipitation extremes at different temporal scales. Extreme precipitation indices such as RX1day, RX5day, and SDII are influenced by slow moving hurricanes over the region. Over 20% of extreme precipitation totals are linked to slow moving hurricanes (Teegavarapu, 2013a,b; Teegavarapu et al., 2013) at several rain gauge stations in south Florida.

The frequency and spatial density of hurricane landfalls are higher in the south and central Florida compared to other regions of the state. This is evident from the analysis of hurricane landfall paths over the region using the data available from ftp://eclipse.ncdc.noaa.gov/pub/ibtracs/v03r05/all/shp/. Approximately double the number of hurricane landfalls is noted in warm phase than those in the cool phase. Links to Atlantic SSTs and hurricane landfalls during AMO phases are already discussed by Teegavarapu et al. (2013). Also, during El Niño upper-level winds over Atlantic basin become unfavorable for tropical hurricane development (Winsberg, 2003; O'Brien et al., 1999). The precipitation characteristics and extremes of Florida are known to be influenced by multiple decadal (i.e., AMO and PDO) and quasidecadal (i.e., NAO) and interannual (i.e., ENSO) oscillations. Independent influences of these oscillations are region-specific and spatially nonuniform.

19. HYDROLOGIC DESIGN USING PRECIPITATION EXTREMES: IMPACTS OF CLIMATE CHANGE

The single most critical input to the hydrologic design used in a single event or continuous modeling approaches is precipitation. Hydrologic designs continue to rely on the assumption of stationarity (Milly et al., 2008). Natural climate variability and future changes in climate that may alter precipitation intensities or storm durations will impact stormwater infrastructure designed under the stationarity assumption. Influence of AMO, a major coupled-oceanic atmospheric oscillation, on extreme precipitation depths in the state of Florida for a given return period was recently reported by Teegavarapu et al. (2013). Urban drainage design practices are now revisited by incorporating climate change factors (Arnbjerg-Nielsen, 2012), analyzing trends in precipitation extremes and their frequencies (De Toffol et al., 2009), evaluating impacts of changing extremes using downscaled precipitation data from GCMs (Grum et al., 2006), and developing frameworks for risk and uncertainty management (Arnbjerg-Nielsen, 2011).

The inabilities of climate change models in reproducing precipitation extremes accurately, and limitations of downscaling models in replicating the spatial and temporal variability of the same are discussed in different studies (Benestad et al., 2008; Teegavarapu, 2010, 2013c; Goly et al., 2014). Highlighting one of the issues related to downscaling, Fowler et al. (2007) conclude that temperature can be downscaled with more skill than precipitation. Evaluation of downscaled precipitation and temperature estimates from multiple GCM models in the state of Florida, USA, by Teegavarapu (2013b) supports a similar conclusion. Emission scenarios and limited skills of multimodel GCM-based projections of future are considered to be the first and second sources of uncertainty, respectively (Jacob and Hurk, 2009). Stainforth et al. (2007) point to many sources of uncertainties in the models, including forcing, initial condition, and climate modeling uncertainties.

Majority of stormwater control and management facilities are designed on the basis of single-event design-storms of a given frequency (ASCE, 1996; Haan et al., 1994; Liew et al., 2013) and with an assumption of stationarity of precipitation time-series. In all single event–based design methods, peak discharges are sought for design with evaporation and evapotranspiration processes considered to be not affecting runoff (ASCE, 1996; Gironas et al., 2009). Future temperature changes will influence runoff volumes that are derived from continuous event-based hydrologic simulation models that consider all abstractions including evaporation and evapotranspiration. In many climate change impact studies, it is assumed that GCM-based climate change scenario-driven simulations and downscaled precipitation data are available. Storm runoff also depends on temperature changes. Single event simulations are usually insensitive to the evaporation rates, and therefore evaporation is typically neglected when a single rainfall event or a synthetic storm is simulated (Gironas et al., 2009). Design storms derived using intensity-duration-frequency (IDF) relationships (Adams and Howard, 1986) are widely employed for the design of stormwater conveyance systems in engineering practice. Watt and Marsalek (2013) suggest that design practices would be well served by adopting a comprehensive approach considering all design storm event characteristics and their sensitivity to climate change and inherent uncertainties in the existing IDF relations. Changes in IDF relationships based on different phases of AMO were recently documented by Teegavarapu et al. (2013) thus confirming the influence of climate variability on regional precipitation extremes with critical implications for hydrologic design.

Traditionally, IDF curve development for hydrologic design relied on the available historical precipitation data under the assumption of stationarity of climate. However, methodologies for development of IDF relationships under changing climate were provided in several recent studies (Peck et al., 2012; Liew et al., 2013; Solaiman and Simonovic, 2011; Zhu et al., 2012). Peck et al. (2012) used a nonparametric weather generator (WG) with shuffling and perturbation mechanisms to generate synthetic rainfall records similar (but not identical) to the observed historic record. In their study, two climate scenarios were selected with lower and upper bounds; the former scenario uses observed rainfall record as WG input, and the latter modifies the observed record using the results of a selected global climate model and then uses the modified rainfall datasets as WG inputs. Liew et al. (2013) used a downscaling-comparison-derivation (DCD) approach to derive IDF curves for present and future climate using the extracted dynamically downscaled data from an ungauged site with simulation data from regional climate model (RCM). Solaiman and Simonovic (2011) used possible realizations of future climate from 29 scenarios derived from downscaled daily precipitation data and nonparametric kernel estimation approach to develop IDF curves for different durations. The daily precipitation data are disaggregated to subdaily durations using a nonparametric nearest neighbor-based disaggregation scheme. As evident from the review of recent literature, IDF curves for future climate can be developed along with upper and lower bounds on the future precipitation intensities and changes due to future climate can be evaluated.

Optimal or sustainable hydrologic design for future climate should answer the following questions:

- Is a deterministic or a stochastic optimization formulation appropriate for optimal hydrologic design?
- How future uncertain precipitation extremes can be addressed in the formulation?
- How to express and quantify multimodel, multiple scenario uncertainties of climate change models?
- How preferences of modelers/hydrologists about possible future extremes can be incorporated in the hydrologic design process?
- How to obtain a sustainable (compromise) climate-change sensitive hydrologic design?

An adaptive approach for stormwater management is to upgrade existing infrastructure with changing IDF relationships over time and evolving temporal precipitation distributions. Alternatively, a sustainable hydrologic design that identifies a compromise between current and future climate-based conditions can be devised (Teegavarapu, 2013c). Preferences of hydrologists, water managers, or design personnel towards future climate change can be included in such a design process. Groves et al. (2008) explored different hypotheses about how different characterizations of climate change uncertainty may affect water managers' beliefs, opinions, choices, and actions. Acknowledging the limitations of the climate change models, hydrologists may develop perceptions toward the accuracy of results from these models. These perceptions can be translated to preferences toward predicted future changes in the main hydrological inputs (Teegavarapu, 2010, 2013b,c). Several studies in the recent past have provided methodologies to develop IDF curves based on GCM simulations and also upper and lower limits of precipitation extremes obtained from downscaled and bias-corrected GCM simulations considering multimodel scenario uncertainties. In a recent study, Teegavarapu (2013c) used these bounds on the precipitation intensities and solved a mathematical programming formulation with an objective of modeling preferences of decision makers toward future precipitation intensities and obtaining a sustainable hydrological design.

20. FUTURE RESEARCH DIRECTIONS

Changes in precipitation extremes and characteristics due to climate change and variability need to be investigated at both regional and global scales. These changes need to be evaluated using a number of statistical techniques available in the literature and those that will be developed in future considering the nature of data. Assessment of stationarity or changes in precipitation time-series is a major challenge considering many issues related to length and quality of the data and changing monitoring networks and lack of metadata in many situations. On the other hand, availability of precipitation data at different temporal and spatial resolutions from radar and satellite-based sources will enhance our understanding of precipitation extremes and also help in evaluation of the same in regions with no rain gauges or monitoring networks with low density of gauges. Future research studies on changes in precipitation extremes and characteristics should focus on:

- Development of new metrics or indices for defining the characteristics and extremes of precipitation time-series.
- Evaluation of long-term changes in precipitation magnitudes, frequency, and intensities mainly due to climate change.

- Understanding how single or multiple coupled oceanic and atmospheric oscillations influence regional precipitation extremes spatially and temporally.
- Mapping the spatial extent of these influences and evaluating how regional hydroclimatology, topography, physiography, and other factors restrain the spatial (extent) and temporal influences of oscillations.
- Evaluation of interactions among multiple hydroclimatological variables under variable climate.
- Assessment of scaling behavior of precipitation extremes in different phases of coupled oceanic and atmospheric oscillations.
- Assessment of rare precipitation extremes confined to specific regions of the world.
- Extensive assessment of climate variability influences on precipitation extremes and characteristics in pristine (i.e., least anthropogenically disturbed) watersheds in different regions of the world.
- Establishment of physical basis for the influences of climate variability and change on precipitation and other variables.
- Development of methodologies for hydrosystems operations impacted by future precipitation variations and hydrologic design under nonstationarity climate.

21. CONCLUSIONS

This chapter provided a review of different methods, indices, and metrics that can be used to evaluate changes in precipitation extremes and characteristics influenced by climate variability and change with more emphasis on the former aspect. Several statistical approaches discussed in Chapter 1 are used to assess changes in precipitation extremes and characteristics. Results from multiple studies are briefly discussed to illustrate the use of different state-of-the-art approaches for understanding spatial and temporal changes in precipitation mainly due to climate variability. Future directions for research are provided to assess changes in precipitation magnitudes and occurrences under changing climate.

REFERENCES

Abramowitz, M., Stegun, I.A., 1972. Handbook of Mathematical Functions. Dover, New York.

Adams, B.J., Howard, C.D.D., 1986. Pathology of design storms. Can. Water Resour. J. 11 (3), 49–55.

Ahrens, B., 2006. Distance in spatial interpolation of daily rain gauge data. Hydrol. Earth Syst. Sci. 10, 197–208.

Arnbjerg-Nielsen, K., 2011. Past, present, and future design of urban drainage systems with focus on Danish experiences. Water Sci. Technol. 63 (3), 527–535.

Arnbjerg-Nielsen, K., 2012. Quantification of climate change effects on extreme precipitation used for high resolution hydrologic design. Urban Water J. 9 (2), 57–65.

ASCE, 1996. Handbook of Hydrology, second ed. American Society of Civil Engineers, New York.

Ashkar, F., 1996. Extreme floods in hydrologic of disasters. In: Singh, V.P. (Ed.), Water Science and Technology Library. Kluwer Academic Publishers, Netherlands.

Ashraf, M., Loftis, J.C., Hubbard, K.G., 1997. Application of geostatisticals to evaluate partial weather station network. Agric. For. Meteorol. 84, 255–271.

Asquith, W.H., 1998. Depth–Duration–Frequency of Precipitation for Texas. Water Resources Investigations Report 98–4044. U.S. Geological Survey, Austin, Texas.

Bardossy, A., Pegram, G., 2014. Infilling missing precipitation records – A comparison of a new copula-based method with other techniques. J. Hydrol. 519, 1162–1170.

BCSD, 2018. Bias Correction and Statistically Downscaled Data. Available online at: http://gdo-dcp.ucllnl.org/downscaled_cmip3_projections/dcpInterface.html#Projections:%20Subset%20Request.

Benestad, R.E., Hanssen-Bauer, I., Chen, D., 2008. Empirical-Statistical Downscaling. World Scientific, Hackensack, New Jersey.

Benson, M.A., 1962. Plotting positions and economics of engineering planning. Proc. ASCE J. Hydraul. Div. 88 (6), 57–71.

Bobee, B., Ashkar, F., 1991. The Gamma Family and Derived Distributions Applied in Hydrology. Water Resources Publications, Littleton, Colorado.

Box, G.E.P., Cox, D.R., 1964. An analysis of transformations. J. R. Stat. Soc. B (26), 211–252.

Brown, M.B., Forsythe, A.B., 1974. Robust tests for the equality of variances. J. Am. Stat. Assoc. 69, 364–367.

Buuren, V.S., 2012. Flexible Imputation of Missing Data. CRC Press, Boca Raton, Florida, USA.

Chang, K.-T., 2006. Introduction to geographic information systems. McGraw Hill, New York.

Changnon, S.A., Kunkel, K.E., 1995. Climate-related fluctuations in Midwestern flooding. J. Water Resour. Plann. Manag. 121, 326–334.

Clark, M.P., Serreze, M.C., McCabe, G.J., 2001. Historical effects of El Niño and La Niña events on the seasonal evolution of the montane snowpack in the Columbia and Colorado river basins. Water Resour. Res. 37 (3), 741–757.

Corder, W.G., Foreman, D.I., 2009. Nonparametric Statistics for Non-statisticians. Wiley, Hoboken.

Cronin, T.M., 2010. Paleoclimates: Understanding Climate Change Past and Present. Columbia University Press, New York.

Curtis, S., 2008. The Atlantic multidecadal oscillation and extreme daily precipitation over the US and Mexico during the hurricane season. Clim. Dyn. 30, 343–351.

De Toffol, S., Laghari, A.N., Rauch, W., 2009. Are extreme rainfall intensities more frequent? Analysis of trends in rainfall patterns relevant to urban drainage systems. Water Sci. Technol. 59 (9), 1769–1776.

Dettinger, M.D., Diaz, H.F., 2000. Global characteristics of streamflow seasonality and variability. J. Hydrometeorol. 1, 289–310.

Dickey, D.A., Fuller, W.A., 1979. Distribution of the estimators for autoregressive time series with a unit root. J. Am. Stat. Assoc. 74, 427–431.

Dingman, S.L., 2011. Physical Hydrology. Prentice Hall, NJ.

Dufek, A.S., Ambrizzi, 2008. Precipitation variability in Sao Paulo State, Brazil. Theor. Appl. Climatol. 93, 167–178.

Edwards, D.C., McKee, T.B., 1997. Characteristics of 20th Century Drought in the United States at Multiple Time Scales. Climatology Rep. 97–2. Department of Atmospheric Science, Colorado State University, Fort Collins, Colorado, 155 p.

El-Sharif, H., Teegavarapu, R.S.V., 2012. Evaluation of spatial interpolation methods for missing precipitation data: preservation of spatial statistics. In: World Environmental and Water Resources Congress 2012, pp. 3822–3832. https://doi.org/10.1061/9780784412312.384.

Enders, C.K., 2010. Applied Missing Data Analysis. The Guilford Press, New York.

Enfield, D.B., Mestas-Nunez, A.M., Trimble, P.J., 2001. The Atlantic multidecadal oscillation and its relation to rainfall and river flows in the continental U.S. Geophys. Res. Lett. 28 (10), 2077–2080.

Fowler, H.J., Blenkinsop, S., Tebaldi, C., 2007. Linking climate change modeling to impacts studies—Recent advances in downscaling techniques for hydrological modeling. Int. J. Climatol. 27, 1547–1578.

Gardner, W.A., Napolitano, A., Paura, L., 2006. Cyclostationarity: half a century of research. Signal Process. 86, 639–697.

Gironas, J., Roesner, L.A., Davis, J., Rossman, L., 2009. Stormwater Management Applications Manual, EPA/600/R-09/077. http://nepis.epa.gov/Adobe/PDF/P100EA7H.pdf.

Glantz, M.H., 2001. Currents of Change: Impacts of El Niño and La Niña on Climate and Society. Cambridge University Press, Cambridge.

Goly, A., Teegavarapu, R.S.V., 2014. Individual and coupled influences of AMO and ENSO on regional precipitation characteristics and extremes. Water Resour. Res. 50 (6), 4686–4079.

Goly, A., Teegavarapu, R.S.V., Mondal, A., 2014. Evaluation of Statistical Downscaling Models for Monthly Precipitation in Florida. Earth Interact. 18 (2014), 1–28.

Goodrich, G.B., Ellis, A.W., 2008. Climatic controls and hydrologic impacts of a recent extreme seasonal precipitation reversal in Arizona. J. Appl. Meteorol. Climatol. 47, 498–508.

Goswami, B.N., Venugopal, V., Sengupta, D., Madhusoodanan, M.S., Xavier, P.K., 2006. Increasing trend of extreme rain events over India in a warming environment. Science 314, 1442–1445.

Groisman, P.Y., Easterling, D.R., 1994. Variability and trends of total precipitation and snowfall over the United States and Canada. J. Clim. 7 (1), 184–205.

Groisman, P.Y., Knight, R.W., Karl, T.R., 2001. Heavy precipitation and high streamflow in the contiguous United States: trends in the twentieth century. Bull. Am. Meteorol. Soc. 82 (2), 219–246.

Groisman, P.Y., Karl, T.R., Easterling, D.R., Knight, R.W., Jamason, P.F., Hennessy, K.J., Suppiah, R., Page, C.M., Wibig, J., Fortuniak, K., Razuvaev, V.N., Douglas, A., Forland, E., Zhai, P.-M., 1999. Changes in the probability of heavy precipitation important indicators of climatic change. Clim. Change 42, 243–283.

Groisman, P.Y., Karl, T.R., Knight, R.W., Stenchikov, G.L., 1994. Changes in snow cover, temperature, and radiative heat balance over the northern hemisphere. J. Clim. 7, 1633–1656.

Groves, D.G., Knopman, D., Lempert, R.J., Berry, H.S., Wainfan, L., 2008. Presenting uncertainty about climate change to water-resource managers. Technical report. RAND Corp.

Grum, M., Jørgensen, A.T., Johansen, R.M., Linde, J.J., 2006. The effect of climate change on urban drainage: an evaluation based on regional climate model simulation. Water Sci. Technol. 54 (6–7), 9–15.

Guillermo, Q., Tabios III, G.Q., Salas, J.D., 1985. A comparative analysis of techniques for spatial interpolation of precipitation. J. Am. water Resour. Assoc. 21 (3), 365–380.
Gujarati, D., 2004. Basic Econometrics. McGraw Hill.
Gujarati, D., 2011. Econometrics by Example. Palgrave MacMilan.
Guttman, N.B., 1994. On the sensitivity of sample L moments to sample size. J. Clim. 7 (6), 1026–1029.
Guttman, N.B., 1999. Accepting the standardized precipitation index: a calculation algorithm. J. Am. Water Resour. Assoc. 35 (2), 311–322.
Haan, C.T., Barfield, B.J., Hayes, J.C., 1994. Design Hydrology and Sedimentology for Small Catchments. Academic Press Inc., San Diego, California.
Haylock, M., Goodess, C., 2004. Inter-annual variability of European extreme winter rainfall and links with mean large-scale circulation. J. Climatol. 24, 759–776.
Hellstrom, C., Malmgren, B.A., 2004. Spatial analysis of extreme precipitation in Sweden 1961–2000. AMBIO A J. Hum. Environ. 33, 187–192.
Helsel, D.R., Hirsch, R.M., 2002. Statistical Methods in Water Resources, Book 4, Hydrologic Analysis and Interpretation, United States Geological Survey.
Hirsch, R.M., Slack, J.R., 1984. Nonparametric trend test for seasonal data with serial dependence. Water Resour. Res. 20 (6), 727–732.
Hodgson, M.E., 1989. Searching methods for rapid grid interpolation. Prof. Geogr. 411, 51–61.
Huntington, T.G., 2006. Evidence for intensification of the global water cycle: review and synthesis. J. Hydrol. 319, 83–95.
IPCC, 2018. http://www.ipcc-data.org/guidelines/pages/scen_selection.html.
IPCC, 2007. Climate change 2007: the physical science basis. In: Solomon, S., Qin, D., Manning, M., et al. (Eds.), Contribution of Working Group I to the Fourth Assessment Report of the Intergovernmental Panel on Climate Change. Cambridge University Press, Cambridge, New York, pp. 1–18.
Iwashima, T., Yamamoto, R., 1993. A statistical analysis of the extremes events: long-term trend of heavy daily precipitation. J. Meteorol. Soc. Jpn. 71, 637–640.
Jacob, D., Hurk, B.V., 2009. Climate change scenarios at the global and local scales. In: Ludwig, F., et al. (Eds.), Climate Change Adaptation in the Water Sector. Earthscan, U.K.
Jain, D., Singh, V.P., 1987. Comparison of Some Flood Frequency Distributions Using Empirical Data. In: Singh, V.P. (Ed.), Hydrologic Frequency Modeling. D. Reidel Publishing Company, The Netherlands, pp. 467–485.
Jarque, C.M., Bera, A.K., 1987. A test for normality of observations and regression residuals. Int. Stat. Rev. 55 (2), 163–172.
Kahya, E., Dracup, J.A., 1993. U.S. streamflow patterns in relation to the El Niño/Southern Oscillation. Water Resour. Res. 29 (8), 2491–2503. https://doi.org/10.1029/93WR00744.
Kajornrit, J., Wong, W.K., Fung, C.C., 2012. Estimation of missing precipitation records using modular artificial neural networks. In: Huang, T., et al. (Eds.), ICONIP 2012, Part IV, LNCS 7666, pp. 52–59.
Karl, R.T., Knight, R.W., Plummer, N., 1995. Trends in high-frequency climate variability in the twentieth century. Nature 377, 217–220.
Karl, T.R., Knight, R.W., 1998. Secular trends of precipitation amount, frequency, and intensity in the United States. Bull. Am. Meteorol. Soc. 79 (2), 231–241.
Karnieli, A., 1990. Application of kriging technique to areal precipitation mapping in Arizona. GeoJournal 22 (4), 391–398.
Keggenhoff, I., Elizbarashvili, M., Amiri-Farahani, A., King, L., 2014. Trends in daily temperature and precipitation extremes over Georgia, 1971–2010. Weather Clim. 4, 75–85.

Khan, S.G., Kuhn, A.R., Ganguly, D.J., Erickson, Ostrouchov, G., 2007. Spatio-temporal variability of daily and weekly precipitation extremes in South America. Water Resour. Res. 43, W11424. https://doi.org/10.1029/2006WR005384.

Kim, U., Kaluarachchi, J., Smakhtin, V.U., 2008. Generation of monthly precipitation under climate change for the Upper Blue Nile River basin, Ethiopia. J. Am. Water Resour. Assoc. 1752–1688.

Kottek, M., Grieser, J., Beck, C., Rudolf, B., Rubel, F., 2006. World Map of KoppenGeiger climate classification updated. Meteorol. Z 15 (3), 259–263.

Kuligowski, R.J., Barros, A.P., 1998. Using artificial neural networks to estimate missing rainfall data. J. Am. Water Resour. Assoc. 34 (6), 1437–1447.

Kundezwewicz, Z.W., Robson, A., 2000. Detecting trend and other changes in hydrological data. In: WMO/TD-No. 1013. UNESCO, WMO, Geneva, Switzerland.

Kunkel, K.E., Andsager, K., Easterling, D., 1999. Long-term trends in extreme precipitation events over the conterminous United States and Canada. J. Clim. 12 (8), 2515–2527.

Kwiatkowski, D., Phillips, P.C.B., Schmidt, P., Shin, Y., 1992. Testing the null hypothesis of stationarity against the alternative of a unit root. J. Econom. 54, 159–178.

Lachniet, M.S., Burns, S.J., Piperno, D.R., Asmerom, Y., Polyak, V.J., Moy, C.M., Christenson, K., 2004. A 1500-year El Niño/Southern Oscillation and rainfall history for the Isthmus of Panama from speleothem calcite. J. Geophys. Res. 109, D20117. https://doi.org/10.1029/2004JD004694.

Landot, T., Sgellari, S., Lima, C., Lall, U., 2008. In-filling Missing Historical Daily Rainfall Data Study. Final Report submitted to South Florida Water Management District. Columbia University, New York, USA.

Latif, M., Barnett, T.P., 1994. Causes of decadal climate variability over the north Pacific and north America. Science 266, 634–637.

Lettenmaier, D.P., Wood, E.F., Wallis, J.R., 1994. Hydro-climatological trends in the continental United States, 1948-88. J. Clim. 7 (4), 586–607.

Levene, H., 1960. Robust tests for equality of variances. In: Olkin, I., Hotelling, H., et al. (Eds.), Contributions to Probability and Statistics: Essays in Honor of Harold Hotelling. Stanford University Press, pp. 278–292.

Liew, S.C., Raghavan, S.V., Liong, S.-Y., 2013. How to construct future IDF curves, under changing climate, for sites with scarce rainfall records? Hydrol. Process. https://doi.org/10.1002/hyp.9839.

Lilliefors, H.W., 1967. On the Kolmogorov-Smirnov test for normality with mean and variance unknown. J. Am. Stat. Assoc. 62, 399–402.

Ling, H.T.M., 2010. Distance coefficients between two lists or sets. Python Pap. 2, 1–31.

Little, R.J.A., Rubin, D.B., 2002. Statistical Analysis with Missing Data. Wiley, New York.

Lloyd, C.D., 2009. Multivariate interpolation of monthly precipitation amount in the United Kingdom. In: Atkinson, P.M., Lloyd, C.D. (Eds.), Geostatistics for Environmental Applications: Proceedings of the Seventh European Conference on Geostatistics for Environmental Applications. Springer, Berlin.

Lohmann, U., 2008. Aerosol effects on precipitation locally and globally. In: Bronnimann, S., Luterbacher, J., Tracy, E., et al. (Eds.), Climate Variability and Extremes during the Past 100 Years, Advances in Global Change Research, vol. 33. Springer, Dordrecht.

Lu, X., 2010. Applying climate information for adaptation decision-making: a guidance and resource document. In: National Communications Support Programme, UNDP.

Lyon, B., Barnston, A.G., 2005. ENSO and the spatial extent of interannual precipitation extremes in tropical land areas. J. Clim. 18, 5095–5109.

Mage, D.T., 1982. An objective graphical method for testing normal distributional assumptions using probability plots. Am. Stat. 36 (2), 116–120.

Manton, M.J., et al., 2001. Trends in extreme daily rainfall and temperature in southeast Asia and the south Pacific: 1961-1998. Int. J. Climatol. 21, 269–284.
Massey, F.J., 1951. The Kolmogorov-Smirnov test for goodness-of-fit. J. Am. Stat. Assoc. 46 (253), 68–78.
McCabe, G.J., 1996. Effects of winter atmospheric circulation on temporal and spatial variability in annual streamflow in the western United States. Hydrol. Sci. J. 41 (6), 873–887.
McCabe, G.J., Palecki, M.A., Betancourt, J.J., 2004. Pacific and Atlantic Ocean influences on multidecadal drought frequency in the United States. Proc. Natl. Acad. Sci. U.S.A. 101 (12), 4136–4141.
McCuen, R.H., 2004. Hydrologic Analysis and Design. Prentice-Hall.
McKee, T.B., Doesken, N.J., Kleist, J., 1993. The relationship of drought frequency and duration to time scale. In: Proceedings of the Eighth Conference on Applied Climatology, Anaheim, California, 17–22 January 1993. American Meteorological Society, Boston, pp. 179–184.
McKee, T.B., Doesken, N.J., Kleist, J., 1995. Drought monitoring with multiple timescales. In: Proceedings of the Ninth Conference on Applied Climatology, Dallas, Texas, 15–20 January 1995. Boston American Meteorological Society, pp. 233–236.
Milly, P.C.D., Betancourt, J., Falkenmark, M., Hirsch, R.M., Kundzewicz, Z.W., Lettenmaier, D.P., Stouffer, R.J., 2008. Stationarity is dead: whither water management. Science 319, 573–574.
Min, S.-K., Zhang, X., Zwiers, F.W., Hegerl, G.C., 2011. Human contribution to more-intense precipitation extremes. Nature 470, 378–381.
Moss, R., Edmonds, J., Hibbard, K., Manning, M., Rose, S., Van Vuuren, D., et al., 2010. The next generation of scenarios for climate change research and assessment. Nature 463, 747–756.
Nicholls, R.J., Wong, P.P., Burkett, V., Woodroffe, C.D., Hay, J., 2008. Climate change and coastal vulnerability assessment—Scenarios for integrated assessment. Sustain. Sci. 3 (1), 89–102.
O'Brien, J.J., Zierden, D.F., Legler, D., Hanson, J.W., Jones, J.W., Smajstrla, A.G., Podesta, G.P., Lenton, D., 1999. El Ni~no, La Ni~na and Florida's Climate: Effects on Agriculture and Forestry, Florida Consortium. Florida State University, Tallahassee.
O'Sullivan, D., Unwin, D.J., 2010. Geographical Information Analysis. John Wiley & Sons, Inc., New Jersey.
Pal, I., Al-Tabbaa, A., 2009. Trends in seasonal precipitation extremes – an indicator of 'climate change' in Kerala, India. J. Hydrol. 367, 62–69.
Panofsky, H.A., Brier, G.W., 1958. Some Applications of Statistics to Meteorology. The Pennsylvania State University Press, 224 p.
Pappas, C., Papalexiou, S.M., Koutsoyiannis, D., 2014. A quick gap filling of missing hydrometeorological data. J. Geophys. Res. Atmos. 1920–1930.
Peck, A., Prodanovic, P., Simonovic, S.P., 2012. Rainfall intensity duration frequency curves under climate change: city of London, Ontario, Canada. Can. Water Resour. J. 37 (2), 177–189.
Phillips, P., Perron, P., 1988. Testing for a unit root in time series regression. Biometrika 75, 335–346.
Presti, R.L., Barca, E., Passarella, G., 2010. A methodology for treating missing data applied to daily rainfall data in the Candelaro River Basin (Italy). Environ. Monit. Assess. 160 (1–4), 1–22.
Regonda, S.K., Rajagopalan, B., Clark, M., Zagona, E., 2006. A multi-model ensemble forecast framework: application to spring seasonal flows in the Gunnison River Basin. Water Resour. Res. 42 (9), 1–14.

Rogers, J.C., Coleman, J.S.M., 2003. Interactions between the Atlantic multidecadal oscillation, El Niño/La Niña, and the PNA in winter Mississippi valley stream flow. Geophys. Res. Lett. 30 (10), 1518. https://doi.org/10.1029/2003GL017216.

Ropelewski, C.F., Halpert, M.S., 1987. Global and regional scale precipitation patterns associated with the El Niño/Southern Oscillation. Mon. Weather Rev. 115, 1606–1626.

Salas, J.D.-J., 1993. Analysis and modeling of hydrological time series. In: Maidment, D.R. (Ed.), Handbook of Hydrology. Mc-Graw-Hill, NY.

Satterthwaite, F.E., 1946. An approximate distribution of estimates of variance components. Biom. Bull. 2, 110–114.

Schneider, N., Cornuelle, B.D., 2005. The forcing of the Pacific decadal oscillation. J. Clim. 18, 4355–4373.

Shepard, D., 1968. A two-dimensional interpolation function for irregularly spaced data. In: Proceedings of the Twenty-Third National Conference of the Association for Computing Machinery, pp. 517–524.

Simanton, J.R., Osborn, H.B., 1980. Reciprocal-distance estimate of point rainfall. J. Hydraul. Eng. Div. 106 (HY7), 1242–1246.

Snedecor, G.W., Cochran, G.W., 1989. Statistical Methods. Iowa State University Press.

Solaiman, A.T., Simonovic, S.P., 2011. Development of Probability Based Intensity-duration-frequency Curves under Climate Change. Report No. 072. University of Western Ontario.

Stainforth, D.A., Allen, M.R., Tredger, E.R., Smith, L.A., 2007. Confidence, uncertainty and decision support relevance in climate predictions. Philos. Trans. R. Soc. 365, 2145–2161.

Suppiah, R., Hennessy, K.J., 1998. Trends in total rainfall, heavy rainfall events, and number of dry events in Australia. Int. J. Climatol. 18 (10), 1141–1164.

Svenson, C., Jones, D., 2010. Review of rainfall frequency estimation methods. J. Flood Risk Manag. 3, 296–313.

Teegavarapu, R.S.V., Goly, A., 2011. Assessment of Spatial and Temporal Variation of Extreme Precipitation Events in Coastal Basins of SFWMD Region. SFWMD final report, 152 p.

Teegavarapu, R.S.V., Goly, A., 2018. Optimal Selection of Predictor Variables in Statistical Downscaling Models of Precipitation. Water. Resour. Manag. 32 (6), 1969–1992.

Teegavarapu, R.S.V., Nayak, A., 2017. Evaluation of long-term trends in extreme precipitation: implications of infilled historical data and temporal-window based analysis. J. Hydrol. 550, 616–634.

Teegavarapu, R.S.V., 2007. Use of universal function approximation in variance dependent surface interpolation technique: an application in Hydrology. J. Hydrol. 332, 16–29.

Teegavarapu, R.S.V., 2010. Modeling climate change uncertainties in water resources management models. Environ. Model. Softw. 25 (10), 1261–1265.

Teegavarapu, R.S.V., 2012. Spatial interpolation using nonlinear mathematical programming models for estimation of missing precipitation records. Hydrol. Sci. J. 57 (3), 383–406.

Teegavarapu, R.S.V., 2013a. Statistical corrections of spatially-interpolated missing precipitation data estimates. Hydrol. Process. 28 (11), 3789–3808.

Teegavarapu, R.S.V., 2013b. Floods in a Changing Climate: Extreme Precipitation. Cambridge University Press, London, UK.

Teegavarapu, R.S.V., 2013c. Climate change-sensitive hydrologic design under uncertain future precipitation extremes. Water Resour. Res. 49 (11), 7804–7814.

Teegavarapu, R.S.V., 2014. Missing precipitation data estimation using optimal proximity metric-based imputation, nearest neighbor classification and cluster-based interpolation methods. Hydrol. Sci. J. 59 (11), 2009–2026.

Teegavarapu, R.S.V., 2016. Climate variability and changes in precipitation extremes and characteristics. In: Kolokytha, E., Oishi, S., Teegavarapu, R. (Eds.), Chapter 1, Sustainable Water Resources Planning and Management under Climate Change. Springer Publication.

Teegavarapu, R.S.V., Goly, A., Obeysekera, J., 2013. Influences of Atlantic multidecadal oscillation phases on regional precipitation extremes. J. Hydrol. 495, 74–93.

Teegavarapu, R.S.V., Aly, A., Pathak, C.S., Ahlquist, J., Fuelberg, H., Hood, J., 2018. Infilling missing precipitation records using variants of spatial interpolation and data-driven methods: use of optimal weighting parameters and nearest neighbor-based corrections. Int. J. Climatol. 38 (2), 776–793.

Teegavarapu, R.S.V., Pathak, C., 2008. Infilling of rain gage records using radar (NEXRAD) data: influence of spatial and temporal variability of rainfall processes. In: Babcock, R.W., Walton, R. (Eds.), Proceedings of World Environmental and Water Resources Conference, Hawaii: ASCE, pp. 1–9. https://doi.org/10.1061/40976(316)406.

Teegavarapu, R.S.V., Nayak, A., Pathak, C., 2011. Assessment of long-term trends in extreme precipitation: implications of in-filled historical data and temporal window-based analysis. In: Proceedings of the 2011 World Environmental and Water Resources Congress. https://doi.org/10.1061/41173(414)419.

Teegavarapu, R., Pathak, C., Chintalapudi, S., 2010. Infilling missing precipitation data using NEXRAD data: use of optimal spatial interpolation and data-driven methods. In: Proceedings of World Environmental and Water Resources Congress, pp. 4659–4667. https://doi.org/10.1061/41114(371)473.

Teegavarapu, R.S.V., Chandramouli, V., 2005. Improved weighting methods, deterministic and stochastic data-driven models for estimation of missing precipitation records. J. Hydrol. 312, 191–206.

Teegavarapu, R.S.V., 2009. Estimation of missing precipitation records integrating surface interpolation techniques and spatio-temporal association rules. J. Hydroinf. 11 (2), 133–146.

Teegavarapu, R.S.V., Tufail, M., Ormsbee, L., 2009. Optimal functional forms for estimation of missing precipitation records. J. Hydrol. 374, 106–115.

Thom, H.C.S., 1966. Some Methods of Climatological Analysis. WMO N. 199. Technical Note number 81., Geneva, p. 53.

Tootle, G.A., Piechota, T.C., Singh, A., 2005. Coupled oceanic-atmospheric variability and U.S. streamflow. Water Resour. Res. https://doi.org/10.1029/2005WR004381.

Tortorelli, R.L., Rea, A., Asquith, W.H., 2005. Depth–duration–frequency of Precipitation for Oklahoma. Water-Resources Investigations Report 99–4232, U.S. Geological Survey.

Trimble, P.J., Obeysekera, J., Cadavid, L., Santee, E.R., 2005. Application of climate outlooks for water management in south Florida. In: Grabrecht, J.D., Piechota, T.C. (Eds.), Climate Variations, Climate Change and Water Resources Engineering. American Society of Civil Engineers (ASCE).

Tung, Y., 1983. Point rainfall estimation for a mountainous region. J. Hydraul. Eng. 109 (10), 1386–1393.

USEPA, 2008. National Water Program Strategy Response to Climate Change. Office of Water (4101M), EPA 800-R-08–001, USA.

Vieux, B.E., 2001. Distributed Hydrologic Modeling Using GIS. Water Science and Technology Library, Kluwer Academic Publishers.

Walsh, P.D., Lawler, D.M., 1981. Rainfall seasonality: description, spatial patterns and change through time. Weather 36, 201–208.

Watt, E., Marsalek, J., 2013. Critical review of the evolution of the design storm event concept. Can. J. Civ. Eng. 40 (2), 105–113.

Wei, T.C., McGuinness, J.L., 1973. Reciprocal distance squared method, a computer technique for estimating area precipitation. Technical Report ARS-Nc-8. In: U.S. Agricultural Research Service, North Central Region, Ohio.

Wentz, F.J.I., Ricciardulli, K., Hilburn, K.K., Mears, C., 2007. How much more rain will global warming bring? Science 317, 233–235.

Westerberg, I., Walther, A., Guerrero, J.-L., Coello, Z., Halldin, S., Xu, C.-Y., Chen, D., Lundin, L.-C., 2009. Precipitation data in a mountainous catchment in Honduras: quality assessment and spatiotemporal characteristics. Theor. Appl. Climatol. 101 (3–4), 381–296.

Winsberg, M.D., 2003. Florida Weather. University Press of Florida, Gainesville, Florida.

WMO (World Meteorological Organization, 2009. Guidelines on Analysis of Extremes in a Changing Climate in Support of Informed Decisions for Adaptation, Climate Data and Monitoring, WCDMP-No. 72, Geneva.

WMO (World Meteorological Organization), 2012. Standard Precipitation Index User Guide, WMO-No. 1090, Geneva.

Xia, Y., Fabian, P., Stohl, A., Winterhalter, M., 1999. Forest climatology: estimation of missing values for Bavaria, Germany. Agric. For. Meteorol. 96, 131–144.

Yue, S., Pilon, P., Cavandias, G., 2002. Power of the Mann-Kendall and Spearman's rho tests for detecting monotonic trends in hydrological series. J. Hydrol. 259 (1–4), 254–271.

Zhu, J., Stone, M.C., Forsee, W., 2012. Analysis of potential impacts of climate change on intensity–duration–frequency (IDF) relationships for six regions in the United States. J. Water Clim. Change 3 (3), 185–196.

FURTHER READING

Akkala, A., Devabhaktuni, V., Kumar, A., 2010. Interpolation techniques and associated software for environmental data. Environ. Prog. Sustain. Energy 29 (2), 134–141.

Kanevski, M., Maignan, M., 2004. Analysis and Modelling of Spatial Environmental Data. EPFL Press, Lausanne, Switzerland.

Teegavarapu, R.S.V., Oishi, S., 2016. Evaluation of Trends and Variability in Sea Level, Temperature and Precipitation of Japan: Links to Climate Variability and Change. Research Center for Urban Safety and Security (RCUSS) report. Kobe University, Japan.

Teegavarapu, R.S.V., 2008. Innovative spatial interpolation methods for estimation of missing precipitation records: concepts and applications. In: Bruthans, J., Kovar, K., Hrkal, Z. (Eds.), Proceedings of HydroPredict-2008. Czech Republic. Association of Hydrogeologists, Prague, pp. 79–82.

Teegavarapu, R.S.V., Goly, A., Viswanathan, C., Behera, P., 2012a. Precipitation extremes and climate change: evaluation using descriptive WMO indices. In: Proceedings of World Environmental and Water Resources Congress 2012: Crossing Boundaries, pp. 1927–1936. https://doi.org/10.1061/9780784412312.193).

Teegavarapu, R.S.V., Meskele, T., Pathak, C., 2012b. Geo-spatial grid-based transformation of precipitation estimates using spatial interpolation methods. Comput. Geosci. 40, 28–41.

CHAPTER 3

Modeling High-Intensity Precipitation for Urban Hydrologic Designs

Chandra R. Rupa, P.P. Mujumdar
Department of Civil Engineering, Indian Institute of Science, Bangalore, India

Contents

Trends and Changes in Hydroclimatic Variables
ISBN 978-0-12-810985-4
https://doi.org/10.1016/B978-0-12-810985-4.00003-7

1. INTRODUCTION

Extreme precipitation events are rare in nature and modeling such events is one of the challenging tasks in hydrology. Quantifying precipitation extremes is important in design and planning of hydrologic infrastructure, especially in urban areas. Estimating the probability of extreme events is difficult in urban areas due to limited records and high spatial variability. Given these constraints in modeling precipitation extremes, however, it is crucial to have reliable estimates of intensities for optimal designing of urban infrastructure. Typically, short-duration high-intensity precipitation events cause heavy flooding in urban areas and these short-duration events are burdened with high uncertainties, necessitating precise quantification. A common measure to appraise extreme events is through return levels for a given return period.

The analysis of extreme precipitation events in urban areas involves two types of modeling: temporal and spatial. In temporal analysis, a single-station data is considered for obtaining return levels, typically, intensity-duration-frequency (IDF) relationships are developed. The temporal changes in return levels, which can be either due to local changes like land use/land cover or due to anthropogenic climate change, are modeled at fine time scales, say 15 min. Precipitation data at such fine time scales is often sparse leading to high uncertainties in modeling return levels. In spatial analysis, variation in precipitation and changes in precipitation patterns are modeled within and surrounding urban areas. A high spatial variability in precipitation is expected in urban areas as the climatology is different compared to its surroundings. Also, it is asserted that urbanization could noticeably modify precipitation patterns affecting the physical properties of precipitation, including intensity, frequency, and accumulated rainfall (Yang et al., 2015). This spatial nonuniformity in precipitation is a localized feature, which cannot be captured by modeling at large-scale. Again, the difficulty in estimation of the spatial distribution of precipitation extremes in urban areas is exacerbated by a sparse observational network, which could lead to a significant uncertainty in the estimates of spatial return levels. Hence, there is a great need to quantify uncertainties in return levels, both spatial and temporal, in urban areas.

A standard methodology for modeling precipitation extremes is to employ an asymptotic distribution to model the extreme events. Based on the model, estimates of parameters are obtained using standard statistical techniques, namely, Maximum Likelihood Estimation (MLE), and further inferences are drawn for hydrologic designs. There are two major concerns as to why the statistical techniques lead to systematic underestimation or overestimation of the probability of disastrous events. First concern is related to the nonavailability of adequate quantity and quality of historical data in developing the asymptotic model. The standard statistical methods rely on sampling distribution, which involves the behavior of a statistic over repeating sampling, and requires sufficiently "large" data to ascertain the correctness of the asymptotic model fitted to the data. The second concern is related to parameter uncertainty, which is often overlooked or ignored most of the times. It is important to quantify these uncertainties and use them in the design decisions and risk analyses. Bayesian approach provides a coherent framework for estimating these uncertainties (Chandra et al., 2015).

In this chapter, modeling precipitation extremes in urban areas using Bayesian analysis are presented. The advantages and disadvantages of employing Bayesian methods are listed. Modeling precipitation extremes at a single station, that is, the temporal modeling is presented in Sections 2 and 3. The uncertainties associated with the return levels due to parameters of fitted model to the observations and due to climate change is quantified using Bayesian method and Reliability Ensemble Averaging (REA) method, respectively. Disaggregation method for obtaining shorter-duration return levels from that of larger durations, using scale invariance theory, is elaborated. Modelling spatial variability, assuming independence as well as considering dependence among the observations using hierarchical Bayesian modeling and max-stable process respectively is discussed in Section 4. The recent developments in both the modeling aspects (Bayesian methods and max-stable processes) and the limitations are discussed in Section 5, and the chapter is concluded with listing the research outcomes in Section 6.

2. METHODS FOR MODELING PRECIPITATION EXTREMES

Statistical modeling of precipitation extremes, which deal with characterizing the tail of the distribution, is a widely studied problem due to its importance in quantifying hydrologic risk. The analysis of extreme events is based on extreme value (EV) and extreme excess (EE) theories, which consider respectively a block maxima (usually annual maxima) and events exceeding a given threshold, resulting in two distribution classes: Generalized Extreme Value (GEV) and Generalized Pareto (GP) distributions. The block maximum series is constructed by considering the block maxima over different years, while the exceedance series (Partial Duration Series, PDS, or the Peaks Over Threshold, POT) is constructed by retaining all excesses above a sufficiently high threshold, u. In this chapter, only the block maxima approach, considering the block as annual, is discussed in the subsequent sections for conciseness.

2.1 Extreme Value Theory and Asymptotic Models

Given a sequence of iid random variables, X_1, X_2, …, with a common distribution, and letting $M_n = \max(X_1, X_2, \ldots)$, if there exists continuous functions $a_n > 0$ and $b_n \in \Re$, the limiting distribution is defined by:

$$\Pr\left(\frac{M_n - b_n}{a_n} \leq y\right) \rightarrow G(y) \quad n \rightarrow \infty$$

The limiting distribution G is nondegenerate and belongs to one of the distributions constituting the family of extreme value distributions depending on the tail behavior. They are the Gumbel ($k = 0$, unbounded light tail), the Frechet ($k > 0$, heavy tail), and the Weibull ($k < 0$, finite upper bounded tail) families. If G exists, then it must be max-stable and the only nondegenerate distribution with this property is the GEV distribution with Cumulative Distribution Function (CDF) given by Eq. (3.1).

$$G(y) = \begin{cases} \exp\left\{-\exp\left[-\dfrac{y-\mu}{\sigma}\right]\right\} & k = 0 \\ \exp\left\{-\left[1 + k\left(\dfrac{y-\mu}{\sigma}\right)\right]^{-1/k}\right\} & k \neq 0, \{x: 1 + k(x-\mu)/\sigma > 0\} \end{cases} \tag{3.1}$$

where μ is location parameter ($\mu \in \Re$), σ the scale parameter ($\sigma > 0$), and k is the shape parameter ($k \in \Re$). The shape parameter, k, defines the tail behavior of the distribution. The return level (z_p) corresponding to the probability of exceedance p (or the return period $T = 1/p$) is given by

$$z_p = F^{-1}(p/\mu, \sigma, k)$$

$$z_p = \begin{cases} \mu - \dfrac{\sigma}{k}\left[1 - \{-\log(1-p)\}^{-k}\right] & \text{for} \quad k \neq 0 \\ \mu - \sigma \log[\log(1-p)] & \text{for} \quad k = 0 \end{cases} \tag{3.2}$$

2.2 Parameter Estimation

Estimates of parameters, $\theta = (\mu,\ \sigma,\ k)^T$, can be obtained by several methods. These include the method of maximum likelihood estimation, the method of probability weighted moments (PWM) or the L-moments, and the method of maximum product spacing (MPS). The L-moments and the MPS method are based on the order statistics and can be used in situations where the MLE does not exist. The L-moments are analogous to conventional moments but can be estimated by linear combinations of the elements of an ordered sample (i.e., by L-statistics). The MPS method is a general estimation method for univariate extremes. The notion of MPS is, on uniform scale after the probability integral transform, to choose parameter values that make the observed data as uniform as possible. This is done by maximizing the geometric mean of the spacing between consecutively ordered observations. The specific details of these two methods are not discussed here for brevity and the MLE method is considered further as it is the prevalent and simpler method in obtaining the estimates of parameters. If $f(.\ |\theta)$ is the probability density function (PDF) of the GEV distribution and n is the total number of maxima for a given time period, the likelihood function is given by

$$L(y|\theta) = \prod_{i=1}^{n} f(y_i|\theta)$$

where θ is the parameter vector. The point estimates of the parameters are obtained by maximizing the log likelihood function, which are used in obtaining the return levels (z_p).

2.3 Bayesian Approach

Though the methods specified in the preceding section are robust and simple in applicability, an uncertainty is associated with the estimates of the parameters, due to uncertainty in observed data (sampling uncertainty). This uncertainty needs to be quantified precisely, which can be achieved by employing Bayesian methods. The basic principle of Bayesian method is to upgrade one's uncertainty about parameter (θ) expressed in terms of prior $\pi(\theta)$, before observing the data to post data availability, using Bayes rule. This is achieved by modifying the prior distribution of parameters due to the information contained in the data, expressed through the likelihood function.

The posterior distribution of parameters $\pi(\theta|y)$ is obtained by multiplying the likelihood function of the GEV distribution, $L(y|\theta)$and the prior distribution of parameters. The choice of prior distribution of parameters is crucial in the Bayesian analysis. The unnormalized posterior density of the parameters is given by

$$\pi(\theta|y) \propto L(y|\theta) \times \pi(\theta) \tag{3.3}$$

where θ is the parameter vector $(\mu, \sigma, k)^T$, $\pi(\theta)$ is the prior distribution of parameters, and $L(y|\theta)$ is the likelihood function equal to, $\prod_{i=1}^{m} f(y_i|\theta)$, m is the number of observations and $f(y_i|\theta)$ is the conditional PDF of annual maximum precipitation given by the parameter vector (θ). Bayesian inference obeys the so called *likelihood principle*, which states "for a given sample of data, any two probability models that have the same likelihood function yield the same inference for θ" (Gelman et al., 2004).

Bayesian inference about the model parameters proceeds by solving Eq. (3.3) and obtaining the posterior distribution of parameters. It is not possible to solve Eq. (3.3) analytically leading to alternative methods, like Markov Chain Monte Carlo (MCMC). If the interest lies in various features of the posterior distribution, this can be approximately studied if one has a large sample θ_1, θ_2, ..., θ_N from the posterior distribution of θ. For example, for a given return period $1/p$, the return level can be obtained from the sequence of θ. MCMC is used for drawing large sample of θ, which involves setting up a Markov chain with posterior distribution as its stationary distribution and then drawing samples using its transition density from the Markov chain, resulting in a set of parameter estimates.

2.4 Estimation of Uncertainty in Parameters

A sampling uncertainty is associated with the point estimates of parameters obtained from any standard statistical methods (MLE, L-moments, and MSE). Delta method provides uncertainty bounds with an assumption that the distribution of return levels obtained from the MLE estimator follows a Gaussian distribution. This method is also applicable for the point estimates obtained using L-moments. The variance of the return level is given by

$$\mathrm{Var}(z_p) \approx \nabla z_p^T V \nabla z_p$$

where V is the variance-covariance matrix of θ and

$$\nabla z_p^T = \left[\frac{\partial z_p}{\partial \mu}, \frac{\partial z_p}{\partial \sigma}, \frac{\partial z_p}{\partial k}\right]$$

evaluated at $\widehat{\theta}$. The variance can be used to obtain the confidence levels of the return levels, under the assumption the distribution of quantiles is Gaussian. However, this assumption may not be valid for return levels at shorter durations. Bayesian approach yields the uncertainty in parameters with credible intervals, obtained from the posterior distribution of parameters.

2.4.1 Example: Bangalore Case Study

The advantage of using the Bayesian method is discussed by considering precipitation data observed at Bangalore City station (IMD station no. 43295), Bangalore. The 15 min precipitation data for the period of 1969–2003 is considered. Annual maximum time series for 15 min, 30 min, 1 h, 3 h, 6 h, 12 h, and 24-h durations are obtained. Fig. 3.1 shows the annual maximum series of 15 min, 30 min, 1 h, and 24-h duration. The dotted line shows linear trend in the time series. The time series is assumed

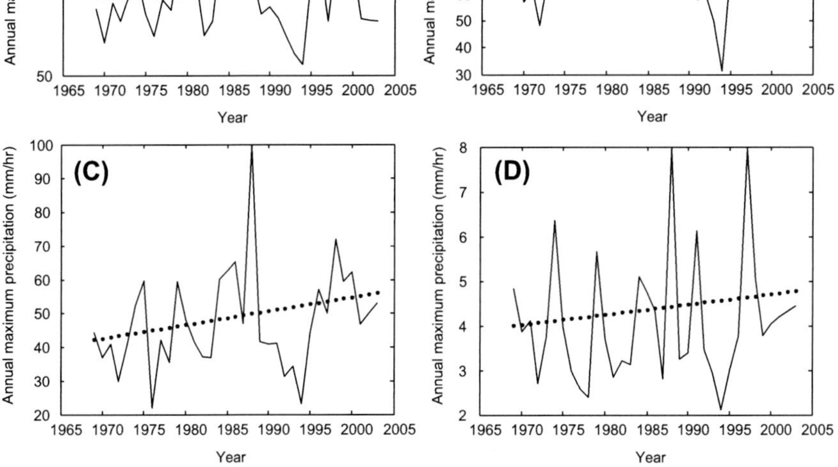

Figure 3.1 Annual maximum precipitation time series for (A) 15 min, (B) 30 min, (C) 1 h and (D) 24 h durations. The *black dotted lines* show the trend line in the extreme precipitation time series.

to be stationary and the estimates of parameters are obtained from both MLE approach and Bayesian analysis. In the Bayesian analysis, a prior distribution needs to be specified for each of the parameters. Non-informative prior distributions are considered for location (μ) and scale (σ) parameter. The location parameter is considered to follow a uniform distribution ($\pi(\mu) \propto 1$) and the scale parameter follows Jeffrey's distribution ($\pi(\sigma) \propto 1/\sigma$ or $\pi(\log\sigma) \propto 1$)). Based on extensive studies on scale parameter by Martins and Stedinger (2001, 2000), beta distribution (B(.)) with hyper parameters $a = 6$ and $b = 9$ is considered. The prior distribution of parameters $\pi(\theta)$ is given by Eq. (3.4).

$$\pi(\mu, \sigma, k) \propto \frac{B(k + 0.5 | a = 6, b = 9)}{\sigma} \tag{3.4}$$

Eq. (3.3) is solved using the method of MCMC by setting up a Markov chain with posterior distribution of parameters, $\pi(\theta|x)$ as its stationary distribution. From the posterior distribution, samples of 20,000 parameter sets (μ, σ, k) are generated and the initial 2000 samples of the chain are discarded, as a burn-in phase. The thinning interval, that is, how frequently to sample from the chain, is varied from 30 to 50. The sets of parameters (μ, σ, k) obtained from the posterior distribution are transformed to obtain return levels using Eq. (3.2) leading to a sample from the corresponding distribution of T year return level.

Table 3.1 gives the estimates of parameters of GEV distribution and the return levels for 10-year return period calculated from (1) MLE method, referred as classical approach and (2) from Bayesian analysis. The table also gives variance in return levels from both the approaches. Fig. 3.2 shows the return levels for 10-year return period along with uncertainty bands (as box plot) for all durations, obtained from Bayesian analysis and the corresponding return levels using MLE method along with 95% confidence limits obtained from Delta method. It can be inferred from Fig. 3.1 and Table 3.1 that the uncertainty in return levels increase as the duration reduces and the uncertainties, which are very high at shorter durations, cannot be estimated accurately by employing classical methods like MLE. The assumption that the return levels follow normal distribution is violated at the shorter durations.

Fig. 3.3A shows the posterior distribution of return levels for 10-year return period and for durations of 15 min, 30 min, 1 h, 3 h, 6 h, 12 h, and 24 h, obtained from the Bayesian approach. Fig. 3.3B shows the

Table 3.1 Parameters of GEV Distribution and the Return Levels (mm/h) for 10-Year Return Period From Classical and Bayesian Approaches. Maximum Likelihood Estimation (MLE) Method Is Used in Classical Analysis

	Classical Approach					Bayesian Approach				
Duration	μ	σ	k	$Z_{T=10}$	Variance of Z_T	Mean of μ	Mean of σ	Mean of k	Mean of $Z_{T=10}$	Variance of Z_T
15 min	87.15	21.10	−0.061	128.48	65.56	84.19	22.74	−0.066	131.8	76.94
30 min	63.50	18.27	−0.158	98.1	32.62	62.94	19.39	−0.129	100.97	51.32
1 h	39.21	12.04	−0.193	61.18	12.44	38.72	12.58	−0.140	63.12	21.34
3 h	16.96	5.64	0.064	30.62	8.71	17.16	6.20	0.001	31.15	7.38
6 h	9.27	2.99	0.086	16.7	2.68	9.38	3.27	0.013	16.86	2.21
12 h	5.13	1.63	0.070	9.09	0.73	5.20	1.78	0.009	9.25	0.65
24 h	2.78	0.86	0.073	4.89	0.21	2.81	0.94	0.006	4.95	0.18

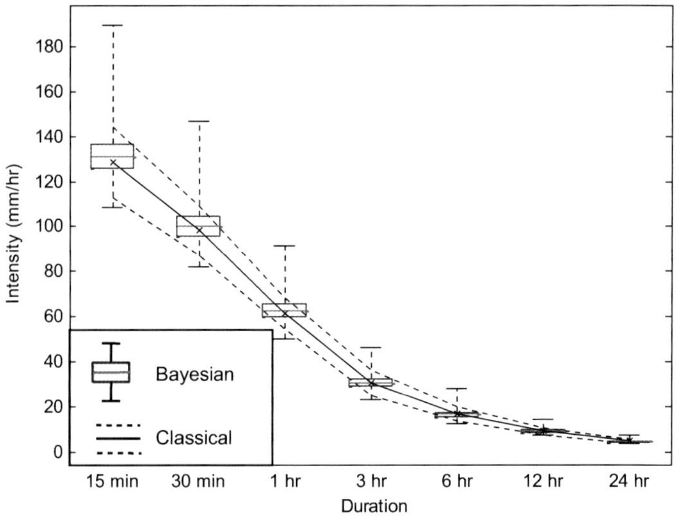

Figure 3.2 Variation in return levels for different durations in comparison with classical approach (historical time line: 1969–2003). Boxplot gives the uncertainty in return levels obtained from Bayesian analysis, IDF derived using classical approach (MLE method) is shown in *solid black line, black dashed lines* gives the uncertainty band using delta method.

posterior distribution of return levels for 15 min duration for different return periods. The return levels obtained from the estimates of parameters using MLE method are marked (*black circles*) in Fig. 3.3. From the figure, it is seen that the return levels obtained from the MLE method shifts from the mode of the posterior distribution of return levels obtained from Bayesian analysis as the duration and the return period reduces. Typically, in urban areas, the infrastructure is built for short-duration precipitation events and for return periods of order 2 year and 5 year. At this durations and return periods, the standard statistical methods like MLE underestimated the return periods.

2.5 Variability, Trends, and Nonstationarity in Extreme Precipitation

The methods for obtaining the return levels in the preceding section assume that the extreme precipitation time series is stationary. However, according to IPCC (2012), climate change is believed to increase the frequency and magnitude of the extreme precipitation events. In addition, to the climate change, rapid modifications in land use/land cover and human

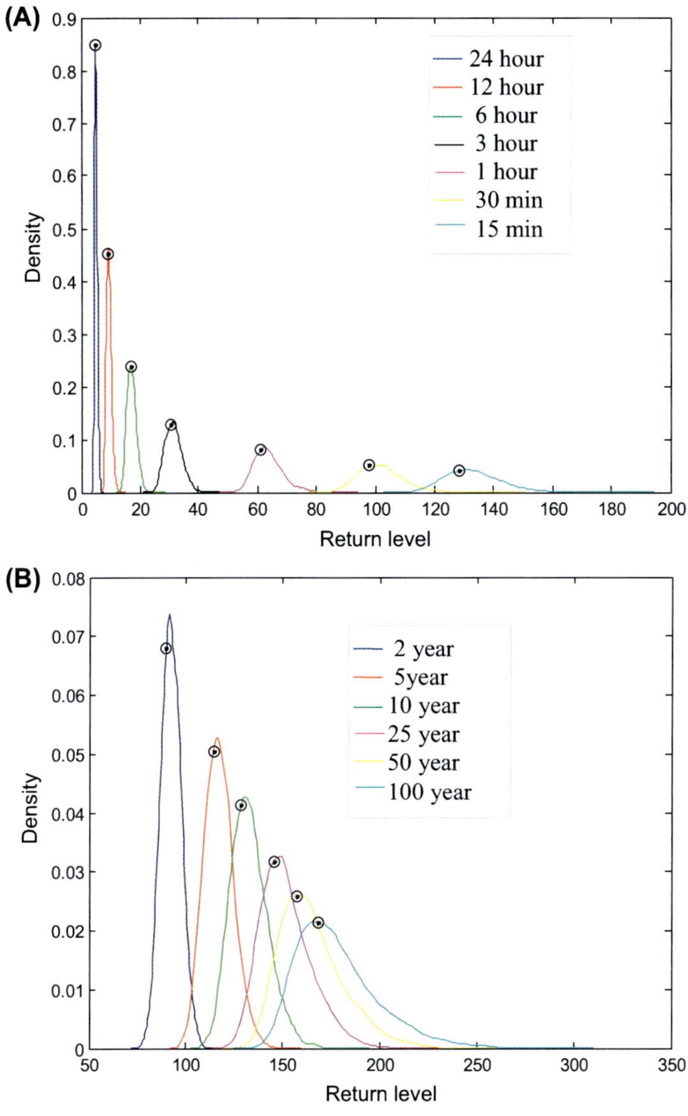

Figure 3.3 (A) Probability density function (PDF) of return levels (mm/h) for 10-year return period for various durations in comparison with return levels from Maximum Likelihood Estimation (MLE) method. Black circle is the return level obtained from MLE method at the corresponding duration. (B) PDF of return levels for 15-min duration at different return periods in comparison with return levels obtained from MLE method (*Black circles*). As the duration and the return period reduce, the return levels obtained from the MLE method shifts away from the mode of the posterior distribution of return levels obtained from Bayesian approach.

interventions augment the changes in extreme precipitation events in urban areas; specifically the short-duration high-intensity precipitation events causing flooding are likely to change. These changes indicate that the past cannot be considered as a reference to model future and the assumption of stationarity is no longer valid in modeling extreme events. In general, there is vast literature on bringing out the issues of nonstationarity in the context of changes in extreme precipitation events, both frequency and intensity, and an increasing number of studies used future-forcing—based scenarios to consider nonstationarity while assessing projected risks of extreme events. Even if the climate change is human induced, which is a common case in urban areas, only low to medium confidence exists in the projection of extreme events (IPCC, 2012). Again, this could be due to the limitations of the modeling or the lack of good quality and sufficient quantity of historical data at urban scales, which is a common scenario while modeling short-duration precipitation events. Furthermore, attributing the changes observed in extremes to climate change are challenging due to high natural variability and poor simulation of human-induced signals by the climate models at regional scales (Mondal and Mujumdar, 2016). Therefore, a holistic and practical approach should be adopted in quantifying changes at local or urban scales. Serinaldi and Kilsby (2015) suggested that non-stationarity analysis should not be based on at-site data, but requires additional information (like borrowing additional information from nearby sites, analyzing hydro-climatic variables describing possible natural and anthropogenic forcings) and Exploratory Data Analysis (EDA). They further concluded that models introduce additional source of uncertainty if the model structure cannot be inferred in a reasonable way and the resulting nonstationary models can provide no practical enhancement and accuracy of the predicted extreme quantiles. Owing to all the issues, nonstationarity detection is a major topic of attention and arriving to a clear conclusion about the existence of trend/periodicities is not a trivial task. Two methods are prevalent in detecting trends: parametric and nonparametric methods. Parametric methods include Phillips-Perron test and Kwiatkowski-Phillips-Schmidt-Shin test. Nonparametric tests include Mann-Kendall, Spearman's ρ, and Cox-Stuart tests (Fatichi et al., 2009). The likelihood ratio test is commonly used to test the suitability of the nonstationary GEV distribution. The other tests are not detailed here for brevity and only the likelihood ratio test is discussed further.

Under nonstationarity, one or more parameters of the distribution under consideration vary with covariates. For GEV distribution, parameters

(μ,σ) are expressed as $\mu(t) = \mu_0 + \mu_1 t$, $\sigma(t) = \sigma_0 + \sigma_1 t$, if the covariate is time. In general, the shape parameter, k, is kept constant as it is difficult to estimate and unlikely to vary with time (Coles, 2001). Let M_1 represent the stationary model with parameters μ, σ, k and M_2 is the nonstationary model with parameters $\mu(t)$, $\sigma(t)$, and k. If $l(M_1)$ and $l(M_2)$ are the maximized log-likelihoods for models M_1 and M_2, respectively, then the null hypothesis of M_1 can be rejected in favor of M_2 at a significance level of α if $2\{l(M_2) - l(M_1)\} > c_\alpha$, where c_α is the $(1 - \alpha)$th quantile of the chi-square distribution with additional number of parameters in M_2 as degrees of freedom. After a detailed EDA and tests, if one decides that a nonstationarity is prevailing, then the Bayesian methods described in the preceding section can be used to obtain the posterior distribution of parameters.

2.5.1 Example: Bangalore Case Study

The precipitation data observed at Bangalore City station is considered for checking nonstationarity. A linearly increasing trend is observed in the extreme precipitation time series for durations 30 min, 1 h, and 6 h (Fig. 3.1). For conciseness, only the location parameter is considered to change with time ($\mu(t) = \mu_0 + \mu_1 t$) and the scale and shape parameters are considered to be stationary. The posterior distributions of parameters are obtained from Bayesian analysis as described in the preceding section. Here, the priors for both stationary and nonstationary models are considered as Gaussian with zero mean and large variance to reflect prior ignorance. The modes of the posterior distributions are used as the parameter estimates for the likelihood ratio test statistic instead of the maximum likelihood estimates to ensure that the test is still valid for Bayesian analysis. At 95% significance level, null hypothesis, that is, stationary model, could not be rejected in favor of nonstationary model for all durations. Table 3.2 gives the values of log likelihood values for both the models and for all durations.

3. STATISTICAL DOWNSCALING OF PRECIPITATION EXTREMES

Obtaining projected IDF relationships for future is essential in calculating anticipated risk and checking resilience of urban infrastructure. Climate change may intensify flood risk, particularly in urban areas due to anticipated increase in extreme precipitation events. Global climate models (GCMs) are the primary tool for understanding how the climate may

Table 3.2 Results of the Likelihood Ratio Test for All Durations

Duration	Maximized Log Likelihood Values: Stationary Model, $l(M_1)$	Maximized Log Likelihood Values: Nonstationary Model, $l(M_2)$	Deviance Statistics $2\{l(M_2) - l(M_1)\}$	c_α
15 min	−156.51	−156.45	0.107	3.84
30 min	−151.44	−151.17	0.534	3.84
1 h	−144.88	−144.12	1.525	3.84
3 h	−130.73	−129.80	1.867	3.84
6 h	−103.10	−102.85	0.485	3.84
12 h	−82.43	−82.39	0.089	3.84
24 h	−59.86	−59.82	0.083	3.84

change in the future. GCM projections indicate that climate change may lead to changes in frequency and intensification of extreme precipitation, which consequently impacts urban drainage and flood risk management decisions. To study the impact of climate change on extreme precipitation in urban areas, downscaling techniques need to be employed as GCMs at local scale are not reliable.

3.1 Motivation for Downscaling

GCMs function at coarser resolutions and they do not provide reliable information on finer scales where hydrological processes typically occur. In particular, GCMs cannot model local circulation patterns leading to hydrological extreme events (Christensen and Christensen, 2003). Downscaling attempts to resolve this scale discrepancy and provides a coherent path for climate change impact assessment at local scales. The downscaling methods are based on the assumption that large-scale weather exhibits a strong influence on local-scale weather but, in general, neglects any reverse effects from local scales on global scales. For obtaining projected precipitation, two approaches to downscaling exist, dynamical and statistical downscaling. Dynamical downscaling layers a regional climate model (RCM) into the GCM to represent the atmospheric physics with a higher grid box resolution in a limited area of interest. Statistical downscaling establishes statistical links between large-scale weather and observed local-scale weather. Statistical downscaling methods can be broadly classified into three categories: weather generators, weather typing, and transfer functions. Further, the downscaled projected precipitation at local scales is at monthly and daily time scales. As the interest lies in obtaining projected short-duration extreme events, a disaggregation from daily time scale to finer time scales (subhourly) is required. Downscaling to urban scale and disaggregation from coarser scale to finer scale is discussed briefly in this section.

3.2 Methods for Downscaling Precipitation Extremes

Extensive research on downscaling methods and applications has been carried out and several reviews of downscaling methods have been published (e.g., Hewitson and Crane, 1996; Zorita and von Storch, 1997; Wilby and Wigley, 1997; Xu, 1999; Hanssen-Bauer et al., 2005). Studies associated with climate change impact on the IDF relationship and using it in urban drainage impact are, however, very few. Maraun et al. (2010) integrated different perspectives on precipitation downscaling from meteorologists, climatologists, statisticians, and hydrologists to identify generic end-user needs. Dynamic downscaling methods, which are based on nesting a fine-scale climate model within GCMs, are computationally expensive, dependent on boundary conditions obtained from GCMs, and lack transferability to local urban scales. Statistical methods are the prevalent techniques for downscaling the GCM outputs to obtain projected extreme precipitation in urban areas. Specifically, weather generator models have been used in numerous studies on climate change impact assessment, where the standard practice is to consider the occurrence and the amount of precipitation separately. A number of studies have been carried out in this area and an extensive review of the models can be found in Wilks and Wilby (1999). An example of downscaling the precipitation to an urban scale using an improved weather generator is presented.

3.2.1 Example: Downscaling Using Improved Weather Generator

Precipitation flux outputs of 26 GCMs for four Representative Concentration Pathways (RCP) scenarios are considered for obtaining the projected precipitation time series for Bangalore City. The downscaling to station scale is performed using an improved weather generator. A set of 85 simulations are considered corresponding to 26 CMIP5 GCMs and four RCP scenarios. The number of GCMs considered for RCP 2.6, 4.5, 6.0, and 8.5 scenarios are 20, 25, 16, and 24, respectively. Future scenarios are constructed for three time slices: 2021–50, 2051–80 and 2071–2100. For each climate model, the grid point that is nearest to the Bangalore City rain gauge station is considered. In case of two or more possible candidate grid points, the grid point with minimum Mean Absolute Percent Error (MAPE) is considered. MAPE is calculated based on the observed and GCM simulated precipitation and return levels. Table 3.3 gives the selected grid points for the GCM model.

The downscaling is performed by modifying the traditional weather generator to obtain both the rainfall occurrence and rainfall amount. For obtaining the precipitation occurrence, states of Markov chain are divided into three states: no-rain, rain of moderate intensity, and rain of high intensity. The high-intensity rainfall events are defined as the events above 0.9th quantile ($\xi > 0.9$), and moderate-intensity rainfall events are the ones below this quantile. Separate distributions are fitted for moderate-intensity and high-intensity rainfall to model the amount of precipitation. It is observed that Gamma distribution is a good fit for both moderate-intensity and high-intensity rainfall days. The three state Gamma (for moderate intensity) + Gamma (for high intensity) distribution has a good ability to simulate extreme precipitation events reproducing the mean and standard deviations. The perturbation factors of the transition probabilities and Gamma distribution parameters obtained as the ratio between future and control period GCMs are applied to historical time period to obtain the projections of precipitation time series.

Fig. 3.4 shows the plot of annual precipitation from various GCMs. It is observed that the models Access1_0, CSIRO_MK3−6−0, HadGEM2−CC, and HadGEM2−EC performed very poorly, while CanESM2, CNRM−CM5, IPSL−CM5A−LR, MIROC5, and MPI−ESM−LR were able to match the observed values reasonably well. The agreement between the model and data is a necessary, but not a sufficient, criterion. Therefore, the underperforming and/or the overperforming GCMs need not be discarded for future analyses. Also, the fact that some models may perform well with respect to the historical data, while the other models may be more accurate in simulating future projections cannot be ruled out. Multiple GCMs, considered for obtaining the projected extremes, introduce uncertainty (model uncertainty), which can be addressed by employing Reliability Ensemble Average (REA) method.

3.3 Uncertainty Due to the Use of Multimodel GCMs

The primary factors of uncertainty is that different GCMs can simulate quite different regional changes even under the same anthropogenic forcing scenario (e.g., Giorgi and Francisco, 2000; Whetton et al., 1996).

Table 3.3 GCM Model With Future Climate Scenarios and Latitude-Longitude Selected for Obtaining Future Projections of Precipitation for Bangalore City

Model Name	Modeling Center	Latitude	Longitude	Abbreviation	Climate Scenarios
ACCESS1_0	Commonwealth Scientific and Industrial Research Organization (CSIRO) and Bureau of Meteorology (BOM), Australia	12.5 N	76.875 E	A1	RCP 4.5, RCP 8.5
BCC-CSM1-1	Beijing Climate Center, China Meteorological Administration	12.55 N	78.75 E	BC1	RCP 2.6, RCP 4.5, RCP 6.0, RCP 8.5
BCC-CSM1-1-M	Beijing Climate Center, China Meteorological Administration	12.89 N	77.6 E	BC1M	RCP 2.6, RCP 4.5, RCP 6.0, RCP 8.5
BNU-ESM	College of Global Change and Earth System Science, Beijing Normal University	12.55 N	78.75 E	BNU	RCP 2.6, RCP 4.5, RCP 8.5
CanESM2	Canadian Centre for Climate Modelling and Analysis	12.55 N	78.7 E	CAN	RCP 2.6, RCP 4.5, RCP 8.5
CCSM4	National Center for Atmospheric Research	12.7225 N	77.5 E	C4	RCP 2.6, RCP 4.5, RCP 6.0, RCP 8.5
CMCC-CMS	Centro Euro-Mediterraneo per I CambiamentiClimatici	12.1 N	76.8 E	CMS	RCP 4.5, RCP 8.5

Continued

Table 3.3 GCM Model With Future Climate Scenarios and Latitude-Longitude Selected for Obtaining Future Projections of Precipitation for Bangalore City—cont'd

Model Name	Modeling Center	Latitude	Longitude	Abbreviation	Climate Scenarios
CNRM-CM5	Centre National de RecherchesMeteorologiques/ Centre Europeen de Recherche et Formation Avancees en CalculScientifique	11.9 N	77.34 E	CN5	RCP 4.5, RCP 8.5
CSIRO-MK3-6-0	Commonwealth Scientific and Industrial Research Organization	12.12 N	76.875 E	CM60	RCP 2.6, RCP 4.5, RCP 6.0, RCP 8.5
FGOALS-G2	LASG, Institute of Atmospheric Physics, Chinese Academy of Sciences and CESS, Tsinghua University	12.55 N	78.75 E	FG2	RCP 2.6, RCP 4.5, RCP 8.5
FGOALS-S2	LASG, Institute of Atmospheric Physics, Chinese Academy of Sciences	12.4 N	75.9 E	FS2	RCP 2.6, RCP 4.5, RCP 6.0, RCP 8.5
GFDL-CM3	NOAA Geophysical Fluid Dynamics Laboratory	13 N	78.75 E	GF3	RCP 2.6, RCP 6.0
GFDL-ESM2G	NOAA Geophysical Fluid Dynamics Laboratory	13.14 N	76.25 E	GF2G	RCP 2.6, RCP 4.5, RCP 6.0, RCP 8.5
GFDL-ESM2M	NOAA Geophysical Fluid Dynamics Laboratory	13.14 N	76.25 E	GF2M	RCP 2.6, RCP 4.5, RCP 6.0, RCP 8.5
GISS-E2-R	NASA Goddard Institute for Space Studies	13 N	76.25 E	GISS	RCP 4.5

HadGEM2-CC	Met Office Hadley Centre (additional HadGEM2-ES realizations contributed by Instituto Nacional de PesquisasEspaciais)	12.5 N	76.87 E	HADC	RCP 4.5, RCP 8.5
HadGEM2-ES	Met Office Hadley Centre (additional HadGEM2-ES realizations contributed by Instituto Nacional de PesquisasEspaciais)	12.5 N	76.875 E	HADE	RCP 2.6, RCP 4.5, RCP 6.0, RCP 8.5
INMCM4	Institute for Numerical Mathematics	12.75 N	78 E	IN4	RCP 4.5, RCP 8.5
IPSL-CM5A-LR	Institut Pierre-Simon Laplace	12.31 N	78.75 E	IPCL	RCP 2.6, RCP 4.5, RCP 6.0, RCP 8.5
IPSL-CM5A-MR	Institut Pierre-Simon Laplace	12.6 N	77.5 E	IPCM	RCP 2.6, RCP 4.5, RCP 6.0, RCP 8.5
MIROC5	Atmosphere and Ocean Research Institute (The University of Tokyo), National Institute for Environmental Studies, and Japan Agency for Marine-Earth Science and Technology	11.9 N	78.75 E	MI5	RCP 2.6, RCP 4.5, RCP 6.0, RCP 8.5
MIROC-ESM-CHEM	Japan Agency for Marine-Earth Science and Technology, Atmosphere and Ocean Research Institute (The University of Tokyo), and National Institute for Environmental Studies	12.55 N	78.75 E	MIEC	RCP 2.6, RCP 4.5, RCP 6.0, RCP 8.5
MIROC-ESM	Japan Agency for Marine-Earth Science and Technology, Atmosphere and Ocean Research Institute (The University of Tokyo), and National Institute for Environmental Studies	12.55 N	78.75 E	MIE	RCP 2.6, RCP 4.5, RCP 6.0, RCP 8.5

Continued

Table 3.3 GCM Model With Future Climate Scenarios and Latitude-Longitude Selected for Obtaining Future Projections of Precipitation for Bangalore City—cont'd

Model Name	Modeling Center	Latitude	Longitude	Abbreviation	Climate Scenarios
MPI-ESM-LR	Max Planck Institute for Meteorology	12.12 N	78.75 E	MPI	RCP 2.6, RCP 4.5, RCP 6.0, RCP 8.5
MRI-CGCM3	Meteorological Research Institute	12.89 N.	77.62 E	MRI	RCP 2.6, RCP 4.5, RCP 6.0, RCP 8.5
NorESM1-M	Norwegian Climate Centre	12.31 N	77.5 E	NEM	RCP 2.6., RCP 4.5, RCP 6.0, RCP 8.5

Reproduced from Chandra, R., Ujjwal, S., Mujumdar, P.P., 2015. Model and parameter uncertainty in IDF relationships under climate change. Adv. Water Resour. 79, 127–139. https://doi.org/10.1016/j.advwatres.2015.02.011.

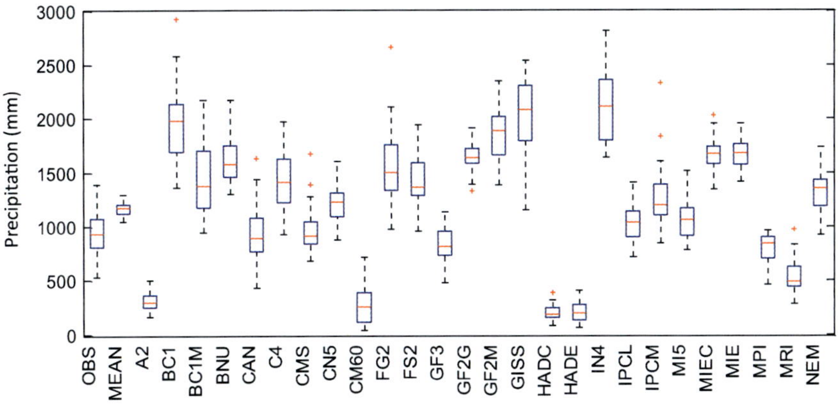

Figure 3.4 Annual precipitation during the control period (1969–98) from 26 GCMs. Mean precipitation across all GCMs (MEAN) and the observed gauge data (OBS) are also shown. *(Reproduced from Chandra, R., Ujjwal, S., Mujumdar, P.P., 2015. Model and parameter uncertainty in IDF relationships under climate change. Adv. Water Resour. 79, 127–139. https://doi.org/10.1016/j.advwatres.2015.02.011.)*

It is highly impossible to ascertain that a particular GCM is most reliable and hence a comprehensive assessment of regional change projections needs to be based on the collective information from the ensemble of GCM simulations. The intermodal or GCM uncertainty resulting from the ensembles of projections generated with multiple GCMs is addressed using REA method.

3.3.1 Reliability Ensemble Average Method

The REA method considers two reliability criteria, one is the model performance criterion, in which the performance of model in reproducing present-day climate is assessed, and the other is the model convergence criterion in which convergence of the simulated changes across models is considered (Giorgi and Mearns, 2003). Typically REA assigns weights to models based both on their performance in the historical period as well as on their ability to provide "good" future projections, the latter being measured with respect to the departure from the ensemble mean. The initial weights are assigned to GCMs on the basis of model performance by taking the CDF deviations between observed precipitation and GCM simulated downscaled precipitation time series for control period (1969–98). The final weights are obtained by taking CDF deviations between individual GCMs for future and weighted mean CDF (derived from weighted projections of multiple GCMs) accounting for the model convergence criterion. The results from all the GCMs can be used without discarding any of the underperformed/overperformed GCMs, since the multimodel projections vary across time and will produce increasingly precise estimates as the number of GCMs increase (Tebaldi and Knutti, 2007).

3.4 Disaggregation of Extreme Precipitation

Disaggregation of precipitation from larger to shorter time scales plays a prominent role in obtaining projected short-duration precipitation extremes. Though numerous deterministic and stochastic disaggregation techniques are developed, cascade models based on scale invariance theory are most widely used for precipitation disaggregation. Scaling implies that the statistical properties of the process observed at various time scales are governed by the same relationship. The distribution equality between two probability distributions at different scales is referred to as "strict sense scaling." For a random variable Z, a function $g(z)$ is said to be scaling if $g(z)$ is proportional to the scaled function $g(\lambda z)$ for all positive values of scale factor λ, that is, if $g(z)$ is scaling then there exists a function $h(\lambda)$ such that $g(\lambda z) \overset{\text{dist}}{=} h(\lambda)g(z)$, where $h(\lambda) = \lambda\beta$, β is called the scaling component. A weaker property, which is usually adopted for assessing the scaling behavior of a process is the equality of raw moments or quantiles, referred to as "wide sense scaling." This implies that the quantiles and raw moments of any order are also scale invariant, that is, $E\left[Z_{\lambda}^{d}\right] = \lambda^{d\ \beta} E\left[Z^{d}\right]$ where d is the order of the moment (Gupta and Waymire, 1990; Burlando and Rosso, 1996). For a simple scaling process, $\beta(d) = d \times \beta$.

For GEV distribution, the NCMs and the relation between statistical properties at two different time scales t and λt for simple scaling are given by:

$$\psi_d = \left(\mu - \frac{\sigma}{k}\right)^d + (-1)^d\left(-\frac{\sigma}{k}\right)^d \Gamma(1 - dk) + d \times \sum_{i=1}^{d-1} (-1)^i \left(-\frac{\sigma}{k}\right)^i \left(\mu - \frac{\sigma}{k}\right)^{d-i} \Gamma(1 - ik)$$

$$\begin{aligned} \mu(\lambda t) &= \lambda^{\beta}\mu(t), \\ \sigma(\lambda t) &= \lambda^{\beta}\sigma(t), \\ k(\lambda t) &= k(t) \quad \text{and} \\ Z_T(\lambda t) &= \lambda^{\beta} Z_T(t) \end{aligned} \tag{3.5}$$

The exponent β is obtained as the slope of log-log plots between NCMs and durations ($\beta = \{\log(\psi_1(\lambda t)) - \log(\psi_1(t))\}/\log(\lambda)$). Plots between the three NCMs versus the scaling exponent are checked to ensure the process follows simple scaling or multiscaling.

3.4.1 Example: Bangalore Case Study

The projected annual maximum precipitation time series for 85 GCM simulations and for the three timelines are obtained from daily precipitation for Bangalore case study. The intermodal or GCM uncertainty resulting from the ensembles of projections generated with multiple GCMs is addressed using REA method. Table 3.4 gives weights of all GCM scenarios and for the three time slices.

The NCMs are calculated for the historical time series of Bangalore case study by using MLE estimates and Bayesian sets of parameters. Log–log plots of the first three precipitation NCMs against duration are obtained for a given set of parameters, and statistics such as mean and quantiles ($q = 0.025$, 0.25, 0.5, 0.75, and 0.975) of the NCMs are plotted. The log-linearity exhibited in the plot indicates the scaling of the precipitation statistical moments with duration for two time intervals: from 15 min to 30 min and from 30 min to 24 h. The linearity of the scaling exponent $\beta(d)$ with the moment order d supports the assumption that the extreme precipitation series for the study area can be described by a simple scaling model. The scaling exponent is obtained from the MLE estimates also. Table 3.5 gives the scaling exponent values for both Classical and Bayesian models. It is observed that the scaling exponent obtained from the classical approach is in good compliance with that of the mean of scaling exponent obtained from Bayesian analysis.

The value of the shape parameter k characterizes the tail behavior of the GEV distribution. Heavy tail behavior implies higher probability of extreme events than that obtained by exponentially tailed distribution (Serinaldi and Kilsby, 2014), which impacts on the designs in hydrologic applications. In general, as the duration reduces, the estimates of k tend to increase (Veneziano and Yoon, 2013). However, in the scale invariance theory, it is considered that the k is constant for all the durations. Therefore, in the present case study, different priors for k are selected, posterior distributions are obtained, and the effect on scaling exponent is studied. The options of various priors and conditions are listed here:

- First option: Beta prior restricted within a range—0.5 and 0.5 according to Eq. (3.4)
- Second option: Since, according to Eq. (3.5), it is common to impose k to be same for all the durations in scaling (e.g., Jiang and Tung, 2015;

Table 3.4 Weights Assigned to GCMs for Four RCP Scenarios and Three Time Lines Using REA Technique

GCM-Model Name	RCP2.6 2021 −2050	RCP2.6 2051 −2080	RCP2.6 2071 −2100	RCP4.5 2021 −2050	RCP4.5 2051 −2080	RCP4.5 2071 −2100	RCP6.0 2021 −2050	RCP6.0 2051 −2080	RCP6.0 2071 −2100	RCP8.5 2021 −2050	RCP8.5 2051 −2080	RCP8.5 2071 −2100
ACCESS1_0	—	—	—	0.024	0.032	0.033	—	—	—	0.025	0.025	0.022
BCC-CSM1-1	0.037	0.049	0.148	0.042	0.033	0.077	0.075	0.073	0.103	0.074	0.090	0.088
BCC-CSM1-1-M	0.056	0.023	0.031	0.035	0.032	0.021	0.033	0.041	0.025	0.021	0.018	0.017
−BNU-ESM	0.042	0.038	0.030	0.019	0.038	0.046	—	—	—	0.051	0.042	0.052
−CanESM2	0.046	0.056	0.098	0.042	0.142	0.047	—	—	—	0.053	0.089	0.089
CCSM4	0.033	0.092	0.113	0.039	0.113	0.068	0.062	0.155	0.114	0.068	0.053	0.057
CMCC-CMS	—	—	—	0.058	0.053	0.035	—	—	—	0.029	0.032	0.041
CNRM-CM5	—	—	—	0.020	0.033	0.029	—	—	—	0.051	0.133	0.068
CSIRO-MK3-6-0	0.059	0.035	0.053	0.035	0.026	0.038	0.081	0.099	0.051	0.049	0.027	0.019
FGOALS-G2	0.032	0.024	0.029	0.056	0.037	0.042	0.029	0.047	0.091	0.036	0.036	0.034
FGOALS-S2	0.021	0.021	0.019	0.049	0.129	0.050				0.039	0.056	0.059
GFDL-ESM2G	0.032	0.020	0.029	0.021	0.030	0.027	0.056	0.060	0.051	0.067	0.033	0.032
GFDL-ESM2M	0.071	0.028	0.022	0.079	0.021	0.028	0.044	0.033	0.040	0.024	0.063	0.076
GISS-E2-R	—	—	—	0.017	0.024	0.030	—	—	—	—	—	—

HadGEM2-CC	—	—	—	0.010	0.032	0.105	—	—	—	0.014	0.025	0.014
HadGEM2-ES	0.015	0.047	0.100	0.071	0.009	0.023	0.152	0.048	0.154	0.038	0.027	0.033
INMCM4	—	—	—	0.027	0.013	0.019	—	—	—	0.019	0.016	0.017
IPSL-CM5A-LR	0.086	0.081	0.056	0.099	0.012	0.019	0.060	0.076	0.031	0.020	0.027	0.022
IPSL-CM5A-MR	0.049	0.038	0.034	0.048	0.035	0.066	0.061	0.067	0.050	0.037	0.028	0.035
MIROC5	0.044	0.031	0.029	0.034	0.024	0.031	0.074	0.057	0.054	0.042	0.035	0.034
MIROC-ESM-CHEM	0.025	0.070	0.021	0.022	0.028	0.018	0.029	0.036	0.026	0.025	0.017	0.017
MIROC-ESM	0.030	0.080	0.028	0.033	0.018	0.022	0.028	0.046	0.037	0.041	0.021	0.021
MPI-ESM-LR	0.128	0.021	0.042	0.038	0.026	0.025	—	—	—	0.023	0.036	0.073
MRI-CGCM3	0.036	0.028	0.028	0.033	0.039	0.071	0.047	0.068	0.064	0.125	0.045	0.044
NorESM1-M	0.117	0.183	0.049	0.047	0.023	0.030	0.118	0.037	0.052	0.027	0.024	0.035
GFDL-CM3	0.040	0.035	0.038	—	—	—	0.053	0.057	0.058	—	—	—

Table 3.5 Scaling Exponent From Classical and Bayesian Analysis With Different Prior Distributions for Shape Parameter k

	Scaling from 24 h to 30 min			Scaling from 30 to 15 min		
Method/Prior	$\beta(1)$	$\beta(2)$	$\beta(3)$	$\beta(1)$	$\beta(2)$	$\beta(3)$
Classical (MLE) Approach	**0.7902**	**1.564**	**2.203**	**0.411**	**0.892**	**1.314**
Bayesian Approach						
$k \sim$ Beta(6,9) $-0.5 < k < 0.5$	0.7897	1.5689	2.332	0.416	0.817	1.215
$k \sim$ Beta(6,9) $-0.5 < k < 0.5$ (Same k for all durations)	0.800	1.598	2.392	0.392	0.756	1.084
$k \sim$ Beta(6,9) $-0.1 < k < 0.1$	0.793	1.580	2.359	0.400	0.778	1.143
$k \sim$ N(0.079, 0.002)	0.796	1.586	2.370	0.397	0.760	1.085
$k \sim$ N(0, 100)	0.787	1.555	2.295	0.409	0.812	1.218

Huard et al., 2010), same value of k (which is k value for 24 h duration) is used for all the durations.

- Third option: prior of k is restricted within a narrow range of $L = -0.1$ and $U = 0.1$ instead of imposing same posterior distribution for all durations.
- Fourth option: Papalexiou and Koutsoyiannis (2013) studied the effect of the sample size on the estimation of the shape parameter (k) and concluded that there is empirical evidence that the shape parameter is not constant but follows approximately a Gaussian distribution with mean $\sim$0.114 and standard deviation $\sim$0.045 as the sample size tends to infinite. Recent study by Serinaldi and Kilsby (2014) shows that the mean value of k tends to be within the range (0.061, 0.097). Gaussian distribution with mean value of 0.079 (average of the range proposed by Serinaldi and Kilsby, 2014) and standard deviation (SD) of 0.045 is considered as prior of k.
- Fifth option: A Gaussian prior with zero mean and large variance as suggested by Coles (2001).

The scaling exponent obtained by considering these options of prior distributions of k is tabulated in Table 3.5. It is observed that there is no significant difference in the mean value of the scaling exponent, from which can be concluded that the assumption of same k for all the durations is in fact valid for GEV distribution as opposed to a case when exceedances above a threshold are considered.

The IDF relationships for future using both the classical and Bayesian approaches are obtained for the four scenarios and for the three time slices. It is observed that the short-duration precipitation return levels increase considerably in future, emphasizing the importance of obtaining the uncertainties associated with short-duration return levels, which can be performed by the Bayesian analysis. It is to be noted that the increase in short-duration return levels is due to the anthropogenic climate change, but not due to the local effects. The range (difference between the maximum and minimum) of the REA averaged return level varies from 35.97 to 49.7 mm/h for RCP 2.6 and 8.5 scenarios for near future (2021−50). For far future (2071−2100), the range varies from 40.3 to 75.7 mm/h for RCP 2.6 and 8.5 scenarios. The uncertainty in REA averaged return level for RCP 8.5 scenario is high, and increases by 53% from near future to far future, while the range increases by 12% for RCP 2.6 scenario. The results indicated that the return level estimates are very sensitive for short-duration time scales and for higher RCP scenarios. The uncertainty in the return level is studied individually for all the 85 GCM runs under consideration and for the three time slices (2021−51, 2051−81, and 2071−2100) using Bayesian and classical approaches. The uncertainty in return levels for 15 min duration, near future (2021−50) results are presented in Fig. 3.5 for 8.5 scenario; the REA averaged return levels and the return levels for historical time period are also shown in the figure. The box plot gives the

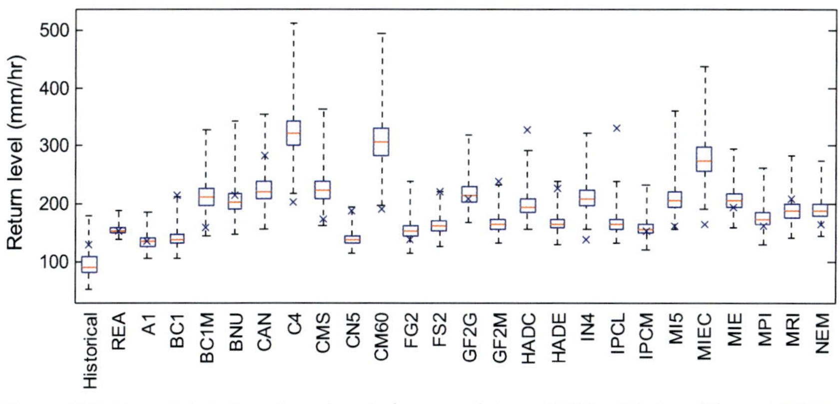

Figure 3.5 Uncertainty in return levels for near future (2021−50) for different GCMs, REA averaged, and the uncertainty in return levels for historical time period (1969−2003) for 15 min duration, for RCP 8.5 Scenario. Boxplot gives the uncertainty in return levels obtained from Bayesian analysis, and the return level obtained from classical approach is shown as "x" on the box plot.

uncertainty in return levels obtained from Bayesian analysis and the return level obtained from classical approach is marked as "x" on the box plot. It is seen from Fig. 3.5 that for RCP 8.5 scenario, most of the GCMs underestimated/overestimated the return levels calculated using classical approach, emphasizing that the uncertainty associated with the return levels is high and cannot be estimated precisely using classical approach, for RCP 8.5. The mean of the REA averaged return levels obtained from Bayesian analyses are approximately equal to that of the return level calculated from classical approach, indicating that the model uncertainty can be addressed using REA method even if classical approach is employed, whereas for accounting the parameter uncertainty, Bayesian method is essential.

4. SPATIAL MODELING OF PRECIPITATION EXTREMES

Modeling extremes at a single station—or, the univariate extremes—are well developed in literature and extending tools used for modeling univariate extremes to model spatial extremes are an active area of research (Cooley et al., 2007). Typically, precipitation measurements are based at specifically located stations, and to obtain the spatial return level maps, interpolation techniques are required. However, in urban areas, the precipitation varies rapidly and the traditional interpolation techniques like kriging and Inverse Distance Weighting (IDW) methods are not capable of capturing this high spatial variation. Also, lack of adequate data due to sparse observational network, spatially, leads to significant uncertainty in the estimates of spatial precipitation. There is, thus, a need to obtain the spatial variation of return levels across the urban areas and quantify the uncertainties precisely. The analysis of spatial extreme data dwells at the intersection of two branches of statistics: extreme value analysis and geostatistics (Cooley et al., 2012). These two topics are introduced briefly in the succeeding subsection before discussing the methods for spatial modeling.

4.1 Multivariate Extremes Theory

Let $X_{i1}, X_{i2} \ldots X_{id}$, $i = 1, \ldots, n$, be a sequence of independent d-dimensional continuous random vectors and let $M_n = (\max_{i=1..n} X_{i1}, \ldots, \max_{i=1\ldots n} X_{id})^T$, and if there exists continuous functions $a_n > 0$ and $b_n \in \Re$,

$$\Pr\left(\frac{M_n - b_n}{a_n} \leq y\right) \rightarrow G(y) \quad n \rightarrow \infty$$

G is nondegenerate, and belongs to class of multivariate max-stable distributions. If certain mixing conditions are met, the iid assumption can be relaxed and the multivariate max-stable distributions continue to work as the possible limiting distributions. The logic for using max-stable is exactly same as for univariate extreme value analysis, where an asymptotic distribution (GEV) is preferred over estimating the distribution of each X and raising to the power n.

Fully parametric representation does not exist for the multivariate max-stable distributions. The univariate marginal distributions must be univariate max-stable, that is, the GEV distribution. For a jth ($j = 1, \ldots, d$) component's marginal, if μ_j, σ_j, and k_j are the location, scale, and shape parameters:

$$\Pr(Y_j \leq y) = G_j(y) = \exp\left\{ - \left[1 + k_j \left(\frac{y - \mu_j}{\sigma_j} \right) \right]^{-1/k_j} \right\}$$

A widely used choice is to assume unit Fréchet margins as accounting for each GEV marginal is cumbersome and also the representations for multivariate max-stable distributions generally assume that the marginals have a common max-stable distribution. Therefore assuming $Z_1, \ldots, Z_d$ has a multivariate max-stable distribution with unit-Fréchet marginal, that is$\Pr(Z_j \leq z) = \exp(-z^{-1})$, then

$$\Pr(Z \leq z) = G^*(z) = \exp\{-V(z)\}$$

where $V(z) = E\left\{ \max_{j=1,\ldots,d} \frac{X(y_i)}{z_j} \right\}$ is called the exponent function, which characterizes the dependence across the elements of random vector Z. The extremal coefficient, θ, is defined as

$$V(z) = \frac{\theta(Y)}{z}, \quad \theta(Y) = E\left\{ \max_{j=1,\ldots,d} X(y_i) \right\}$$

For a bivariate case, in the spatial context, the extremal coefficient function is $\theta : h \rightarrow E[\max\{X(y), X(y+h)\}]$, which takes values in [1,2], where the lower bound corresponds to perfect dependence and the upper bound to independence.

$$\begin{aligned}\Pr\{Z(y+h) \leq z | Z(y) \leq z\} &= \Pr\{Z(y+h) \leq z\}^{\theta(h)-1} \\ &= \begin{cases} 1 & \text{perfect dependence} \\ \Pr\{Z(y+h) \leq z\} & \text{independence} \end{cases}\end{aligned}$$

4.1.1 Extremal Coefficient

The d-dimensional extremal coefficient can be implicitly defined as $\Pr(X_1 < x, \ldots, X_d < x)$ for any x and transforming the marginals of X to obtain Z and due to the homogeneity property of V, the probability can be written as

$$\Pr(Z_1 \le z, \ldots, Z_d \le z) = \exp\left\{\frac{V(1, \ldots, 1)}{z}\right\} \tag{3.6}$$

with $\theta_d = V(1, \ldots, 1)$. This quantity, θ_d, is extremal coefficient of observations and varies from $\theta_d = 1$ when observations are fully dependent among locations to $\theta_d = d$ when they are independent. In case of fitting the models for accounting dependency, empirical estimators for pairs of sites with the fitted versions from models are compared and therefore, for a bivariate case, $\theta_d = 2$. It is possible to extend Eq. (3.6) to be a distance-based dependence metric as $\theta(h) = \Pr(Z(s+h) \le z,\ Z(s) \le z)$. The distance-based dependence measures have been primarily estimated via madogram. The madogram is a first-order semivariogram, which requires the first moment to be finite which is not always the case in modeling dependency in extremes. Cooley et al. (2006) proposed the F-madogram, which first transforms the random variable by applying the CDF and is finite for any distribution, is given by:

$$\nu(h) = \frac{1}{2} E|G\{X(s)\} - G\{X(s+h)\}|$$

The F-madogram's values range from 0 to 1/6, corresponding to complete dependence and independence respectively. The F-madogram is related to the extremal coefficient, as

$$\theta(h) = \frac{1 + 2\nu(h)}{1 - 2\nu(h)}$$

Let $x_i(s)$, $i = 1, \ldots, n$, be iid replicates of a max-stable process with marginal distribution G observed at a finite set of locations, then the sample F-madogram is given by

$$\widehat{\nu}(h) = \frac{1}{n} \sum_{i=1}^{n} \frac{1}{2|N(h)|} \sum_{(s,s') \in N(h)} |G\{X_i(s)\} - G\{X_i(s')\}| \tag{3.7}$$

where N(h) denotes the number of pairs (s, s') separated by distance h. Eq. (3.7) assumes that the marginal distribution G is known. When the marginal distribution is not known, Naveau et al. (2009) defined the λ-madogram, which is related to the Pickands' dependence function to completely describe the bivariate dependence.

4.2 Geostatistics

Let S be a study region; if s denotes a location in the study region and if $Y(s)$ denotes a stochastic process, field of geostatistics provides a framework for modeling, predicting, and/or interpolating $Y(s)$. A basic model for $Y(s)$ can be formulated as $Y(s) = a(s) + e(s)$, where $a(s)$ is the mean function and $e(s)$ is a zero mean stochastic process, which accounts for any behavior not captured by mean. The mean function, $a(s)$, is generally considered to be a regression relation given by $Q(s)^T\alpha$, where $Q(s)$ is a vector of covariate information at location s and α is the vector of regression coefficients.

The process e is considered to be second-order stationary and isotropic in a simple geostatistical model, which is characterized by its covariance function (also called as semivariogram):

$$\gamma(h) = \frac{1}{2}\mathrm{Var}[e(s+h) - e(s)] = \frac{1}{2}E[\{e(s+h) - e(s)\}^2]$$

The sample variogram is given by:

$$\widehat{\gamma}(h) = \frac{1}{2|N(h)|}\sum_{(s,s')\in H(h)}\{e(s) - e(s')\}^2 \tag{3.8}$$

A variety of models are available in literature for covariance function (Banerjee et al., 2004) such as linear, exponential, Gaussian, spherical etc. After selecting a parametric family for the semivariogram (or variogram), the estimates of model parameters of stochastic process are obtained.

It is to be noted that in geostatistics model, e is stationary and Gaussian and marginal distribution can only vary with $a(s)$ capturing the spatial variation, that is, after considering the regional effects with a, the dependence in e is characterized by variogram, which focuses on the central tendencies (as it is based on first and second moments) but not on the distribution's tail. Though the primary goal of geostatistic analysis is spatial prediction/interpolation by considering an estimated variogram, modeling extreme data is not feasible considering standard geostatistics, thus motivating to develop proficient models which can handle extreme data along with dependencies.

4.3 Hierarchical Bayesian Method

Two major applications in spatial modeling are: first, to make inference at locations where there are no data, that is, to perform spatial interpolation and second, to borrow strength across locations when estimating marginal parameters. Hierarchical models perform both the applications in a simpler way. These models are extensively used in literature in describing the relation between observations (or parameters of fitted model to observations) and a complex process that generates them. To explain spatial variation of data, hierarchical models assume that the observations or the parameters of fitted model to observations are driven by a latent process. Such models are termed as latent variable models or latent process models.

The Bayesian Hierarchical Model (BHM) is formed in three layers (or levels). The base layer or the data layer models the data, by pooling the block maxima from all the stations over all the years. The second layer is the process layer, where the latent process that drives the extreme precipitation in the region is modeled. The third layer is the prior layer, in which the prior distributions of the parameters that control the latent process are defined. Given the data (Y) at a particular station, following Bayes' rule, the unnormalized posterior density of the parameters is given by Eq. (3.3), the hierarchical model is formulated by introducing a process layer in Eq. (3.3), given by:

$$\pi(\theta|Y) \propto L(Y|\theta_1) \times \pi_1(\theta_1|\theta_2) \times \pi_2(\theta_2) \tag{3.9}$$

$\theta = (\theta_1, \theta_2)$, where θ_1 are the parameters in the data layer (parameters of the GEV distribution) and θ_2 are the parameters in the process layer (parameters of latent spatial process). Each term in Eq. (3.9) is associated with each layer (base, process, and prior layers) of the hierarchical model. $L(Y|\theta_1)$ is the likelihood function, $\pi_1(\theta_1|\theta_2)$ describes the distribution of the latent process, and $\pi_2(\theta_2)$ is prior distribution of the parameters of the latent process.

4.3.1 Data Layer

In the data layer, extreme precipitation values for all years and from all stations are pooled to form the likelihood function. Let $y_{j,i}$ be the jth recorded extreme precipitation at location i. n_i is the number of block maximum precipitation values recorded at station i and S is the number of stations. μ_i and σ_i are the location and scale parameters at station i and the shape parameter k is considered to be constant at all the stations. The scale parameter σ_i is reparameterized as $\phi_i = \log(\sigma_i)$ as $\sigma_i > 0$. The likelihood function is given by

$$L(y|\theta_1) = \prod_{i=1}^{S}\prod_{j=1}^{n_i}\frac{1}{\sigma_i}\left[1 + k\left(\frac{y_{j,i} - \mu_i}{\sigma_i}\right)\right]^{(-1/k)-1}$$

$$\exp\left\{ - \left[1 + k\left(\frac{y_{j,i} - \mu_i}{\sigma_i}\right)\right]^{(-1/k)}\right\}$$

4.3.2 Process Layer

The second layer in hierarchy is the process layer, where the spatial latent process is characterized by relating the parameters of the GEV model with the geography and climatology of the region. Typically, the GEV parameters are considered to follow an independent Gaussian processes with mean as a linear function of the covariates and variance as a function of distance between the stations, usually a covariance model. For example, if $\mu \sim GP\left(m_\mu, \Sigma_\mu\right)$

$$m_\mu = \alpha_{\mu,0} + \alpha_{\mu,1} \times \text{cov}_1 + \alpha_{\mu,2} \times \text{cov}_2$$

where cov is the geographic or climatologic covariate such as elevation, mean precipitation etc. α_μs are referred as mean parameters or the regression parameters. Σ_μ is a parametric function, considered to define variogram. Several functions for variogram are available in literature (Banerjee et al., 2004) such as linear, exponential, Gaussian, spherical etc. Eq. (3.10) corresponds to an exponential variogram.

$$\Sigma_\mu = \beta_{\mu,0} \exp\left(- \beta_{\mu,1} \times \|h\|\right) \tag{3.10}$$

where h is the geographic distance between stations. $\beta_{\mu,0}$ and $\beta_{\mu,1}$ represents the sill and decay (1/range) parameters of the variogram, respectively.

4.3.3 Prior Layer

Priors to parameters of the process layer that characterize the spatial latent process are assigned in the third layer. These parameters are assumed to be independent of each other in this layer. For example, if the latent process is a Gaussian process, for mean parameters of the Gaussian process (α_μ, α_σ, α_k) no prior information is available about GEV parameters' relation to the covariates; therefore, uninformative priors are considered as a safe choice, which allows proper posteriors even after choosing an improper prior (Banerjee et al., 2004). For the parameters of the variogram (β_μ, β_σ, β_k), setting priors is complicated as improper priors may generally result in improper posteriors (Banerjee et al., 2004). Therefore, one needs to choose informative priors describing the spatial structure of GEV parameters.

4.3.4 Model Inference and Return Level Maps

As in the case of single-station analysis, since the conditional distributions of parameters (Eq. 3.3) do not exist in closed form in general, random sampling of parameters is required. MCMC methods are used for sampling, where Gibbs sampling is one of the most popular approaches. A Metropolis-Hastings (MH) step is to be included within each Gibbs sampler as extreme value distributions do not have conjugate priors. Samples can be drawn from the conditional distributions of parameters from the independent Gaussian processes for each state of the Markov chain given its current state. These parameter sets define the return level for a given return period (Eq. 3.2). Note that as the parameters can also be obtained for ungauged locations, the return levels can be obtained at the ungauged locations. This is the major strength of the latent process approach. This sampling output can be exploited to obtain point-wise means and other uncertainty measures and quantiles, which are used for producing the return level maps for the study region.

4.3.4.1 Example: Bangalore Case Study

Bangalore City and its surrounding areas between 77.2°E to 78.2°E and 12.5°N to 13.25°N are considered for the study. The elevation varies between 620 and 1082 m over an area of 1300 km^2 approximately within the Bangalore City boundary. Daily precipitation data is observed at 19 Telemetric Rain Gauge (TRG) stations, maintained by India Meteorological Department (IMD) in and around Bangalore City. This daily precipitation data is considered in developing BHM. In the spatial analysis, the modeling of precipitation extremes is limited to daily time scale and a similar procedure discussed can be adopted for short-duration precipitation time series. Fig. 3.6A shows the study region with Bangalore City boundary. The precipitation stations considered in the hierarchical Bayesian analysis are marked in the figure. The least distance between the stations is 2 km (between Nelamangala and Hessarghatta) and the farthest distance is 52 km (between Devanahalli and Kanakapura).

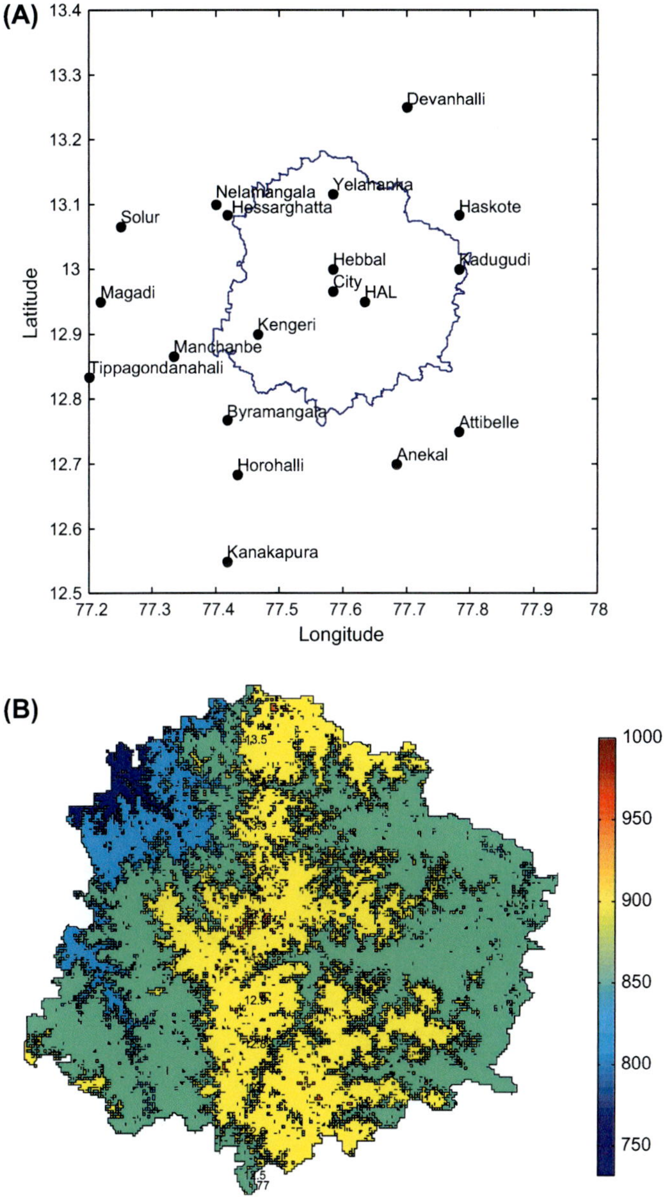

Figure 3.6 (A) Bangalore City boundary marked with 19 Telemetric Rain Gauge stations considered in the developing spatial models analysing extreme precipitation and (B) Map showing elevation (in meters) above mean sea level within Bangalore City boundary.

A detailed exploratory analysis of the extreme precipitation data observed in the study region is performed. The extremal coefficient is estimated for the common time period. Fig. 3.8 shows the extremal coefficient, both for the sample and the binned estimates, for the annual maximum precipitation time series. For the binned estimates, the extremal coefficient takes on values between 1.68 and 1.84, which indicates that the amount of dependence between the observations is low even at short distances. Also it is observed that the spatial variation of precipitation is highly nonuniform in the study area and its surroundings, making the assumption of spatial independence less critical in this study region.

4.3.4.1.1 Covariates The parameters of the GEV distribution follow a latent process assumed to be Gaussian in this case with mean as a function of geographic and climatologic covariates. The covariates or the explanatory variables selected should have important information describing the latent spatial process. The extreme precipitation may be related to mean summer, monsoon, and annual precipitation and the mean diurnal temperature range, that is, difference between daily maximum and minimum temperature. Therefore, other than the geographic covariates—latitude, longitude, and elevation—the mean precipitation and the mean diurnal temperature range are considered as covariates in the study. The list of covariates selected is given in Table 3.6.

The covariate information should be available over the entire region to obtain the spatial map of return levels and not just at the TRG locations.

Table 3.6 List of Covariates in the Bayesian Hierarchical Model

S.No.	Covariate	Short Form
Geographic		
1	Elevation	Elev
2	Latitude	Lat
3	Longitude	Lon
Climatologic		
4	Mean Annual Precipitation	MAP
5	Mean Summer Precipitation	MSP
6	Mean Monsoon Precipitation	MMP
7	Mean Annual Diurnal Temperature Range	DAT
8	Mean Summer Diurnal Temperature Range	DST
9	Mean Monsoon Diurnal Temperature Range	DMT

Elevation is obtained from the 30 m resolution digital elevation model from the Advanced Spaceborne Thermal Emission and Reflection Radiometer (ASTER) Global Digital Elevation Model (GDEM) by National Aeronautic and Space Administration (NASA) (http://reverb.echo.nasa.gov/reverb/). Fig. 3.6B gives the variation in elevation within the Bangalore City boundary. The daily precipitation is obtained from the gridded dataset (0.25° × 0.25°) prepared by IMD by interpolation of precipitation records measured at approximately 6900 rain gauge stations over India, for the period 1901–2010. Three different spatial covariates—mean annual precipitation, mean summer precipitation (March–May), and the mean monsoon precipitation (June–September)—are obtained from this dataset to capture the spatial variability in precipitation extreme patterns. Similarly, the diurnal temperatures are obtained from IMD gridded dataset of resolution 1° × 1° resulting from an interpolation of temperature records measured at approximately 550 observatories over India, available for the period 1969–2005. As in the case of mean precipitation, three different spatial covariates—mean annual, summer (March–May), and monsoon (June–September) diurnal temperature range—are obtained from this dataset.

4.3.4.1.2 Return Level Maps The spatial variation of the return levels is modeled through the variation of the GEV parameters (μ, σ, and k), determined from the BHM. Both location and scale parameters are modeled as two independent Gaussian processes with mean as $E(\varphi(s)) = m_\varphi(s) = f(\alpha_\varphi, \text{covariates})$, where $\varphi = (\mu, \phi)$. The function, $f(.)$, is considered as a linear function of the covariates (listed in Table 3.6); for example, $m_\varphi(s) = \alpha_{\varphi,0} + \alpha_{\varphi,1} \times \text{elev} + \alpha_{\varphi,2} \times \text{MSP}_x$. Only the first-order regression equation with single covariates and second-order with two covariates are considered in this example, to avoid complexity. The variance of the Gaussian process is considered as exponential variogram model, following Eq. (3.10). In the prior layer, the parameters that characterize the spatial latent process are assigned a prior distribution. For regression parameters, α_φ uninformative priors are considered, that is, the parameters are set to follow uniform distribution $(\alpha_\varphi \sim \text{Unif}(-\infty, +\infty))$. The sill and range represent the spatial structure of the parameters μ andσ, and therefore, priors are set for $\beta_{\varphi,0}$ and $\beta_{\varphi,1}$ based on the maximum and minimum range and sill of the variogram model. This is done by fitting a GEV distribution to each station independently, obtaining the MLE estimates of μ and σ and fitting an empirical variogram (Eq. 3.10) to μ's and σ's. Based on the fitted empirical variogram, the range is set to cover the maximum range of the exponential variogram model and sill is chosen so that it covers a wide envelope around the exponential variogram. The resulting priors for variance function, for location parameter, μ, are $\beta_{\mu,0} \sim \text{Unif}(2, 35)$, $\beta_{\mu,1} \sim \text{Unif}(16, 100)$ and for scale parameter σ are $\beta_{\sigma,0} \sim \text{Unif}(0.005, 0.05)$ and $\beta_{\sigma,1} \sim \text{Unif}(5, 50)$. Using the geographic and climatologic covariates, 45 Bayesian hierarchical models are formed and MCMC sampling is performed to generate random sets of parameters of the GEV and the Gaussian processes.

4.3.4.1.3 Sampling Using Markov Chain Monte Carlo The MCMC algorithm is used to obtain draws from the posterior distribution of the parameters (Eq. 3.9). MCMC involves setting up a Markov chain with posterior distribution of parameters as its stationary distribution and then drawing samples using a transition density from this Markov chain. Metropolis–Hastings (MH) algorithm within a Gibbs sampler (Gelman et al., 2004) is used to update each parameter of the model. The sampling of parameters μ, σ, and k are partitioned while applying Gibbs sampler, that is, the location parameter μ for all the stations is first sampled, then scale parameter σ for all the stations is sampled, and finally the shape parameter k is updated. The MH step is used to update the parameter set by drawing a value from a candidate density with an acceptance rate. A similar approach as suggested by Cooley et al. (2007), that is, obtaining suitable candidate density to update parameters using the information from the maximum likelihood estimates, to update both the location and scale parameters is considered to speed up the convergence. According to the asymptotic properties of MLEs, if $\widehat{\varphi}$ represents the MLEs then $\widehat{\varphi} \sim \varphi + \omega$, where $\varphi \sim \text{MVN}(m_\varphi, \Sigma_\varphi)$, m_φ is the mean and Σ_φ is the covariance matrix obtained from Eq. (3.10) and $\omega \sim \text{MVN}(0, \Phi^{-1})$, Φ is the Fisher information matrix assuming φ and ω are independent. Therefore the conditional distribution (Cooley et al., 2007) used as the candidate density in MH step is

$$\varphi \big| \widehat{\varphi} = \text{MVN}\left(m_\varphi + \Sigma_\varphi(\Phi^{-1} + \Sigma_\varphi)(\widehat{\varphi} - m_\varphi), \Sigma_\varphi - \Sigma_\varphi(\Phi^{-1} + \Sigma_\varphi)^{-1}\Sigma_\varphi\right)$$

For updating regression parameters (α_φ), sill and range parameters ($\beta_{\varphi,0}$ and $1/\beta_{\varphi,1}$), and shape parameter k, random walks are considered as MH candidate densities. The parameters are updated sequentially starting with location parameter μ and the mean and covariance parameters of μ, scale parameter, σ and the mean and covariance parameters of σ and then the shape parameter k.

A set of about two lakh samples is obtained using MCMC and the initial 10,000 samples are discarded, as a burn-in phase and subsequently thinned (i.e., how frequently to sample from the chain) by 30 to obtain uncorrelated chains. The acceptance rate achieved was between 20% and 30%, which is an acceptable range for high-dimensional models. Several types of diagnostic plots were used to check the convergence properties of the MCMC chains. Subsequently, from the GEV parameters(μ, σ, k), spatial return levels for the study region are obtained using Eq. (3.2). Pointwise means and pointwise quantiles are used for producing the return level maps for the study region.

The Deviance Information Criteria (DIC) is used to select the best hierarchical model. Lower the DIC, better is the score; based on this criterion, the model considering two covariates, elevation and Mean Monsoon Precipitation (MMP), is chosen as the best-fit model for annual maximum precipitation. It is observed that the covariates for the second best-fit model are DTA along with the elevation. The best model, with the lowest DIC score, is considered for further analysis in obtaining the return levels maps for Bangalore City. Fig. 3.7 shows the mean location parameter, scale parameter, and return level for a return period of 10 years for the best-fit model for annual maximum precipitation along with the lower and upper bounds of the 95% credible intervals. It is observed from the figure that return levels and the associated uncertainty on the western part of the Bangalore City are higher compared to the other parts for annual maxima.

4.4 Modeling Spatial Dependence

The hierarchical models typically assume conditional independence, which could be feasible while modeling urban precipitation extremes. In case of high dependence in the data, which need to be modeled, max-stable process models are appropriate for capturing the spatial dependencies. Though the copula approaches provide an alternative framework to model dependence structure of a multivariate distribution with known marginals, these are not discussed here for brevity.

4.4.1 Max-Stable Processes

Let X_1, X_2 ... be a sequence of independent copies of a stochastic process, and if there exists continuous functions $a_n > 0$ and $b_n \in \Re$ such that

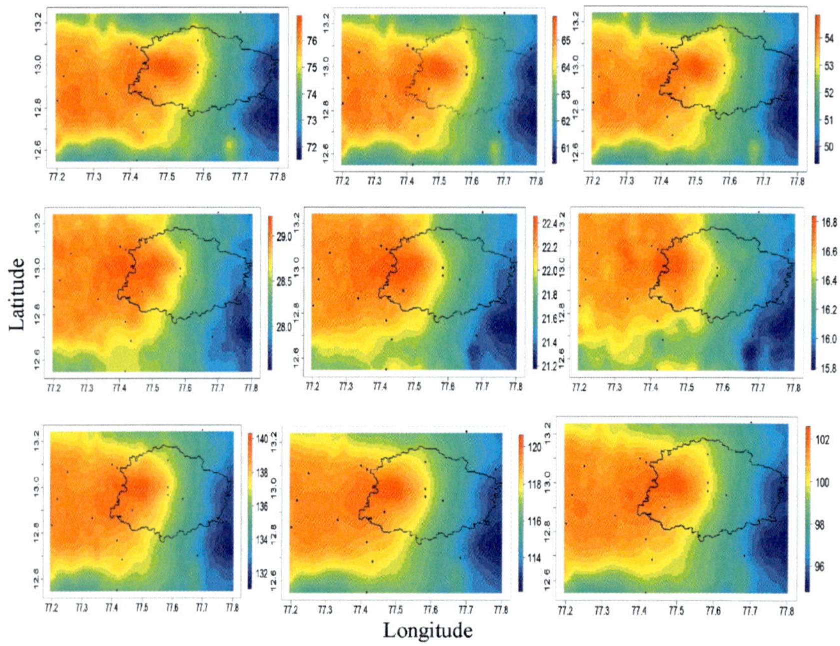

Figure 3.7 Spatial maps of means of location (top row), scale (bottom row) parameters, and return levels for 10-year return period (bottom row) for the selected hierarchical model. The left and right most columns show the upper and lower bounds of the 95% pointwise credible intervals, and the middle column shows the mean values. All units in mm.

$$\left(\frac{\max_{i=1,\ldots n} X_i(s) - b_n(s)}{a_n(s)}\right) \rightarrow Y(s) \quad n \rightarrow \infty$$

which has a nondegenerate distribution, then $Y(s)$ is a max-stable process. The definition of max-stable distributions and the max-stable processes do not have a conceptual construction. de Haan (1984) first proposed a spectral representation of extremal process or the max-stable process. The processes can be described by two components, a Poisson process Π on $(0, \infty)$ with intensity $d\zeta/\zeta^2$ and $\{X(s)\}$ is a stochastic process. If $\{X_i(s)\}$ is independent realizations of the stochastic process, $X(s)$, with $E[X(s)] = 1$ and let $\zeta_i \in \Pi$ be points of the Poisson process, then

$$Z(s) = \max_{i \geq 1} \zeta_i X_i(s) \quad s \in S,$$

is a max-stable process with unit Fréchet margins. It can be noted that the points of the Poisson process play the role of the radius, while the stochastic process the role of angle. It is typically considered as a rainfall storm impacting a region $s \in S$ which has an overall intensity ζ and spatial extent driven by $X(s)$, that is, $\zeta X(s)$ corresponds to the amount of rain for this storm at the location s and the max-stable processes correspond to pointwise maxima over an indefinite number of storms. This spectral characterization is not unique, different choices of the process $X_i(s)$ lead to different classes of max-stable processes.

4.4.2 Inference

To fit max-stable process to data, there is a lack of closed form expressions for the likelihood. The joint likelihood can be only written in low-dimensional settings of order $d < 2$ for all the max-stable processes, except for Smith model, which can be written for $d < 3$. Padoan et al. (2010) developed a likelihood-based approach to fit max-stable processes by substituting composite likelihood instead of joint likelihood. The composite likelihood is a weighted sum of log-likelihoods for each event, which is maximized to obtain the estimates of parameters, termed as Maximum Composite Likelihood Estimators (MCLE). Pairwise likelihood is an example of composite likelihood, given by

$$l_P(\psi; z) = \sum_{k=1}^{n} \sum_{j=1}^{d-1} \sum_{j'=j+1}^{d} \log f\{z_k(s_j), z_k(s_{j'}); \psi\}$$

where ψ is a spatial dependence parameter. Each term $f\{z_k(s_j), z_k(s_{j'}); \psi\}$ is a bivariate density function based on observations at locations j and j'. The two inner summations add up all unique pairs and the outer summation adds over the n iid replicates. MCLE obtained by maximizing the pairwise likelihood is consistent and asymptotically normal. The variance matrix of the MCLE $\widehat{\psi}$ is estimated by an information sandwich of the form $V(\widehat{\psi}) = J^{-1}(\widehat{\psi})H(\widehat{\psi})J^{-1}(\widehat{\psi})$, where $J^{-1}(\widehat{\psi})$ is the observed information matrix and $H(\widehat{\psi})$ is the estimated variance of the score contributions corresponding to the composite log likelihood l_p. The model selection, when different covariates are considered, is effected by minimizing the Composite Likelihood Information Criteria (CLIC), which has analogous properties to those of Akaike's information criterion (AIC) and Takeuchi's information criterion (TIC).

4.4.2.1 Example: Bangalore Case Study

The daily precipitation data from the 19 TRGs in and around Bangalore city is considered to model the spatial dependencies considering max-stable process. Though the exploratory analysis, the data shows independence; the extremal coefficient is between 1.68 and 1.84, which indicates there is some amount of dependence in observations (Fig. 3.8). The dependence is modeled using the max-stable process. *SpatialExtremes* package, developed by Ribatet Mathieu in R is used to model the max-stable process. The same covariates (Table 3.6) are considered in developing the dependency model and 45 models are formed considering the different combinations of covariates (first-order regression equation with single covariates and second-order regression equation with two covariates). TIC is considered to choose the best model. Different max-stable process models are fit to the data and with different covariates. It is observed that Extremal-t and Brown-Resnick models fit the pair-wise estimates of the extremal coefficient (Fig. 3.8) reasonably. Extremal-t max-stable model is considered and based on the minimum TIC scores, MMP along with the latitude perform better as covariates. The return levels maps for Bangalore City considering the dependence are obtained from the selected model. Fig. 3.9 shows the location parameter, scale parameter, and return level for a return period of 10 years along with the lower and upper bounds of the 95% confidence intervals. It is observed from Figs. 3.7 and 3.9 that return levels obtained

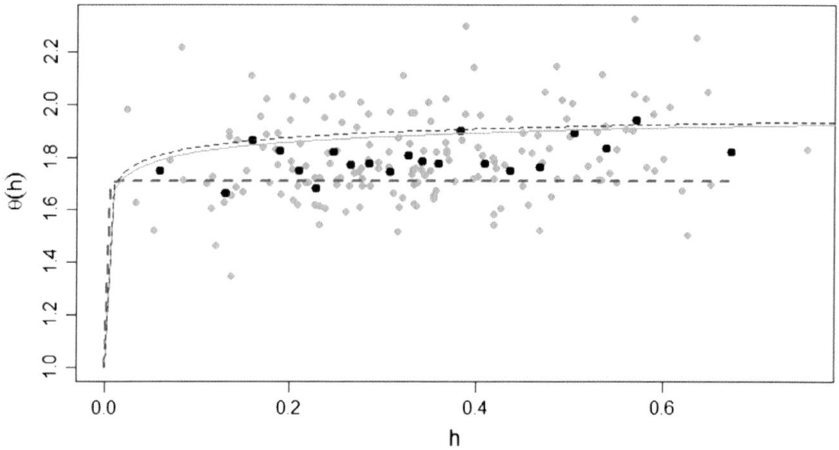

Figure 3.8 F-madogram estimates of the pairwise extremal coefficients (grey), binned estimates (black), and the fitted extremal coefficient functions for different max-stable models: Schatler – *dashed line,* Brown-Resnick – *dotted line,* Extremal-t – *solid line.*

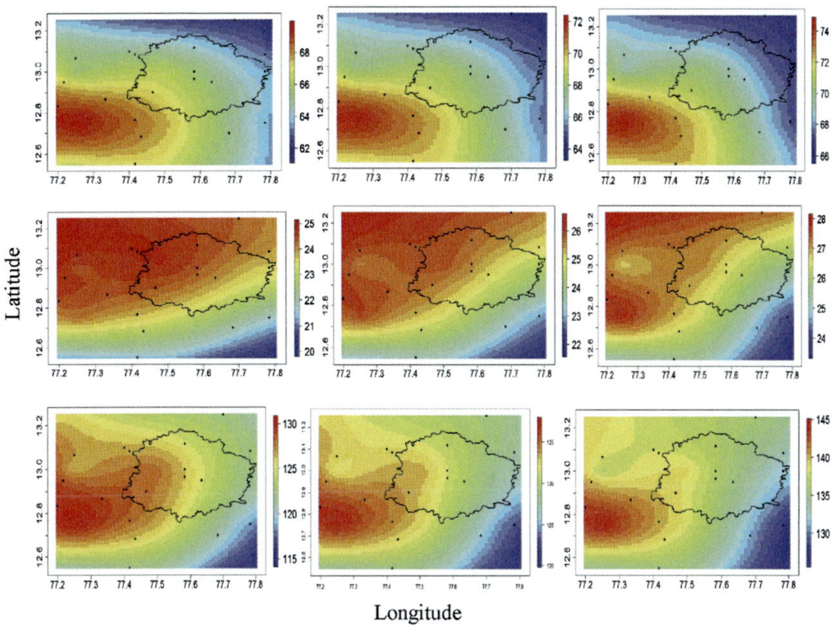

Figure 3.9 Spatial maps of means of location (top row), scale (bottom row) parameters, and return levels for 10-year return period (bottom row) for the selected extremal-t model. The middle column shows the estimates and the left and right most columns show upper and lower bounds of the 95% confidence intervals. All units in mm.

from max-stable process are nearly equal to that of the BHM. The pattern of the return levels is similar for both the models. Therefore, it can be concluded that spatial dependence not accounted in the observations is of little consequence in the BHM and the latent process models can be used in obtaining the spatial return levels for Bangalore City, considering the simplicity in formulating the model.

5. DISCUSSION

The Bayesian methods and models discussed in the preceding sections have advantages as well as disadvantages. Though Bayesian methods are proven to be advantageous in quantifying uncertainties, there are a few disadvantages with these methods. The first constraint with the Bayesian methods is the choice of prior distribution for parameters. Though a well-documented literature is available on selecting priors, this becomes a concern in the spatial analysis when choosing the prior distributions for variogram parameters. However, it is to be noted that as the sample size (number of observations) increases, the effect of prior on the posterior distribution of parameters is negligible. In urban areas, when modeling short-duration high-intensity precipitation, the historical observations are sparse in general, necessitating choosing the prior judicially. In most of the cases, the choice of prior could be informative, based on the physical process causing precipitation. The second limitation with the Bayesian methods is that they are computationally demanding. Again, this limitation is not of major concern in single-station analysis, but in modeling extreme precipitation spatially (BHMs/max-stable processes), computations become highly challenging. The extreme precipitation analysis considering a single station is quite well developed but the spatial extreme precipitation analysis is young and extending tools for modeling univariate extremes to model spatial extremes is still developing. In urban areas, precipitation is expected to vary spatially in a rapid manner and have different climatology when compared to the other areas, introducing spatial nonuniformity in precipitation. This spatial variation is a localized phenomenon which cannot be captured by modeling spatial variation in a large scale. Also, the availability of spatial data in urban areas is in general course leading to high uncertainties in estimating the spatial return levels maps of extreme precipitation.

Bayesian hierarchical model gives spatial distribution of parameters using stochastic processes which enable consideration of complex variation of parameters over space. The advantages of employing BHM are its flexibility and its simplicity in prediction at unobserved locations. These models allow us to add further layers of variation, if required, though the computations are arduous. Also, prediction at ungauged sites is straightforward using the simulation from conditional distributions of Gaussian process for each state of the chain, from which one can infer properties at the ungauged sites and it is straightforward to obtain measures of uncertainty for quantities of interest, like return levels. In general, it is observed that the uncertainty in return levels reduced when data from all the stations was pooled in a hierarchical model. But, the BHMs have disadvantages as well in modeling the precipitation extremes spatially. Major limitation could be the dependence in the extreme precipitation data at adjacent sites which violates the assumption of conditional independence. This assumption may be unrealistic and lead to underestimation of specific events when forecasted, but the method is still applicable for computation of marginal properties of extreme distributions, like return levels (Davison et al., 2012; Cooley and Sain, 2012).

The alternatives to BHMs are the max-stable models, where they capture the local behavior of spatial extremes appropriately considering dependence. Each modeling strategy has its own advantages and disadvantages (Turkman et al., 2010). Overcoming the disadvantage of assuming conditional independence in hierarchical models, copula, and max-stable models capture the residual dependence correctly, but need considerable effort in building up the model (Cooley and Sain, 2012). When the objective is to model extreme precipitation instead of marginal properties at local scales, capturing the correct dependence structure between the extremes is crucial and max-stable processes are suitable. In the case study presented, there is little dependence in the extreme precipitation observations and employing BHMs, which are flexible tools in modeling, could be a viable alternative to the max-stable processes. Therefore, the choice of the modeling strategy is dependent on the specific dataset, purpose of the study, and the benefits of using each of the modeling methods. A sensible choice of modeling strategy depending on the requirements could be carefully weighted.

Another alternative to model dependence is the copula-based approach. However, study by Davison et al. (2012) suggests that frequentist copula model does not adequately explain the local dependence structure of extreme precipitation data at 51 weather stations in Switzerland. There are other methods suggested to reduce the computational burden of using pairwise likelihood methods for very large spatial datasets, for example, downweighing or removing the likelihood contributions from sites which are far apart (Bevilacqua et al., 2012; Padoan et al., 2010). Approximate Bayesian Methods (ABCs) are viable solutions in situations where the likelihood function either cannot be evaluated numerically or written in closed form. It is to be noted that, in practice, the dependence often decreases as the number of rare events increases, which could be due to inadequacies in the classical formulation. Wadsworth and Tawn (2012) developed models for near-independence spatial extremes data.

6. CONCLUSION

Temporal and spatial modeling of extreme precipitation in urban areas is a major challenge due to sparse data availability and huge spatial nonuniformity in precipitation. High uncertainties are associated with the short-duration precipitation events, which need to be modeled and further to be considered in the design and risk analyses. Bayesian methods provide a comprehensive way to quantify the uncertainties in the precipitation data. The following are some of the inferences from the modeling of extreme precipitation, spatially and temporally.

- High uncertainties in GEV parameters and return levels are associated at shorter durations. While obtaining the IDF relationships, the uncertainty in return levels for short durations could not be assessed correctly using MLE method, raising an important question about the authenticity of standard statistical estimates. Bayesian approach is a simple and viable alternative to estimate the uncertainties associated with the sampling uncertainty.
- The uncertainties due to the use of multiple GCM models can be handled using REA method. It is observed that the uncertainty introduced due to choice of GCM is smaller than that due to parameter uncertainty. The uncertainty in return levels is large for RCP 8.5 scenario compared to the other scenarios as most of the GCMs underestimated/overestimated the return levels in the Bangalore City case study. Quantification of uncertainties, both model and parameter, is important to obtain reliable projected IDF relationships in urban areas.
- It is important to understand the spatial behavior of extreme precipitation over urban areas for effective infrastructure design and operation, which can be modeled using BHMs. This model allows us to obtain estimates of spatial return levels including those at ungauged locations. Elevation and mean annual precipitation are the predominant covariates for modeling extreme annual precipitation for Bangalore City. The uncertainty in return levels varies considerably across the city.
- Methods like kriging and IDW for estimating spatial variability of extreme precipitation are not suitable in urban areas due to high spatial variation in precipitation within short distances. BHMs overcome this disadvantage by obtaining spatial distribution of parameters using stochastic processes which enable consideration of complex variation of parameters over space. Also, by pooling the data from all the stations in a hierarchical model and adding covariates in obtaining the estimates of parameters reduced the uncertainty in return levels when compared to that from the single station analysis.
- When the extreme precipitation is modeled considering max-stable process, latitude and mean monsoon precipitation are the best covariates for the case study. The return levels obtained considering max-stable process, including dependence in the annual maximum observations,

are in comparison with the return levels obtained from BHMs. Also the spatial patterns obtained from both the models are identical. The choice of modeling strategy can be based on specific dataset and purpose of the study.

Lastly, it could be concluded that Bayesian methods provide the finest approach for modeling extreme precipitation in urban areas. Turkman et al. (2010) argued that the Bayesian models are not max-stable, but they are mixtures of max-stable processes, compatible with the asymptotic theory due to the spatially varying parameters.

REFERENCES

Banerjee, S., Carlin, B.P., Gelfand, A.E., 2004. Hierarchical Modeling and Analysis for Spatial Data. Chapman and Hall, Boca Raton, FL.

Bevilacqua, M., C.Gaetan, J.M., Porcu, E., 2012. Estimating space and space—time covariance functions: a weighted composite likelihood approach. J. Am. Stat. Assoc. 107, 497. https://doi.org/10.1080/01621459.2011.646928.

Burlando, P., Rosso, R., 1996. Scaling and multiscaling models of depth-duration-frequency curves for storm precipitation. J. Hydrol. 187, 45—64.

Chandra, R., Ujjwal, S., Mujumdar, P.P., 2015. Model and parameter uncertainty in IDF relationships under climate change. Adv. Water Resour. 79, 127—139. https://doi.org/10.1016/j.advwatres.2015.02.011.

Christensen, J.H., Christensen, O.B., 2003. Climate modelling: severe summertime flooding in Europe. Nature 421, 805—806.

Coles, S., 2001. An Introduction to Statistical Modelling of Extreme Values. Springer, London.

Cooley, D., Sain, S.R., 2012. Discussion of "statistical modeling of spatial extremes" by A. C. Davison, S. A. Padoan and M. Ribatet. Stat. Sci. 27 (2), 187—188. https://doi.org/10.1214/12-STS376A. Main article doi: 10.1214/11-STS376.

Cooley, D., Cisewski, J., Erhardt, R.J., Jeon, S., Mannshardt, E., Omolo, B.O., Sun, Y., 2012. A survey of spatial extremes: measuring spatial dependence and modeling spatial effects.

Cooley, D., Nychka, D., Naveau, P., 2007. Bayesian spatial modeling of extreme precipitation return levels. J. Am. Stat. Assoc. 102, 824—840. https://doi.org/10.1198/016214506000000780.

Cooley, D., Naveau, P., Poncet, P., 2006. Variograms for Spatial Max-stable Random Fields. In: Lect. Notes Statist., vol. 187. Springer, New York, pp. 373—390.

Davison, A.C., Padoan, S., Ribatet, M., 2012. Statistical modeling of spatial extremes (with Discussion). Stat. Sci. 27 (2), 161—186. https://doi.org/10.1214/11-STS376.

de Haan, L., 1984. A spectral representation for max-stable processes. Ann. Prob. 12, 1194—2004.

Fatichi, S., Barbosa, M., Caporali, E., Silva, M.E., 2009. Deterministic versus stochastic trends: detection and challenges. Water Resour. Res. 114, D18121. https://doi.org/10.1029/2009JD011960.

Gelman, A., Carlin, J.B., Stern, H.S., Rubin, D.B., 2004. Bayesian data analysis: texts in statistical science. Chapman & Hall/CRC Press, London.

Giorgi, F., Mearns, L.O., 2003. Probability of regional climate change calculated using the Reliability Ensemble Averaging (REA) method. Geophys. Res. Lett. 30 (12), 1629. https://doi.org/10.1029/2003GL017130.

Giorgi, F., Francisco, R., 2000. Evaluating uncertainties in the prediction of regional climate change. Geophys. Res. Lett. 27, 1295–1298.

Gupta, V.K., Waymire, E., 1990. Multiscaling properties of spatial rainfall and river flow distributions. J. Geophys. Res. 95 (D3), 1999–2009.

Hanssen-Bauer, I., Achberger, C., Benestad, R., Chen, D., Forland, E., 2005. Statistical downscaling of climate scenarios over Scandinavia. Clim. Res. 29 (3), 255–268.

Hewitson, B., Crane, R., 1996. Climate downscaling: techniques and application. Clim. Res. 7, 85–95.

Huard, D., Mailhot, A., Duchesne, S., 2010. Bayesian estimation of intensity-duration-frequency curves and of the return period associated to a given rainfall event. Stoch. Environ. Res. Risk Assess. 2, 337–347.

IPCC, 2012. Summary for policymakers. In: Field, C.B., Barros, V., Stocker, T.F., Qin, D., Dokken, D.J., Ebi, K.L., Midgley, P.M. (Eds.), Managing the Risks of Extreme Events and Disasters to Advance Climate Change Adaptation. A Special Report of Working Groups I and II of the Intergovernmental Panel on Climate Change. Cambridge University Press, Cambridge, UK/New York, pp. 1–19.

Jiang, P., Tung, Y.-K., 2015. Incorporating daily rainfalls to derive rainfall DDF relationships at ungauged sites in Hong Kong and quantifying their uncertainty. Stoch. Environ. Res. Risk Assess. 29, 45–62. https://doi.org/10.1007/s00477-014-0915-2.

Maraun, D., Wetterhall, F., Ireson, A.M., Chandler, R.E., Kendon, E.J., Widmann, M., Brienen, S., Rust, H.W., Sauter, T., Themeßl, M., Venema, V.K.C., Chun, K.P., Goodess, C.M., Jones, R.G., Onof, C., Vrac, M., Thiele-Eich, I., 2010. Precipitation downscaling under climate change: recent developments to bridge the gap between dynamical models and the end user. Rev. Geophys. 48 https://doi.org/10.1029/2009RG000314, 8755–1209.

Martins, E.S., Stedinger, J.R., 2000. Generalized maximum likelihood generalized extreme value quantile estimators for hydrologic data. Water Resour. Res. 28 (3), 737–744.

Martins, E.S., Stedinger, J.R., 2001. Historical information in a generalized maximum likelihood framework with partial duration and annual maximum series. Water Resour. Res. 37 (10), 2559–2568.

Mondal, A., Mujumdar, P.P., 2016. Hydrologic extremes under climate change: non-stationarity and uncertainty. In: Teegavarapu, R.S.V., Kolokytha, E., Satoru, O. (Eds.), Sustainable Water Resources Planning and Management under Climate Change. Springer, Springer Singapore. https://doi.org/10.1007/978-981-10-2051-3_2. ISBN: 978-981-10-2049-0 (Print) 978-981-10-2051-3 (Online).

Naveau, P., Guillou, A., Cooley, D., Diebolt, J., 2009. Modeling pairwise dependence of maxima in space. Biometrika 96, 1–17.

Padoan, S.A., Ribatet, M., Sisson, S.A., 2010. Likelihood-based inference for max-stable processes. J. Am. Stat. Assoc. 105, 263–277.

Papalexiou, S.M., Koutsoyiannis, D., 2013. Battle of extreme value distributions: a global survey on extreme daily rainfall. Water Resour. Res. 49, 187–201. https://doi.org/10.1029/2012WR012557.

Serinaldi, F., Kilsby, C.G., 2015. Stationarity is undead: uncertainty dominates the distribution of extremes. Adv. Water Resour. 77, 17–36.

Serinaldi, F., Kilsby, C.G., 2014. Rainfall extremes: toward reconciliation after the battle of distributions. Water Resour. Res. 50, 336–352. https://doi.org/10.1002/2013WR014211.

Tebaldi, C., Knutti, R., 2007. The use of the multi-model ensemble in probabilistic climate projections. Philos. Trans. R. Soc. A 365. https://doi.org/10.1098/rsta.2007.2076.

Turkman, K.F., Turkman, M.A.A., Pereira, J.M., 2010. Asymptotic models and inference for extremes of spatio-temporal data. Extremes 13, 375–397.

Veneziano, D., Yoon, S., 2013. Rainfall extremes, excesses, and intensity-duration-frequency curves: a unified asymptotic framework and new nonasymptotic results based on multifractal measures. Water Resour. Res. 49, 4320–4334. https://doi.org/10.1002/wrcr.20352.

Wadsworth, J.L., Tawn, J.A., 2012. Dependence modelling for spatial extremes. Biometrika 99 (2), 253–272. https://doi.org/10.1093/biomet/asr080.

Whetton, P.H., England, M.H., O'Farrell, S.P., Watterson, I.G., Pittock, A.B., 1996. Global comparison of the regional rainfall results of enhanced greenhouse coupled and mixed layer ocean experiments: implications for climate change scenario development. Climatic Change 33, 497–519.

Wilby, R.L., Wigley, T.M.L., 1997. Downscaling general circulation model output: a review of methods and limitations. Prog. Phys. Geogr. 21, 530–548.

Wilks, D.S., Wilby, R.L., 1999. The weather generation game: a review of stochastic weather models. Prog. Phys. Geogr. 23 (3), 329–357.

Xu, C., 1999. From GCMs to river flow: a review of downscaling methods and hydrologic modelling approaches. Prog. Phys. Geogr. 23 (2), 229–249.

Yang, L., Tian, F., Niyogi, D., 2015. A need to revisit hydrologic responses to urbanization by incorporating the feedback on spatial rainfall patterns. Urban Clim 12, 128–140.

Zorita, E., von Storch, H., 1997. A Survey of Statistical Downscaling Techniques. GKSS Rep., 97/E/20, GKSS Res. Cent., Geesthacht, Germany.

CHAPTER 4

Introduction to Physical Scaling: A Model Aimed to Bridge the Gap Between Statistical and Dynamic Downscaling Approaches

Abhishek Gaur[1], Slobodan P. Simonovic[2]
[1]Construction Research Centre, National Research Council Canada, Ottawa, ON, Canada;
[2]Facility for Intelligent Decision Support, Western University, London, ON, Canada

Contents

Trends and Changes in Hydroclimatic Variables
ISBN 978-0-12-810985-4
https://doi.org/10.1016/B978-0-12-810985-4.00004-9

1. INTRODUCTION

In their pioneering work, Lowry (1977) identified "background climate," "effects of local landscape," and "effects of local urbanization" as three main components shaping climate at any location. Although first proposed 40 years ago, this fundamental and simple theoretical framework forms the basis of several detailed and complex climate models used currently like the global climate models (GCMs). GCMs simulate complex large-scale biogeophysical and chemical processes, as well as their interactions with geophysical characteristics like land-use and topography to model global climate system. They are very useful mathematical tools that have been used extensively to study the impacts of historically observed and future projected greenhouse gas emissions on climate.

Very high computational demand is associated with GCM simulations even though they simulate climate at a very coarse spatial resolution of 3—5 degrees (~330—~550 km). To provide a rough estimate, at 3 degrees spatial resolution a GCM performs computations at 7200 grids across the globe each day. For simulating long-term climate projections for 250 years (as typically projections are made from 1850 to 2100), it will be performing the calculations ~700 million times to produce one set of climatic projections. Due to such high computational demand associated with GCM runs, climate simulations cannot be performed at very high resolutions, which will produce more realistic estimates of climate but will also be impractical. As such, at 3–5 degrees spatial resolution, GCM outputs are unsuitable for local and regional-scale risk assessment studies. Additionally, climatic effects of subgrid-scale geophysical features like land-cover and elevation are averaged out over a large area in the data associated with a GCM grid.

To estimate higher resolution climate projections from lower resolution GCM projections, a step referred to as downscaling is performed. Two broad classes of downscaling methods have been used in the past: statistical and dynamic. In the statistical downscaling methods, higher resolution climate data are estimated by developing statistical relationships between low resolution climatic or atmospheric data and locally observed data. The developed relationships are then used to estimate downscaled future climatic projections. A review of the state-of-the-art statistical downscaling techniques used in previous climate change impact assessment studies is provided in Section 2.2 of this chapter. In dynamic downscaling methods, local-scale climate is estimated by coupling a mesoscale higher resolution model or regional climate model (RCM) with a GCM. Boundary conditions generated from the GCMs are used as inputs into a RCM to estimate higher resolution climate taking into consideration biogeophysical interactions occurring at the land–atmosphere interface within the grid. Thus in dynamic downscaling methods, climate variability within a GCM grid is modeled taking into consideration geophysical variability present within a GCM grid-cell. On the other hand, in existing statistical downscaling methods same is modeled based on the location-specific differences in statistical relationships between low-resolution GCM and observed local climate.

The statistical relationships derived when performing statistical downscaling can be different for different locations within a GCM grid because of the differences in "background climate," "effects of local landscape," and "effects of local urbanization" as discussed in Lowry (1977). In previous studies, changes in "background climate" between different subgrid locations have been considered when downscaling GCM data. However, the contribution of land-cover and topography has largely been ignored. To model changes in statistical relationships between large-scale and local-scale climate at a location more accurately, land-cover and elevation characteristics of the location should be also considered when performing statistical downscaling of future GCM projections. Another important benefit of performing statistical downscaling, considering geophysical characteristics of a location, is that the developed relationships can then be used in geophysical environments that are drastically different from the present. Typically in statistical downscaling, relationships are derived for a location over the current time-period, and the relationships are used directly to model future projections at that location without considering possible future shifts in geophysical characteristics of the location which can change the nature of statistical relationships developed on the current time-period. When making decadal to century-scale predictions, considerable changes in geophysical characteristics at local to regional scales can be expected. It is therefore important to develop statistical downscaling methods that can model the effects of changes in geophysical characteristics in future climatic projections.

Physical scaling (SP) downscaling model is one of the first steps toward this direction. In this chapter we introduce an original model, perform validation of the model in current climate, compare the performance of the model with other state-of-the-art models, demonstrate an example application of the model, as well as explain R codes that can be used to apply the model at any location across the globe. The chapter is organized as follows. A review of literature related to SP model is provided in Section 2. This is followed by an introduction of SP method in Section 3. The advantages of SP method over other state-of-the-art statistical downscaling methods are presented in Section 4. Results from the validation of SP method are discussed in Section 5. A case study of SP method where it is used to quantify land-cover change–induced hydroclimatic changes in four catchments located in southern Saskatchewan region of Canada is presented in Section 6. The R code developed to formulate SP model and used to downscale future GCM projections is presented and discussed in Section 7. This is followed by a summary of conclusions and future work in Section 8.

2. LITERATURE REVIEW

In this section we provide theoretical background on concepts related to SP model. Most of the discussion provided in this section revolves around the downscaling of two important climate variables: precipitation and temperature, and how that can be performed more realistically using SP model.

First a brief introduction to the phenomenon of climate change and land-use land-cover (LULC) change is provided in Section 2.1. Then a review of statistical downscaling methods used in the past is presented in Section 2.2. This is followed by a discussion on the mechanisms and observation of the evidence of the effects of land-cover and elevation on local climate in Section 2.3. Finally a review of previous studies which have quantified the impact of future land-cover change on hydroclimatic variables is provided in Section 2.4.

2.1 Climate Change and LandUse Land-Cover Change

In this subsection, a brief introduction to climate change and its impacts on the past and future climates are presented. This is followed by a discussion of the historically observed and future projected LULC trends across the globe.

2.1.1 Climate Change

Climate change is referred to as the phenomenon of a persistent and detectable long-term change in climate conditions of a location. It is undeniable that the global climate has changed drastically over the past 60 years or so. Observational evidences of drastic shifts in climatic patterns have been observed across the globe. Global temperatures have been found to increase consistently with a mean increase of ~0.85°C rise noted over the period 1880–2012. Other hydroclimatic changes include an increase in global sea levels, decrease in snow-cover and sea-ice content, changes in precipitation and flow patterns, changes in circulation patterns, etc. (Stocker et al., 2013). Greenhouse gases emitted as a result of rapid human and industrial growth have been found to be the most prominent reason behind this change. Greenhouse gases emitted in the atmosphere hinder longwave radiations reflected by the earth's surface to travel back into the space thereby increasing temperatures near and around the earth's surface. Indeed, continual emission of greenhouse gases as a result of continuous industrial and human growth has resulted in considerable changes in the earth's radiation budget. It is estimated that earth is receiving more energy from the sun than it is releasing since at least about 1970 (Stocker et al., 2013). A change in the earth's radiation budget affects a range of atmospheric and climatic variables as evident in the observational findings discussed before.

Observed trends of increases in greenhouse gas concentration are expected to continue over the 21st century. To study the possible range of future emissions and their implications on climate, future greenhouse gas emission scenarios have been developed by different integrated assessment modeling (IAM) groups operating across the globe. Out of all the proposed future greenhouse gas trajectories, Van Vuuren et al. (2011) identified four representative concentration pathways (RCPs): RCP2.6, RCP4.5, RCP6, and RCP8.5 that correspond to an end of the 21st century radiative forcing of 2.6, 4.5, 6, and 8.5 W/m^2, respectively. A description of the four RCPs, their respective modeling groups, and related references are provided in Table 4.1. One or more of the selected RCPs are typically chosen to perform climate change impact assessment studies. In the Fifth Assessment Report (AR5) published by the Intergovernmental Panel for Climate Change (IPCC) (Stocker et al., 2013), future climate projections made by a range of GCMs, corresponding to the above specified four RCPs, have been presented and discussed.

Future climatic and hydrologic regimes are expected to be shaped by the greenhouse gas emission trajectory that is experienced. Temperature is expected to increase for all RCPs although the magnitude of change varies among different RCPs. The projected global mean increases in temperature between 1986–2005 and 2081–2100 time-periods as simulated by different GCMs are expected to be in the range of 0.3–1.7°C, 1.1–2.6°C, 1.4–3.1°C, and 2.6–4.8°C for RCP2.6, RCP4.5, RCP6, and RCP8.5,

Table 4.1 The Four Representative Concentration Pathways (RCPs) Chosen and Described in Van Vuuren et al. (2011)

RCP	Modelling Group	Description	Reference
RCP2.6	IMAGE	Rising radiative forcing pathway leading to 8.5 W/m^2 by 2100.	Van Vuuren et al. (2006, 2007)
RCP4.5	GCAM	Stabilization without overshoot pathway to 4.5 W/m^2 at stabilization after 2100.	Clarke et al. (2007), Smith and Wigley (2006) and Wise et al. (2009)
RCP6	AIM	Stabilization without overshoot pathway to 6 W/m^2 at stabilization after 2100.	Fujino et al. (2006) and Hijioka et al. (2008)
RCP8.5	MESSAGE	Peak in radiative forcing at ~3 W/m^2 before 2100 and then decline to 2.6 W/m^2 by 2100.	Riahi et al. (2007)

respectively (Collins et al., 2013). Future mean precipitation is expected to increase with rising temperatures. At a global scale mean precipitation is projected to increase at a rate of 1% $°C^{-1}$ to 3% $°C^{-1}$ for all RCPs except RCP2.6 in which case the rate of change in precipitation is associated with a considerable uncertainty among different GCMs (0.5% $°C^{-1}$ to 4% $°C^{-1}$). It should be noted that considerable regional and distributional variations are associated with the global temperature and precipitation changes outlined above (Stocker et al., 2013). Temperature and precipitation extremes are projected to change more drastically in the future than the means. For instance, it is projected that a one in 20-year maximum temperature event will at least become a one in 10-year event at most places across the globe; however, at many places it will become an annual or one in 2-year event. Similarly, for precipitation, a clear tendency for an increase in precipitation extremes with temperatures at a rate of roughly 5% $°C^{-1}$ to 10% $°C^{-1}$ is projected (Stocker et al., 2013; Collins et al., 2013).

2.1.2 LULC Change

Earth has experienced significant LULC change with human development and growth so much so that carbon emissions owing to land use change have been found comparable to that due to fossil fuel emissions over the period 1850–1990 (Houghton, 1999). Studies have found that during the last 300 years more than 50% of the land surface has been affected by land-use change. A rapid clearing of around 25% of the forested lands and a rapid growth of agricultural land to around 30% of the total land surface has also been observed (Turner et al., 1990; Hurtt et al., 2006).

With increasing population and agricultural and industrial growth, LULC change is projected to increase across the globe over the 21st century. The degree of change however is projected to vary depending on the manner in which future society growth occurs. Harmonized land-use data is a 0.5° × 0.5° continuous annual estimate of land-use for the period 1500–2100 and is prepared by merging historical reconstructions of land-use with future estimates corresponding to different RCPs as provided by the modeling groups listed in Table 4.1. An analysis of harmonized land-use datasets produced by Hurtt et al. (2011) suggests that future LULC change is highly dependent on the emission scenario that future society follows while developing. An analysis of future projected land-use indicated significant change among all land-use classes. Most consistently the total area of secondary land (i.e., land which had been influenced by human activity in the past) is projected to increase under all RCPs. On the contrary, few land-use classes such as total agricultural land can change by +13% to −24% depending on the choice of RCP.

In summary, rapid changes in LULC have been observed in the past 300 years or so. It is expected that this trend will continue in the future; however, the magnitude of change will depend on the manner in which human development progresses in the future.

2.2 Statistical Downscaling

Statistical downscaling models aim to estimate regional to local-scale climate from low spatial resolution GCM climate data using statistical techniques. Here we provide an overview on some of the models that have been used in the past to perform statistical downscaling of GCM projections. A discussion on other relevant aspects of statistical downscaling such as inherent assumptions and commonly used predictors is also presented.

2.2.1 Assumptions

All statistical downscaling methods must adhere to following assumptions (Hewitson and Crane, 1996; Giorgi et al., 2001; Wilby et al., 2004; Benestad et al., 2008). Firstly, there must be a strong relationship between the predictor and predictant variables chosen in the method. In statistical terms, the predictor variable should have high degree of covariance and similar time structure to that of the predictant variable for it to be selected for prediction. Secondly, chosen predictor variables must be adequately simulated by the GCMs. This is very important because if important physical and statistical aspects of the predictor variable are not accurately simulated by the GCMs, then downscaling will only result in a more precise estimate of the erroneous predictor variables. Third assumption is that chosen predictors must be able to represent the climate change signal in the predictant variable adequately. For instance, sea level pressure is considered a very good predictor for observed temperature. However, if it is used alone to predict temperatures it can lead to the prediction of unrealistically low temperature estimates (Benestad et al., 2008). Therefore when performing downscaling, it is important to identify a set of predictor variables which can model the predictant variable with acceptable accuracy. Lastly and most importantly, the relationships derived are assumed to be time-invariant, that is, the relationships do not change over time. The validity of this assumption is very critical to obtaining reliable future estimates of climate in drastically changed future environments. Although this assumption cannot be tested explicitly, several studies have tested the robustness of the statistical relationships by calibrating and validating the methods on contrasting climatic conditions. For instance, Wilks (1999) tested the robustness of a statistical downscaling model towards downscaling precipitation by calibrating the method on wet days and validating it on dry days and vice versa. These tests do not guarantee superior method skill in future climates but models that perform well in such can be considered preferable for making downscaled future climatic projections than the ones that do not perform well.

2.2.2 Common Predictors

The predictor variables considered for downscaling in the past have been guided by the availability of long time-series of atmospheric variables in the calibration and prediction timelines. Atmospheric data for calibration is generally extracted from the reanalysis products and for the prediction period is obtained from the GCMs. The most relevant set of predictors are chosen by accessing the relationship between the predictor variables and observed climate data. Predictors that are found to be consistently correlated with the observed climate are selected for prediction.

The choice of variables has also been found to depend on the temporal resolution of the predictant variable. For instance, monthly mean predictant variables have been found to be linked to upper level atmospheric and circulation variables (like geopotential height and sea level pressure) while daily predictant variables, which are more variable and skewed, are likely also dependent on variables linked to horizontal flux and convergence of moisture (like specific humidity, winds, and vorticity) (Cavazos and Hewitson, 2005; Schoof, 2012). Another important factor that can influence the choice of predictor variables in statistical downscaling is the spatial, temporal, and distributional scale at which the relationship between predictors and observed climate are accessed. Gaur and Simonovic (2015) identified this as one of the three prominent sources of uncertainty in climate change impact assessment studies. They demonstrated that GCM-based climatic and atmospheric variables differ considerably with the spatial, temporal, and distributional scale of analysis. Masson and Knutti (2011) and Räisänen and Ylhäisi (2011) have discussed the variability of GCM skill over different spatial scales and have suggested an optimal spatial scale at which GCMs perform best. Evaluation of predictors at different spatial, temporal, and distributional scales is therefore very important and should be considered when selecting optimal predictor variables for statistical downscaling of GCM projections.

2.2.3 Statistical Downscaling Models

Statistical downscaling methods used in the past can be classified as being either PerfectProg or model output statistics (MOS) models (Fowler et al., 2007; Schoof, 2013). In PerfectProg models, the relationship between predictor and predictant variables are derived by using observed or reanalysis based predictors. On the other hand, in MOS techniques GCM data is used directly for downscaling. The raw GCM data is sometimes bias-corrected (Teutschbein and Seibert, 2012) before it is used for downscaling in the MOS models. The PerfectProg models can be further classified into three categories of models: (1) regression-based, (2) weather pattern-based, and (3) weather generator-based methods (Wilby and Wigley, 1997).

Several MOS methods have been used in the past to estimate higher resolution downscaled climatic projections from raw GCM projections. The most popular method under this category is the bias correction and spatial disaggregation (BCSD) technique (Salathé, 2003; Wood et al., 2004). In this method the raw GCM data are first bias-corrected with reference to gridded observed climate data. The bias-corrected data is then used to obtain local-scale spatial anomaly pattern which is then used to modify future bias-corrected data. The method has been used extensively to downscale GCM projections in many climate change impact assessment studies (Salathé, 2005; Widmann et al., 2003; Sharma et al., 2007 among others).

Among the PerfectProg models, regression-based statistical downscaling models are perhaps the most popular. In regression-based downscaling methods a regression relationship is developed between predictor and predictant variables. The predictor and predictant variables have previously been linked using functions such as standard ordinary least squares (Wilby et al., 2002), quantile regression (Friederichs and Hense, 2007), partial least squares regression (Bergant and Kajfez-Bogataj, 2005), generalized linear modeling (Fealy and Sweeney, 2007), etc. In some studies, techniques such as singular value decomposition and canonical correspondence analysis (CCA) have also been used to identify relationships between the predictor fields (Hertig and Jacobeit, 2008; Huth, 2002). Studies have also employed artificial neural networks (ANN) and machine learning algorithms such as genetic programming (Coulibaly, 2004), support vector machines (Tripathi et al., 2006), and relevance vector machines (Ghosh and Mujumdar, 2008) to link predictor and predictant variables.

One of the most popular regression-based downscaling techniques is statistical downScaling model (SDSM). Downscaling by SDSM method

(Wilby et al., 2002) is performed by analyzing a range of atmospheric variables available from the reanalysis and GCM databases. In SDSM a set of 25 candidate predictor variables have been suggested for initial screening (Wilby et al., 2002). Out of the candidate predictor variables, a set of most relevant atmospheric predictor variables which are highly correlated with locally observed climate variable of interest are chosen for prediction. This step is referred to as the initial screening process. Multiple linear regression relationship between the selected large-scale atmospheric variables (predictors) and locally observed climate (predictant) data is then formulated over the calibration period. The calibrated regression relationship is thereafter used to predict climate variable of interest over the validation period. Another regression-based precipitation downscaling model is generalized linear modeling (GLM)—based model. In GLM model, first relevant set of atmospheric predictors are identified for prediction using an approach similar to that used in the SDSM model. Precipitation occurrence is modeled from the chosen predictor variables using a logistic regression model and precipitation magnitudes on wet days are modeled using generalized additive model (GAM). The GLM model has been used extensively to perform precipitation downscaling across the globe (Fealy and Sweeney, 2007; Kigobe et al., 2011; Beecham et al., 2014; Lee et al., 2011; Manzanas et al., 2015).

Weather pattern—based downscaling methods are based on the principle that observed climatic variations at the local scale are largely shaped by synoptic climatic patterns. The most common method under this category is referred to as the analog method (Zorita and von Storch, 1999). In this method historical records are searched for synoptic patterns matching the GCMs. The observed climate corresponding to those patterns is then considered as the downscaled GCM data. Several studies have employed analog methods to downscale GCM data (Timbal and Jones, 2008; Timbal et al., 2008; Li and Smith, 2009; Chiew et al., 2010; Frost et al., 2011; Hope et al., 2010). Another popular method in this category is the nonhomogeneous hidden Markov model (Hughes and Guttorp, 1994), which has been widely used to downscale GCM precipitation data (Robertson et al., 2004, 2007; Bellone et al., 2000; Hughes et al., 1999; Samuels et al., 2009; Liu et al., 2011; Kwon et al., 2011). In this method precipitation occurrence probabilities and wet -day precipitation amounts at a location are linked to the classes of synoptic scale atmospheric fields such as geopotential height. The applicability of weather pattern—based downscaling methods is highly dependent on the availability of long observational climatic and atmospheric records so that patterns found in the GCM data can be associated with an observational synoptic scale pattern.

Weather generators are stochastic models that can be used to produce long synthetic replicates of observed climate time-series. A wide range of parametric, nonparametric, and semiparametric weather generators have been used in climate change impact assessment studies in the past (Wilks, 2012, 2010; Dibike and Coulibaly, 2005; Charles et al., 2007; Kwon et al., 2011; Eum and Simonovic, 2012; Hashmi et al., 2011; Richardson, 1981; Soltani and Hoogenboom, 2003; Kuchar, 2004; Craigmile and Guttorp, 2011; Apipattanavis et al., 2007; Semenov and Barrow, 1997; Sharif and Burn, 2006, 2007; Lee et al., 2012; King et al., 2012, 2016; Srivastav and Simonovic, 2015). Downscaling by weather generator technique is performed by first obtaining scaled historical climate data, which is regarded as equivalent to future projections without internal climate variability component. Scaled historical data has been obtained using several approaches in the past. One of the approaches is by estimating change factors using historical and future GCM modeled climate data (Anandhi et al., 2011; Gaur and Simonovic, 2015). Change factors can be calculated for various temporal scales (daily, monthly, seasonal, yearly), can have different mathematical formulations (additive, multiplicative), and can be associated with different quantiles of the distribution (mean, median, or other quantiles). The scaled historical climate is then used as input into a weather generator to obtain future climate projections with internal climate variability. This is done by generating several runs of historical scaled data such that key statistical properties of each run is similar to the input climatic time-series with some perturbation component which incorporates internal climate variability component to the input time-series. In some studies, weather generator parameters have been modified in place of using change factors to include the climate change signal in the weather generator downscaled outputs (Schoof et al., 2007, 2010).

2.3 Statistical Versus Dynamic Downscaling of GCM Projections

Both statistical and dynamic downscaling approaches have strengths and limitations. Statistical downscaling methods present a way to downscale GCM projections quickly and with minimal computational requirements. For this reason they are desirable for uncertainty assessment in future climate projections. They are also particularly useful in cases where location-specific downscaled projections are desired. Dynamic downscaling approaches can provide downscaled projections at a finer resolution grid-scale than the GCMs (typically at 10–20 km spatial resolution), but location-specific projections still need to be extracted from them using some spatial interpolation technique. This can be problematic in regions that have complex, varying topography or land-cover within a downscaled grid-cell. Statistical downscaling approaches are also very useful for real-time applications. They have been found to produce excellent results especially in regions where observed data are abundant and are considered very suitable for downscaling near-term climate predictions since the relationships derived in the historical timeline are considered to be time-invariant in statistical downscaling. This assumption is practically valid in near-term (5–10 years) projections. Finally, only large-scale changes in climate can be accounted for when performing statistical downscaling as regional geophysical characteristics are not considered in these models.

On the other hand, physically based dynamic downscaling methods are suitable for both near-term and long-term predictions. In other words, they can model climate in drastically changed climatic and geophysical environments. They are particularly useful in modeling climate projections in regions where observed data is limited. Dynamic downscaling methods are computationally expensive and use boundary conditions that are extracted from the GCMs. Therefore, bias associated with the boundary conditions is also transferred to the dynamically downscaled climate projections. Statistical downscaling models, on the other hand, offer an opportunity to bias-correct GCM data before they are downscaled.

As evident from above discussion, both statistical and dynamic downscaling methods have areas of strengths and weaknesses. An ideal downscaling model would combine the strengths of both these categories of methods and minimize the weaknesses.

2.4 Effects of Land-Cover and Elevation on Climate

Both land-cover and elevation are known to influence climate both from observational studies and physical sciences. Some of these concepts and observational evidences have been summarized in this section.

2.4.1 Land-Cover and Climate

Regional and local-scale climate, which are most relevant for catchment-scale hydrologic assessment, are shaped not only by large-scale atmospheric processes but also by its physical characteristics like land-cover and topography (Lowry, 1977; Oke, 1982). Observational studies have found detectable influence of both land-cover (Pielke et al., 2006) and elevation (Pepin et al., 2015), which are also supported by the laws of atmospheric and physical sciences.

The influence of land-cover on temperature has been particularly well explored in the case of urban areas. Urban areas are characterized by low albedo and heat absorbing surfaces because of which their near-surface heat transfer is led by an enhanced sensible heat flux and reduced latent heat flux as compared to the nonurban areas. Additionally, urban areas are unable to dissipate heat absorbed during the daytime primarily due to a reduced sky-view factor in the canopy layer. This phenomenon is well documented in the scientific literature and is commonly termed as the urban heat island (UHI) effect (Voogt and Oke, 2003; Oke, 1982). UHI effect has been recorded for many cities across the globe including Colombo, Sri Lanka (Emmanuel and Johansson, 2006), San Juan, Puerto Rico (Velazquez-Lozada et al., 2006), Singapore (Wong and Yu, 2005), Ouagadougou, Burkina Faso (Offerle et al., 2005), Gaborone, Botswana (Jonsson, 2004), Phoenix, Arizona (Fast et al., 2005), Addis Ababa, Ethiopia (Abebe and Magento, 2016), Toronto, Canada (Wang et al., 2015), Noida, India (Kikon et al., 2016), among others.

The effect of land-cover on temperature has also been reported for other land-cover classes. For example, Ge (2010) found agricultural regions to be 2.3°C cooler than the surrounding grassland areas in the growing season during the daytime and 1.61°C warmer than the surroundings after the harvest. Roy et al. (2007) investigated the effect of massive agricultural green revolution on temperature in India using both monthly observational climatic data and regional climate models. They found that during the growing season agriculture can considerably reduce temperatures in the surrounding areas. Boucher et al. (2004) performed a global-scale study to investigate the impact of agricultural growth on atmospheric water vapor and climate. They found a cooling of up to ~0.8K over irrigated areas together with an increase in global mean radiative forcing in the same period due to an enhanced water vapor concentration due to irrigation. McPherson et al. (2004) investigated the impact of winter wheat crops on climate over Oklahoma and found considerable cooling in the growing season and warming after the crops have been harvested. Similar cooling of irrigated areas of Great Plains over nonirrigated areas is also found by Mahmood et al. (2006). Another study has noted a local cooling effect as a result of intensive horticulture due to increase in surface albedo in south-eastern Spain (Campra et al., 2008). Studies have also found decrease in temperatures due to afforestation/deforestation, which results in changes in the surface albedo, leaf area index, roughness lengths, and rooting depths (Sellers, 1992; Jackson et al., 1996; Pitman, 2003; Fahey and Jackson, 1997; Xue and Shukla, 1996; Nosetto et al., 2005). Li et al. (2017) evaluated contribution of historical land-cover change to changes in extreme temperature over Eurasia. They compared temperature simulations obtained from National Centre for Atmospheric Research's Community Atmosphere Model Version 5.0 (NCAR CAM5.0) (Neale et al., 2012) forced with land-covers corresponding to years 1850 and 2000. Results indicated that historically over Eurasia land-cover change has led to changes in extreme indices of both minimum and maximum temperatures; however, the magnitude of change varies significantly among different regions. Other observational studies further corroborate to a relationship between regional land-cover and temperatures (Qiao et al., 2014; Fall et al., 2010a,b; Hale et al., 2006, 2008; Lawrence and Chase, 2010).

Several observational studies have also noted the influence of land-cover on precipitation, especially for urban areas. Li et al. (2011) studied urban signature in strong and weak precipitation events over the Pearl River Delta (PRD) in China using tropical rainfall measuring mission (TRMM) satellite precipitation data. They found that over and around the urban regions "strong" precipitation events have increased with urbanization while "weak" precipitation events have decreased. They also found strengthening of the precipitation intensity, a decrease in rainfall frequency, and an increase in convective rainfall and afternoon precipitation events over and around the urban areas. Similar findings were reported by De and Rao (2004). They analyzed rainfall trends of several Indian megacities (with population more than 1 million) such as New Delhi, Kolkata, Mumbai, and Chennai between 1901 and 2000 and found statistically significant increasing trends in annual and monsoon precipitations. A decreasing trend was also found for a few cities. They found more pronounced increases in precipitation during the period 1951–2000 when rapid industrial development took place over the selected urban locations. Rao et al. (2004) performed a similar analysis on precipitation trends for the duration 1901–2000 and found similar statistically significant increasing trends for the cities analyzed. Kishtawal et al. (2010) analyzed mean and extreme rainfall trends of urban locations within India using observed as well as remotely sensed TRMM precipitation data and identified an increasing trend linked to the pace of urbanization of the cities. Further urban locations were found to have more possibility of witnessing an extreme precipitation event than the surrounding nonurban area. Several other studies have also found evidences of land-cover linkage with rainfall (Kug and Ahn, 2013; Schlüzen et al., 2010; Halfon et al., 2009; Fujibe et al., 2009; Mote et al., 2007; Diem and Mote, 2005; Inoue and Kimura, 2004; Dixon and Mote, 2003; Shepherd et al., 2002). There are three hypotheses as to how urban areas can impact regional precipitation distribution: (1) by modifying the thermodynamic processes such as energy balance and urban heat island induced circulation within and around the city, (2) by causing winds to converge over and downwind of the cities due to roughness of the city elements, and (3) by effecting cloud microphysical processes due to the presence of large amounts of aerosol in the urban air (Han et al., 2014).

The influence of agricultural development has also been found to influence rainfall. Segal et al. (1998) adopted a modelling approach for investigating this and found an increase in precipitation with increase in irrigated agricultural areas over North America. An increase in agricultural areas in the Indian subcontinent has been found to decrease the monsoon rainfall (Lee et al., 2009; Niyogi et al., 2010). This is considered to happen as a result of reduced temperatures over the agricultural areas due to irrigation. This in turn diminishes the land–sea temperature contrast in agricultural areas, which is one of the driving factors for monsoon rainfall in India. A detectable effect of afforestation/reforestation on precipitation has been found and has been found to be dependent on geographical location, regional atmospheric characteristics, and the extent of afforested-reforested area (Xue and Shukla, 1996; Pitman and Narisma, 2005; Pielke et al., 2006 and references therein).

It has also been found that different sections within a city can have different climates based on the geophysical characteristics of their neighborhoods. For example, urban green areas such as parks are generally cooler than their surrounding built-up areas, and can produce temperature differences of up to 7°C, a phenomenon termed as "park cool island" (Oke et al., 1989; Barradas, 1991; Spronken-Smith and Oke, 1998; Jansson et al., 2007; Chang et al., 2007). The cooling effect of these vegetated areas has also been found to extend into the surrounding urban areas as well (Jauregui, 1991; Chen and Wong, 2006; Wong and Yu, 2005; Jonsson, 2004). For instance, after analyzing temperatures at seven gauging stations in Curitiba, Brazil, Kruger and Givoni (2007) found significant thermal effects of neighboring land-cover cells on the temperature recorded at those stations. Sun (2011) analyzed temperatures for two streets in Taichung city, Taiwan, and found them to be significantly correlated with the surrounding green ratio and building ratio in the night. Yokobori and Ohta (2009) too found temperatures in the Tokyo metropolitan area, Japan, to decrease with the total amount of vegetated area present in the neighborhood areas. Yan et al. (2014) found the air temperature in Beijing, China, to be significantly affected by the manmade and tree cover surface composition of the city. They also analyzed this relationship at different spatial scales (from 20 to 300 m in radius) and found that the relationships varied with the spatial scale considered.

Above studies provide evidence that local-scale climate is not only affected by the land-cover properties of the location of interest but also by the land-cover composition of the areas surrounding the location. Further, the effect of neighborhood land-cover composition is also dependent on the spatial scale at which those relationships are analyzed.

2.4.2 Elevation and Climate

The relationship between elevation and temperature is well understood in meteorology. The rate of decrease of temperature with elevation is referred to as the lapse rate. An approximate lapse-rate value of 6.5°C per 1000 m in the troposphere is considered as standard in many meteorological applications (Lydolph, 1985). The relationship between elevation and precipitation is more uncertain than the temperature with some studies finding consistent relationships between the two variables while some studies finding variable relationships. For example, Puvaneswaran and Smithson (1991) found both increasing and decreasing trends while analyzing precipitation-elevation relationships across Sri Lanka and termed the relationship to be complex. On the other hand using geographically weighted regression (GWR), Brunsdon et al. (2001) found a definite relationship between elevation and precipitation over the Great Britain. They highlighted the importance of considering GWR while studying these relationships as they vary in space.

Sensitivity of different regions to climate change has also been found to be elevation-dependent. In general, higher elevation regions are considered to be more susceptible to climate change than the lower elevation regions. Yan and Liu (2014), for instance, found that higher elevation areas (>2000 masl) in the Tibetan Plateau show a higher rate of warming than other lower elevation regions surrounding it. Giorgi et al. (1996) performed modeling experiments using a regional climate model and compared present-day and doubled CO_2 experiment modeled climates over the Alpine region. They obtained more pronounced warming over higher elevation regions than the lower elevation regions from their experiments.

Li et al. (2012) analyzed the variation of temperature and precipitation extremes across southwestern China over the period 1961–2008 and found significant links between elevation and changes in diurnal temperature range, frost days, ice days, cold night frequency and cold day frequency, consecutive dry days, consecutive wet days, wet-day precipitation, and the number of heavy precipitation days in the region during this period. Historical and future projected trends in minimum temperature were analyzed with reference to elevation in the Tibetan plateau by Liu et al. (2009) and greater vulnerability of higher elevation regions over the lower elevation regions was demonstrated. A more detailed review of the observational evidence and plausible operating mechanisms that lead to these elevation-dependent responses to greenhouse gases is provided in Pepin et al. (2015). Climate modification brought due to snow-albedo feedback, more frequent cloud cover, and water vapor–related radiative feedbacks are considered as possible mechanisms for a higher warming rate in the higher elevation regions. It is also recognized that the conclusions made for elevation-dependent changes in climate variables are not well understood because of less data availability at higher elevation regions (Pepin et al., 2015).

The literature review presented above provides observational and physical evidence that land-cover and elevation and their distribution across a region influences its climate. Under rapidly changing projected future geophysical conditions (Hurtt et al., 2011) it is therefore incessant to consider changes in geophysical characteristics when making future climate projections.

2.5 Previous Assessments of Future Land-Cover Change Impact on Climate

The impact of future projected land-cover change on climate has been studied using GCMs at regional to global scales. The typical methodology has been to compare climate projections obtained from the GCMs with and without considering future land-cover change. For example, Hua et al. (2015) analyzed future temperature projections obtained with and without considering LULC change from a GCM: CanESM2 under two RCPs: RCP2.6 and RCP8.5 and found small magnitudes of change at a global scale; however, changes of upto 0.1°C were obtained at regional scales. Quesada et al. (2017) investigated the impact of future LULC change on monsoonal precipitation by analyzing future projections under RCP8.5

from five earth system models (ESMs) and found significant changes in four (out of eight) of those regions. A unique methodology to identify land-cover change–induced climatic changes is proposed by Kumar et al. (2013). They compared climatic projections made for two neighboring sites: one which has projected to experience land-cover change in future and another which is projected to remain unchanged in future. From their analysis using 14 GCMs under RCP8.5, they found substantial land-cover change–induced summer warming in parts of North America and Eurasia.

Studies have also evaluated land-cover change effects on climate at local scales. Malyshev et al. (2015) analyzed the effect of local-scale land-cover heterogeneity in the geophysical fluid dynamics laboratory (GFDL) ESM and found significant effects on climate. Georgescu et al. (2013) simulated mid- and end-of-century temperature for the Arizona city using the weather research and forecast (WRF) modeling system. They found that the projections are highly sensitive to the scenario-based land-cover trajectory. Both mid- and end-century temperature estimates were found to be strongly dependent on the built environment and future emission pathways in the Sun Corridor expansion projections. Chen and Frauenfeld (2015) downscaled future temperature and precipitation projections made by Community ESM under RCP4.5 using WRF modeling system over China. They obtained significant effects of future-projected urbanization on both temperature and precipitation across China. Argueso et al. (2014) used the WRF modeling system to downscale CSIRO MK3.5 GCM outputs to 2 Km grid-scale. They simulated the present (1990–2009) and future (2040–59) climates for Sydney area and concluded that coupling of future urbanization effects and climate change will significantly affect the local climatology. They projected more intense changes in minimum temperatures than in maximum temperatures, particularly in winter and spring season when urban effects will contribute almost equally toward temperature change as the changes in global emissions. Kusaka et al. (2012) used dynamically downscaled projections from three GCMs MIROC3.2-medres, MRI-CGCM2.3.2a, and CSIRO-Mk3.0 climate models to access future (2070s) summertime (August) temperature for three largest urban settlements in Japan: Tokyo, Osaka, and Nagoya. They used WRF model to get high resolution temperature distributions in the cities and found that the city temperatures will increase by 2–3°C in future. Also the impact from UHI effect was found to be comparable to that from greenhouse gas emissions. Hamdi et al. (2014) used a high resolution limited area model ARPEGE-IFS coupled with Town Energy Balance model to estimate changes in UHI magnitudes as projected by the ARPEGE-Climat GCM in the timeslice 2071–2100. The model was run at 4 km spatial resolution for SRES A1B scenario and it was found from the inline run (where TEB was activated for regional as well as urban runs) that the intensity of daytime UHI will decrease by 0.2–0.24°C. Also, strong UHI events were projected to decrease in the future by 1°C. McCarthy et al. (2012) used the latest version of the Hadley Centre Regional Climate model HadRM3 at 25-km resolution coupled to a simple urban land-surface scheme (Best et al., 2006) to assess the sensitivity of UK urban climates to large-scale greenhouse gas–induced climate change, local forcing from urban land use, and anthropogenic heat flux resulting from energy use in the urban areas. Adachi et al. (2012) also calculated future UHI intensities for Tokyo city by using five future projections from climate models downscaled using the TERC-RAMS regional model. After comparing the results obtained with and without incorporating urban effects, they concluded that the temperature change between 1990s and 2070s because of greenhouse gas emissions is projected to be around 2°C while due to land-cover changes is projected to be around 0.5°C. Several other studies have also evaluated local-scale changes in climate due to future projected urbanization (Zong-Ci et al., 2013; Zhao et al., 2013; Cao et al., 2016; Bounoua et al., 2015; Zhang et al., 2016).

2.6 Conclusions From Literature Review

In this section we first presented theoretical background about climate change and land-cover change, and discussed how they have considerably affected earth's climate in the past and are expected to continue doing so in the future. From the discussion provided, it is clear that earth has experienced extensive climate and land-cover over the 20th century and this trend is expected to continue in the 21st century. An introduction to statistical downscaling with a discussion on inherent assumptions, criteria for the selection of predictors, and a review of the state-of-the-art statistical downscaling models used in the past is presented next. From the discussion it can be concluded that a variety of PerfectProg and MOS models have been used in the past to statistically downscale GCM projections. They are built on a set of statistical assumptions. The choice of predictor variables have primarily been guided by the availability of atmospheric and climatic data with the climate models although other minor factors also guide the predictor selection. A summary of the strengths and weaknesses of statistical and dynamic downscaling models is provided next. The discussion provides evidence that both dynamic and statistical downscaling models can be improved in order to make the downscaling process both realistic and practical. A detailed discussion on the observational evidence of linkage between climate and geophysical characteristics of a region is presented next. The discussion provides ample evidence of that land-cover and elevation effects of temperature and precipitation patterns from local to regional scales. Finally, previous studies that have projected future land-cover change—driven climatic changes have been reviewed. It is seen that all such studies have adopted a dynamic downscaling approach for doing so.

The discussion provided in this section clearly highlights that under the backdrop of rapidly changing climatic and geophysical characteristics, the need to develop a new statistical downscaling model that can consider geophysical characteristics of a region like land-cover and elevation when performing downscaling is particularly high. The new model should be fast and robust like traditional statistical downscaling models but should also account for geophysical characteristics of a region as done in dynamic downscaling. The physical scaling model, which is envisioned to have these qualities, is presented in the following section.

3. PHYSICAL SCALING MODEL OF DOWNSCALING

It has been highlighted in previous sections that statistical downscaling methods used currently do not explicitly account for geophysical characteristics of the region of study. This is one of the major drawbacks of statistical downscaling because of which geophysical changes in a region cannot be modeled using statistical downscaling methods. In SP method (Gaur and Simonovic, 2016a,b) this is achieved by including additional geophysical covariates representing land-cover and elevation distribution of the region. The predictant (observed local-scale climate) and predictor (large-scale climate and geophysical covariate) variables are linked using a generalized additive model (GAM) regression relationship. In GAM regression, the predictant variable is connected to smoothed predictor variables using a link function. The smoothing is generally performed using nonparametric algorithms. The regression function is totally nonparametric in nature and hence is suitable for modeling a range of climate variables including precipitation and temperature. A more detailed description of SP model and its extensions is provided below.

3.1 SP Model

SP model approach to downscaling temperature (surface or air) can be mathematically expressed as:

$$g(T_{obs}) = B_0 + f_1(T_{\text{mod}}) + f_2(E_p) + f_3(LC_p) \tag{4.1}$$

where T denotes temperature, E denotes elevation (masl) of the reference pixel, LC denotes categorical land-cover variable associated with the reference pixel, and B denotes regression parameters. Subscript *obs* and mod denote if the climatic data is observed or model-based, respectively. T_{mod} denotes large scale "background" climate data obtained by bilinear interpolation of GCM temperature data at the reference pixel. Subscript p indicates that the data used is a pixel-scale data. Variables g and f represent the link and smoothing functions, respectively. In this study smooth functions are fit using penalized likelihood maximization algorithm. The penalized likelihood maximization algorithm is a variant of maximum likelihood estimation algorithm and applies a tradeoff between model fit wiggliness and goodness of fit by incorporating a penalty function (Wood, 2000).

In the case of precipitation, method involves a two-step process of predicting precipitation occurrence using a logistic regression model and a wet-day precipitation amounts model using a GAM regression model. SP method approach to downscale precipitation (Gaur and Simonovic, 2018) can be mathematically expressed as:

$$\ln\left(\frac{P_{obs}}{1 - P_{obs}}\right) = B_0 + B_1 P_{\text{mod}} + B_2 E_p + B_3 LC_p \tag{4.2}$$

$$g(P_{obs,wet}) = B_0 + f_1(P_{\text{mod},wet}) + f_2(E_p) + f_3(LC_p) \tag{4.3}$$

where notations have previously defined meanings. Additionally, subscript *wet* denotes values on wet days only (i.e., days with more than 0.1 mm of precipitation).

3.2 SP With Surrounding Pixel Information (SPS) Model

SPS method is an extension to SP method where land-cover and elevation properties of the reference, as well as neighborhood, pixels are incorporated into the method formulation (Gaur and Simonovic, 2016b). SPS method for downscaling air temperature can be mathematically expressed as:

$$\begin{aligned} g(T_{obs}) = B_0 + f_1(T_{\text{mod}}) + f_2(E_p) + f_3(LC_p) + \\ f_4(Fr_{W,s}) + \cdots + f_{17}(Fr_{BSV,s}) + f_{18}(R_{E,s}) \end{aligned} \tag{4.4}$$

where, symbols have similar meanings as explained above. Predictors $Fr_{W,s}$, $\ldots Fr_{BSV,s}$ represent the fraction of total area surrounding the reference pixel that is occupied by *Water, ...Barren and Sparsely Vegetated* land-cover classes, respectively. The value of predictors: $Fr_{W,s}$, ...$Fr_{BSV,s}$ is between 0 and 1 and they add up across all neighborhood land-cover classes to give a value of 1. Neighborhood elevation information is incorporated by including a predictor $R_{E,s}$, which represents the ratio between reference pixel elevation and mean elevation of pixels surrounding the reference pixel at a specific neighborhood scale, *s*.

For precipitation, SPS downscaling method involves two steps of forming a precipitation occurrence and wet-day precipitation amounts model (Gaur and Simonovic, 2018). The two steps involved in SPS method can be mathematically expressed as:

$$\ln\left(\frac{P_{obs}}{1-P_{obs}}\right) = B_0 + B_1P_{\text{mod}} + B_2E_p + B_3LC_p + B_4Fr_{W,s} + \cdots + B_{17}Fr_{BSV,s} + B_{18}R_{E,s} \tag{4.5}$$

$$g(P_{obs,wet}) = B_0 + f_1(P_{\text{mod},wet}) + f_2(E_p) + f_3(LC_p) + f_4(Fr_{W,s}) + \ldots. + f_{17}(Fr_{BSV,s}) + f_{18}(R_{E,s}) \tag{4.6}$$

where, symbols have same meanings as explained before.

The choice of spatial scale within which neighborhood geophysical characteristics are analyzed is critical to SPS method application and performance. Four neighborhood scales (represented as *s* in Eqs. 4.4—4.6): 3x3, 5x5, 7x7, and 9x9 have been chosen before to define neighborhood characteristics (Gaur and Simonovic, 2016b). The selected neighborhood scales are shown in Fig. 4.1 in darker shades of grey for neighborhood scales: 33, 5x5, 7x7, and 9x9, respectively, while the reference pixel is shown in black. Larger neighborhood scales are considered to be inclusive of smaller spatial scales, for instance, neighborhood scale 5x5 encompasses pixels corresponding to neighborhood scale 3x3 and additional darker shaded pixels specific to 5x5 neighborhood scale.

4. ADVANTAGES OF SP AND SPS METHODS OVER OTHER TRADITIONAL STATISTICAL DOWNSCALING MODELS

SP and SPS methods described in this chapter have following advantages over the traditional downscaling methods:

- They can account for geophysical characteristics of the region of interest when used to perform downscaling. As explained before, elevation and land-cover properties of the location of interest, as well as its neighborhoods, can be included into the SP and SPS method definition.

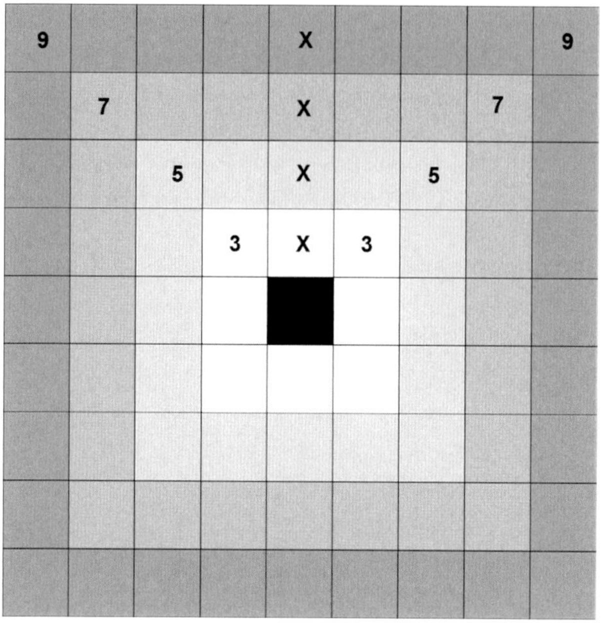

Figure 4.1 Neighborhood scales considered in this study. Increasing neighborhood scales are shown in progressively darker shades of grey. The reference pixel is shown in black.

The inclusion of additional geophysical predictors in the method can produce more accurate results in regions where geophysical characteristics play an important role in shaping local climate.

- The method also provides an opportunity to investigate climatic implications of historically observed and/or future projected geophysical changes for any region. Such analysis is possible with the SP and SPS methods because they include geophysical characteristics of the region of interest into the method definition.
- The method builds on a delicately balanced compromise between statistical and dynamic downscaling approaches and attempts to build on the positives of both downscaling approaches. It provides local-scale climate estimates consistent with the geophysical characteristics of a region and is able to do it using minimal computational and time resources.
- The method provides an opportunity to explore uncertainty in future projected climatic changes under projected geophysical changes. Since the method is fast and computationally inexpensive, it can be used in conjunction with data from multiple GCMs and RCPs to access the uncertainty associated with such changes.

- The method can also be used in real-time applications for downscaling GCM projections, as well as to model climatic effects of changes in geophysical characteristics of a region. Preparation of web-tools (Neset et al., 2016) and serious gaming tools (Savic et al., 2016) for instructional, educational, and entertainment purposes can be one such real-time application of the method.
- The method has been developed and implemented using publicly available datasets (Friedl et al., 2010; Wan, 2006; Mesinger et al., 2006; Taylor et al., 2012) and an open source programming language R (R Development Core Team, 2008). So it can be implemented conveniently and free of charge to obtain downscaled GCM projections for any region across the globe.

5. VALIDATION OF SP AND SPS MODELS

In this section the analysis results and conclusions obtained from extensive assessment of SP and SPS models towards downscaling North American regional reanalysis (NARR) surface temperature, air temperature, and precipitation are presented. A comparison of the performance of SP and SPS models with other state-of-the-art statistical downscaling models is also presented and discussed.

5.1 Study Region

The validation of SP and SPS models is performed at the southern Saskatchewan region of Canada. The physiographic setting of the study region is shown in Fig. 4.2. The study region is representative of the Canadian prairies and is characterized by diverse topography and land-cover. Two major urban centers located in this region are: Saskatoon and Regina. They are shown as black dots in Fig. 4.2. It experiences a continental climate. Large fluctuations in temperature are observed owing to its landlocked position. Significant spatial variability in precipitation is also observed. The region receives almost two-thirds of its precipitation during the summer season, which usually occurs due to large-scale convective and cyclonic systems. Snow cover plays a critical role in shaping the hydrometeorology of this region as it stays snow-covered almost 6 months a year.

5.2 Data Used

List of data used for performing model validation is provided below. Data sources and other specifications have also been summarized.

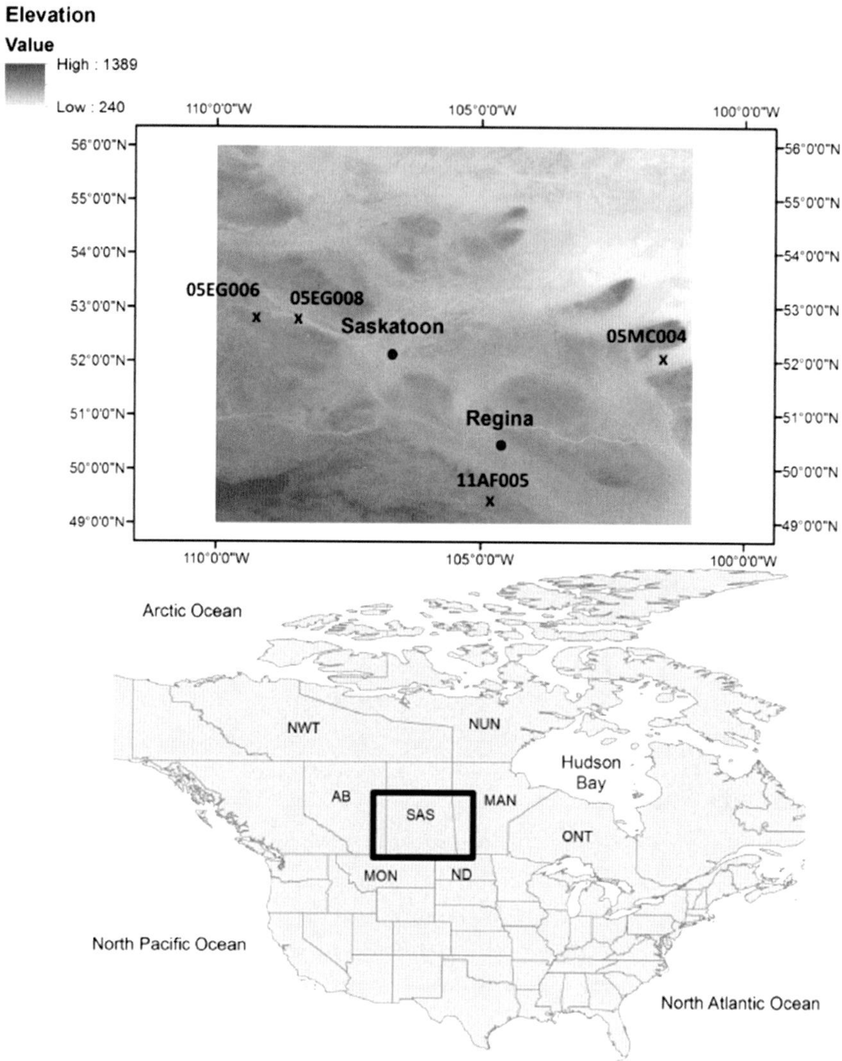

Figure 4.2 The location of southern Saskatchewan region, which is chosen as the study region in Gaur and Simonovic (2016a,b, 2018).

5.2.1 Remotely Sensed Surface Temperature

MODIS surface temperature level 3 Terra (MOD11A1) and Aqua (MYD11A1) products are used in this study (Wan, 2006). Both day-time: Terra-Day (TD), Aqua-Day (AD) and night-time: Terra-Night (TN), Aqua-Night (AN) data are collected for the period 2006–13. These data are available at an approximate spatial resolution of 1 × 1 km. Quality assessment files of the data are assessed and only reliable data are selected for analysis. It was found that a higher percentage of reliable data is available for snow-free months (April to October) than for the snow-covered months (November to March). The reason behind this can be that more cloud-free conditions occur during the summers than in the winters facilitating the sensing of reliable surface temperature data. The distribution of data is also found uneven within a day with higher percentage of data available in the night-time than in the daytime. Overall the data is found to be evenly distributed across the study period.

5.2.2 Remotely Sensed Land-Cover

MODIS recorded level 3 annual land-cover product (MCD12Q1) in University of Maryland (UMD) classification scheme (Friedl et al., 2010) is used in this study. In UMD classification scheme, land-cover is classified into 14 different classes. These land-cover classes as well as the classification codes used to refer them in this study are listed in Table 4.2. The land-cover dataset is available at a 500 m spatial resolution. Land-cover for the period 2006–12 is considered.

5.2.3 Remotely Sensed Elevation

The National Aeronautics and Space Administration (NASA) Shuttle Radar Topographic Mission (SRTM) elevation product is used. This data has a spatial resolution of 90 m.

Table 4.2 Land-Cover Classes as Identified in the UMD Classification System

Classification Code	UMD Classes
W	Water
ENF	Evergreen Needleleaf Forest
EBF	Evergreen Broadleaf Forest
DNF	Deciduous Needleleaf Forest
DBF	Deciduous Broadleaf Forest
MF	Mixed Forest
CS	Closed Shrublands
OS	Open Shrublands
WS	Woody Savannas
S	Savannas
G	Grasslands
C	Croplands
UB	Urban and Built-up
BSV	Barren or Sparsely Vegetated

Classification code used in downscaling models to represent each land-cover class is also provided.

5.2.4 Reanalysis Surface Temperature

North American Regional Reanalysis (NARR) 3-hourly surface (skin) temperature data for the period 2006–13 are used for downscaling. Daily air temperature and precipitation rate data are also used for downscaling. These data are produced by running the NCEP Eta model together with the Regional Data Assimilation System (Mesinger et al., 2006) and have an approximate spatial resolution of 32 km.

5.2.5 Gauged Daily Air Temperature and Precipitation Data

Daily precipitation data gauged at 52 locations within the study region over the period 2006–13 is acquired from the Environment Canada. The data can be accessed at: http://climate.weather.gc.ca/. The list of gauging stations selected for study is provided in Table 4.3. Using MODIS land-cover data it is found that these gauging stations are associated with UMD land-cover classes: S, OS, G, DNF, UB and C.

5.3 Surface Temperature Downscaling

SP model was evaluated for its performance in downscaling NARR surface temperature data over the period 2006–13 in Gaur and Simonovic (2016a) by adopting a cross-validation approach. The model was formulated for each month and for each timeline: AD, TD, AN, and TN using data from reliable MODIS surface temperature and land-cover pixels distributed across the study region. To adjudge if a particular MODIS pixel was reliable, its quality assessment files were accessed and pixels associated with an error of magnitude <1 K were adjudged reliable for model calibration. Similarly, land-cover pixels that were specified as "good quality" in land-cover quality assessment files were chosen for analysis.

SP model considers observed (or remotely sensed in this case) temperature data as predictant variable and elevation, land-cover, and model-derived climate as predictor variables. A GAM model with these predictor and predictant variables was formulated. The method was thereafter calibrated on randomly chosen 90% of the data and validated on the rest 10% of the data. A total of 100 such calibration and validation sets were generated for each month and each timeline giving a total of 4800 (100 sets × 4 timelines × 12 months) calibration and validation sets for model evaluation.

Model performance was evaluated for each timeline—AD, TD, AN, and TN—and model performance was found satisfactory in all timelines. The average root mean squared error (RMSE) across all models was found to be 0.05, 0.02, 0.02, and 0.03 K for timelines AD, TD, AN, and TN,

Table 4.3 List of Stations With Data Considered in This Study

S.No.	Station Name	Latitude	Longitude	Elevation
1	Rosetown East	51.57	−107.92	586.00
2	Last Mountain Cs	51.42	−105.25	497.00
3	Bratt's Lake Climate	50.20	−104.71	580.00
4	Wynyard (Aut)	51.77	−104.20	560.10
5	Nipawin	53.33	−104.00	371.90
6	Assiniboia Airport	49.73	−105.95	725.50
7	Hudson Bay(Aut)	52.82	−102.32	358.10
8	Pilger	52.42	−105.15	552.00
9	Prince Albert A	53.22	−105.67	428.20
10	Outlook Pfra	51.48	−107.05	541.00
11	North Battleford	52.77	−108.25	548.00
12	Coronach Spc	49.05	−105.48	756.00
13	Watrous East	51.67	−105.40	525.60
14	Melfort	52.82	−104.60	490.00
15	Elbow Cs	51.13	−106.58	595.00
16	Kindersley A	51.52	−109.18	693.70
17	Meadow Lake A	54.13	−108.52	480.70
18	North Battleford Rcs	52.77	−108.26	548.00
19	Yorkton	51.26	−102.46	498.40
20	Eastend Cypress (Aut)	49.44	−108.99	1059.00
21	Spiritwood West	53.37	−107.55	584.30
22	Yorkton	51.26	−102.46	498.30
23	La Ronge A	55.15	−105.27	379.20
24	Regina Int'l A	50.43	−104.67	577.60
25	Regina Rcs	50.43	−104.67	577.30
26	Saskatoon Intl A	52.17	−106.70	504.10
27	Val Marie Southeast	49.06	−107.59	796.00
28	Broadview	50.37	−102.57	599.80
29	Estevan	49.22	−102.97	580.60
30	Estevan A	49.22	−102.97	580.30
31	Indian Head Cda	50.55	−103.65	579.10
32	Loon Lake Rcs	54.02	−109.14	545.60
33	Lucky Lake	50.95	−107.15	664.70
34	Meadow Lake	54.13	−108.52	481.00
35	Moose Jaw A	50.33	−105.57	576.70
36	Moose Jaw Cs	50.33	−105.54	577.00
37	Nipawin	53.33	−104.02	371.90
38	Nipawin	53.33	−104.01	371.90
39	North Battleford	52.77	−108.24	548.30
40	Rockglen (Aut)	49.17	−105.98	917.00
41	Saskatoon Rcs	52.17	−106.72	504.10

Continued

Table 4.3 List of Stations With Data Considered in This Study—cont'd

S.No.	Station Name	Latitude	Longitude	Elevation
42	Scott Cda	52.36	−108.83	659.60
43	Swift Current	50.29	−107.69	816.90
44	Swift Current A	50.30	−107.68	816.90
45	Swift Current Cda	50.27	−107.73	825.00
46	Weyburn	49.70	−103.80	588.60
47	Yorkton	51.26	−102.46	498.30
48	Cypress Hills Park	49.64	−109.51	1259.00
49	Jimmy Lake Awos	54.91	−109.96	637.10
50	Leader Airport	50.91	−109.50	675.50
51	Mankota	49.10	−107.02	830.00
52	Maple Creek	49.90	−109.47	766.70

respectively. Additionally, superior model performances were obtained in snow-free months than in the snow-covered months, and in the night-time than in the daytime. Further the method was found to perform better in the lower elevation and flat areas of the study region than in areas with complex topography. Based on the obtained results, the method was considered suitable for the downscaling of GCM surface temperature projections across the southern Saskatchewan region.

5.4 Air Temperature Downscaling

SP and SPS models were assessed for their skill in downscaling NARR air temperature data in Gaur and Simonovic (2016b). The assessment was performed over the period 2006−12. A large ensemble of SP and SPS models were evaluated to identify most appropriate functional form, approach, and spatial scale for the application of SP and SPS models.

Two approaches, direct and indirect, for downscaling NARR air temperatures were tested. In the direct approach, NARR air temperature data were used directly to obtain local air temperature whereas in the indirect approach, NARR surface temperature are first used to estimate downscaled surface temperature, which are then used to predict downscaled air temperature using a model that estimates air temperature from surface temperature (referred to as ST→AT model hereafter). Three different functional forms were used to link predictor and predictant variables. The total ensemble of models evaluated in Gaur and Simonovic (2016b) are summarized in Table 4.4. The method, functional form used to link predictor and predictant variables, and predictors used in the ST → AT model, for each model are also summarized.

Table 4.4 Models Evaluated for Air Temperature Downscaling Skill in Gaur and Simonovic (2016b)

S.No.	Approach	Model	Method	Functional Form	Predictors (ST → AT Model)
1	Direct	SP-lm	SP	LR	–
2		SP-qr	SP	QR	–
3		SP-gam	SP	GAM	–
4		SPS3x3-lm	SPS	LR	–
5		SPS5x5-lm	SPS	LR	–
6		SPS7x7-lm	SPS	LR	–
7		SPS9x9-lm	SPS	LR	–
8	Indirect	SP-lm-ST	SP	LR	ST
9		SP-qr-ST	SP	QR	ST
10		SP-gam-ST	SP	GAM	ST
11		SP-lm-ST-LC	SP	LR	ST, LC
12		SP-qr-ST-LC	SP	QR	ST, LC
13		SP-gam-ST-LC	SP	GAM	ST, LC
14		SP-lm-ST-LC-AVs	SP	LR	ST, LC, AVs
15		SP-qr-ST-LC-AVs	SP	QR	ST, LC, AVs
16		SP-gam-ST-LC-AVs	SP	GAM	ST, LC, AVs
17		SPS3x3-lm-ST-LC-AVs	SPS	LR	ST, LC, AVs
18		SPS5x5-lm-ST-LC-AVs	SPS	LR	ST, LC, AVs
19		SPS7x7-lm-ST-LC-AVs	SPS	LR	ST, LC, AVs
20		SPS9x9-lm-ST-LC-AVs	SPS	LR	ST, LC, AVs

The downscaling skills of models listed in Table 4.4 were also compared with the skill of a state-of-the-art statistical downscaling method: SDSM. The performances of SDSM and SP models toward downscaling NARR air temperature data were accessed by using leave-one-out cross-validation technique (Stone, 1974) over 52 gauging stations listed in Table 4.4. In this technique, validation was performed at a reference gauging station while calibration was performed on the remaining gauging stations. This process was repeated until validation was performed at all gauging stations located within the study region. Downscaling methods were evaluated based on two metrics: root mean squared error (RMSE) and Pearson correlation coefficient (r). Additionally, Akaike Information Criteria (AIC) was used to compare the quality of SP methods. AIC statistic (provided in Eq. 4.7) quantifies a regression model quality by meriting it based on the goodness of fit of the regression relationship and demeriting it based on its increased complexity.

$$\mathrm{AIC} = 2k - 2\ln(L) \tag{4.7}$$

where AIC represents AIC statistic value, k represents the number of estimated parameters in a model, and L represents the maximum value of likelihood function for the model.

Downscaling of NARR temperature data by SDSM method was performed by first screening large-scale atmospheric variables for prediction. In Gaur and Simonovic (2016b) the atmospheric variables listed in Table 4.5 were considered for the initial screening process. The choice of atmospheric variables was made taking into account the recommended large-scale atmospheric variables in Wilby et al. (2002) and the availability of atmospheric data in the NARR product. Next, monthly correlations between large-scale atmospheric variables and temperature across all temperature recording stations are examined for each timeline. Results indicated that the correlation between large-scale atmospheric variables and observed

Table 4.5 NARR-Derived Atmospheric Variables Considered for Screening in SDSM Method

S.No.	Short-Name	Long-Name
1	dlwrf	Downward Longwave Radiation Flux
2	dswrf	Downward Shortwave Radiation Flux
3	uswrf	Upward Shortwave Radiation Flux
4	ulwrf	Upward Longwave Radiation Flux
5	cape	Convective Available Potential Energy
6	tke	Total Kinetic Energy
7	hgt500hpa	Geopotential Height at 500 hpa
8	hgt850hpa	Geopotential Height at 850 hpa
9	hgt1000hpa	Geopotential Height at 1000 hpa
10	shum500hpa	Specific Humidity at 500 hpa
11	shum850hpa	Specific Humidity at 850 hpa
12	shum1000hpa	Specific Humidity at 1000 hpa
13	hcdc	High Cloud Area Fraction
14	mcdc	Medium Cloud Area Fraction
15	lcdc	Low Cloud Area Fraction
16	prate	Precipitation Rate
17	prmsl	Mean Sea Level Pressure
18	rhum	Relative Humidity
19	wnd	Wind Speed

temperature depends on the atmospheric variable, month of the year, and time of the day being analyzed. Based on the obtained correlation values and personal judgment, three atmospheric variables, shum1000hpa, hgt850hpa and ulwrf, were selected to model local temperature using SDSM model.

Before performing downscaling by SP methods (listed in Table 4.3), they were tested for the presence of multicollinearity in predictor variables. Variance inflation factors (VIFs) associated with different predictor variables were calculated and their values are found within an agreeable range of 1–2.5 indicating acceptable multicollinearity among the predictor variables (Rogerson, 2001; Pan and Jackson, 2008). From leave-one-out cross-validation of SDSM and SP methods, at more than 50% of all gauging stations the performances of SP methods considered under direct approach were found superior than the SDSM method in terms of both RMSE and correlation. On the other hand, the performance of SP methods considered under indirect approach was found inferior to SDSM method at more than 50% of the stations. A summary of RMSE values obtained for different models assessed is presented in Fig. 4.3. It can be noted from the figure that the lowest RMSE values are associated with direct SP downscaling methods, followed by SDSM, and followed by indirect SP methods.

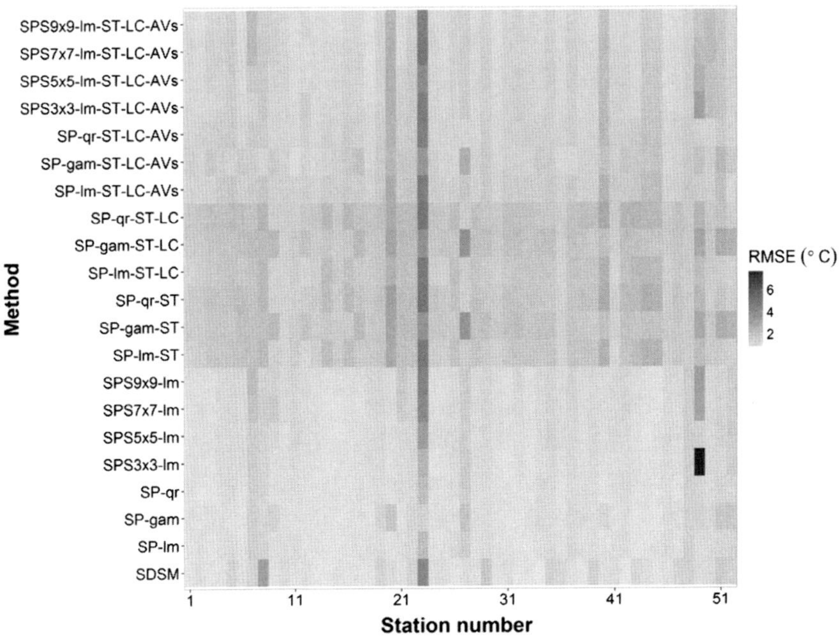

Figure 4.3 RMSE obtained from leave-one-out cross-validation procedure on the 52 stations on which SP and SPS methods are evaluated for air temperature downscaling skill.

Based on the average RMSE across all gauging stations, methods considered under direct SP approach were found to show almost similar performances (RMSE ~ 0.8°C). SDSM method was obtained as the next best performing method (RMSE = 0.9°C). The performances of indirect methods were found to be inferior to both direct SP and SDSM methods. The performance of methods considering atmospheric variables as predictors in the ST → AT method were found to be similar (RMSE ~ 1.1°C), as well as superior to the methods that do not account for them (RMSE ~ 1.5°C). Similar method performances were also observed in terms of correlation.

Relative performance of methods at each gauging station was also accessed. Three best performing methods based on RMSE and correlation were identified at each gauging station and overall composition of first, second and third ranked methods across all gauging stations were analyzed. Among SDSM and SP methods, SP-gam was found to be the most consistent first and second ranked method while SP-lm was obtained as the most consistent third ranked method. The results obtained in Gaur and Simonovic (2016b) suggest that direct SP methods tend to outperform SDSM method while the latter outperforms indirect SP methods in terms of the RMSE and correlation.

The variation of SDSM and SP method performances across different months was also analyzed. In general, superior model performances are obtained in the summer months (April—September) than in the winter months (October—March). This is expected in the case of SP method because the quality of remotely sensed data is known to be higher in cloud and snow-free conditions which usually occur during the summer months.

The qualities of SP methods considered under direct and indirect approaches were further compared in terms of AIC. Data across all gauging stations was used to estimate AIC associated with each SP and SPS model. Among the models considered under direct approach, method SPS5x5-lm was obtained to be of the highest quality while SP-qr was obtained as the lowest quality model. Among the models considered under the indirect approach, method SP-gam-ST-LC-AVs was found to be of the highest quality while SP-qr-ST was obtained to be of the lowest quality. The qualities of SPS models were found to be superior to SP methods in both direct and indirect approaches with method SPS5x5-lm demonstrating highest method quality among SPS methods. The results from this analysis suggested 5 × 5 (area within 6.25 km^2 from the reference location) to be the best spatial scale at which neighborhood information should be included in the SPS models. Among the methods under indirect approach, methods considering atmospheric variables in the ST → AT models were found to have a higher quality than the other methods.

To summarize, the results obtained in Gaur and Simonovic (2016b) demonstrated that air temperature downscaling skill of direct SP and SPS models is superior to the SDSM model while the skill of indirect SP and SPS models is inferior to the SDSM model for the study region analyzed in the study. Between SP and SPS models, latter have higher quality in the regression relationship than the SP method. A comparison of results obtained at different neighborhood scales suggest 5×5 to be the optimum scale at which maximum downscaling skill is demonstrated by the SPS method.

5.5 Precipitation Downscaling

SP and SPS methods were evaluated for their skill in downscaling NARR precipitation over the period 2006–12 in Gaur and Simonovic (2018). The downscaling skill of SP and SPS methods were also compared with the skills of two state-of-the-art precipitation downscaling models: SDSM and GLM, described before in Section 2.2.3. Since SP and GLM-based models both employ GAM as the regression function, the same is used to build relationship between low resolution atmospheric variables and locally observed climatic data in the SDSM model. This is done to maintain regression function consistency among all models being evaluated in this study so that an unbiased evaluation of downscaling methodologies can be performed.

In Gaur and Simonovic (2018) the selection of predictors for SDSM and GLM models was made by analyzing monthly correlations between atmospheric variables and locally observed precipitation data at all gauging stations located within the study region. In this study, specific humidity (shum), high cloud area fraction (hcdc), medium cloud area fraction (mcdc), low cloud area fraction (mcdc), air temperature (air), and geopotential height (hgt) were tested for their correlation with observed precipitation across 57 precipitation gauging stations located in the study region. Highest correlations were obtained in the case of cloud-cover variables: high cloud area fraction (hcdc), medium cloud area fraction (mcdc), and low cloud area fraction (lcdc). Therefore they were selected as atmospheric predictor variables for performing downscaling by SDSM.sig and GLM.sig models (SDSM and GLM models with only significant or relevant atmospheric predictors). Another version of SDSM and GLM models which considers all atmospheric variables listed above for downscaling were also considered

Table 4.6 Models Evaluated in Gaur and Simonovic (2018)

S.No.	Model Name	Model Name (Short)	Predictors
1	SP	M1	P, LC, E
2	SP_LC	M2	P, E
3	SP_LC_elev	M3	P
4	SPS3x3	M4	P, LC, E, NLC_{3x3}, NE_{3x3}
5	SPS5x5	M5	P, LC, E, NLC_{5x5}, NE_{5x5}
6	SPS7x7	M6	P, LC, E, NLC_{7x7}, NE_{7x7}
7	SPS9x9	M7	P, LC, E, NLC_{9x9}, NE_{9x9}
8	SDSM	M8	wnd, rhum, prmsl, lcdc, shum1000hpa,
9	GLM	M9	mcdc, hcdc, air, shum850hpa, shum500hpa, hgt1000hpa, hgt850hpa, hgt500hpa
10	SDSM.sig	M10	lcdc, mcdc, hcdc
11	GLM.sig	M11	

and referred to as "SDSM" and "GLM" models in the study. The list of models evaluated in Gaur and Simonovic (2018) is provided in Table 4.6.

Two different tests of robustness were performed: (1) Test for temporal robustness (TR) and (2) Test for spatial robustness (SR). In the temporal robustness test, downscaling models were calibrated over the period 2006–10 and validated over the period 2011–13. On the other hand, in the spatial robustness test the downscaling models were calibrated across 29 stations located across the study region and validated across the rest of the gauging stations. The downscaled precipitation were evaluated on the basis of the downscaling model's ability to simulate following seven precipitation-based indices: (1) Spearman correlation coefficient between model simulated and observed data (sp.cor), (2) fraction of dry days, that is, fraction of days with less than 0.1 mm of rainfall (ddays), (3) maximum precipitation intensity (ppt.max), (4) mean wet day precipitation (ppt.wet), (5) total number of 1-day precipitation events (p1d), (6) total number of 2–4 day precipitation events (p2 to 4d) and (7) total number of 5 or more day precipitation events (p5d).

While calibrating SP, SPS, and GLM models, probability predictions made in the occurrence model are associated with an occurrence (1) or no-occurrence (0) value using a threshold value such that:

$$f(p) = \begin{Bmatrix} 1 \text{ if } p >= p_{\text{threshold}} \\ 0 \text{ if } p < p_{\text{threshold}} \end{Bmatrix} \tag{4.8}$$

where p denotes predicted probabilities as obtained from the occurrence model and $p_{threshold}$ denotes the threshold probability value chosen for analysis. In Gaur and Simonovic (2018), a series of $p_{threshold}$ values ranging between 0 and 1 were tested to select a threshold probability value that provides maximum prediction accuracy to the SP method and GLM-based models.

The variation of model efficiencies with probability threshold values for SP method—based models and GLM models for both snow-cover states and robustness tests were analyzed. Twenty-one probability threshold values evenly spaced between 0 and 1 at a spacing of 0.05 are considered for analysis. Model efficiency was calculated by evaluating the percentage of total data length correctly predicted by the calibrated model on validation time-series. It was found that occurrence model performance for SP and GLM-based models varied significantly with the choice of probability threshold value. Further, minor variations in model efficiency were also observed with differences in snow-cover state, robustness test, and downscaling model considered. Optimal threshold value for each model, snow-cover state, and robustness test combination was used for making prediction from these models. A summary of these optimal threshold values is presented in Table 4.7. It is noticed that threshold values for GLM models are higher than the threshold values of SP method—based models.

The calibrated models were used to downscale NARR precipitation grid data across the validation period (for TR test) and validation stations (for SR test). The performance of models in terms of average (across all

Table 4.7 Optimum Probability Threshold Values for Different Models for Snow-free (sf) and Snow-Covered (sc) Months, and for TR (SR) tests as Obtained in Gaur and Simonovic (2018)

	Snow	
Model	**sf**	**sc**
GLM	0.55 (0.55)	0.5 (0.6)
GLM.sig	0.55 (0.5)	0.45(0.55)
SP	0.35 (0.35)	0.3 (0.45)
SP-LC	0.35 (0.35)	0.35 (0.45)
SP-LC-elev	0.35 (0.35)	0.3 (0.45)
SPS3x3	0.4 (0.35)	0.3 (0.45)
SPS5x5	0.4 (0.4)	0.4 (0.4)
SPS7x7	0.4 (0.4)	0.4 (0.45)
SPS9x9	0.4 (0.35)	0.4 (0.5)

experiments and snow-cover states) rank correlation between modeled and observed data is shown in Table 4.8. Best performing model in terms of average performance is highlighted in grey. From the results it is clear that SP method–based models majorly outperform both SDSM and GLM models in terms of correlation. From the results obtained in Gaur and Simonovic (2018), it was also found that the performance of SP models was better in the snow-free months ($rho_{avg.} = 0.5$) as compared to snow-covered months ($rho_{avg.} = 0.4$) and better in the TR test ($rho_{avg.} = 0.5$) than the SR test ($rho_{avg.} = 0.4$). Following SP method–based models, SDSM model is found to perform best, followed by GLM model. Further the performance of SDSM.sig and GLM.sig models is found to be inferior than the SDSM and GLM models.

SP models perform best in simulating the fraction of total number of dry days in the validation time-series as evident from Table 4.8. The performance of SP models was again found to be better in snow-free months ($RMSE_{avg.} = 0.1$) than the snow-covered months ($RMSE_{avg.} = 0.5$) and in the TR test ($RMSE_{avg.} = 0.11$) than in the SR test ($RMSE_{avg.} = 0.12$). Among other models, GLM model was also found to perform well ($RMSE_{avg.} = 0.11$) followed by SDSM ($RMSE_{avg.} = 0.5$). Again the performance of SDSM.sig and GLM.sig models toward simulating dry day fraction was found to be inferior than the SDSM and GLM models.

Maximum precipitation intensity was not simulated satisfactorily by all three types of models considered in this study. This can be seen from Table 4.8 where biases associated with maximum precipitation values are presented. SP model was found to perform best followed by SP_LC_elev model. Among the three types of models, SP method–based models are found to perform best, followed by GLM and then by SDSM model. SP model performance was found to be significantly better in snow-free months ($RMSE_{avg.} = 54$ mm) than the snow-covered months ($RMSE_{avg.} = 90$ mm) and in the TR test ($RMSE_{avg.} = 70$ mm) than SR test ($RMSE_{avg.} = 78$ mm). Further the performance of SDSM.sig and GLM.sig models towards simulating maximum precipitation intensity was found to be inferior than SDSM and GLM models.

Wet-day mean precipitation was simulated reasonably well by SP method–based models (Table 4.8). Among the three types of models considered, GLM model was found to perform best ($RMSE_{avg.} = 0.8$ mm), followed by SP method ($RMSE_{avg.} = 1$ mm), and then by SDSM ($RMSE_{avg.} = 2.8$ mm). In the case of wet-day mean precipitation, SP model performance was found to be better in the snow-covered months

Table 4.8 Spearman Correlation Coefficient Between Model Simulated and Observed Precipitation, and RMSE Associated With Dry Day Fraction, Maximum Precipitation, Mean Wet-Day Precipitation, Total 1-Day Precipitation Events, Total 2–4 Days Precipitation Events, and Total >=5 Day Precipitation Events, for Models Considered in This Study

Statistic	M1	M2	M3	M4	M5	M6	M7	M8	M9	M10	M11
Spearman correlation coefficient	0.44	0.44	0.44	0.43	0.43	0.42	0.42	0.43	0.40	0.41	0.36
RMSE-dry day fraction	0.09	0.11	0.10	0.10	0.12	0.13	0.12	0.48	0.13	0.55	0.14
RMSE- maximum precipitation	71	71	71	71	71	71	71	82	75	85	83
RMSE-mean wet-day precipitation	0.87	1.00	0.95	1.01	1.06	1.12	1.07	2.69	0.95	2.9	1.03
RMSE-total 1-day precipitation events	65	78	59	173	231	269	278	1615	181	1957	139
RMSE-total 2–4-day precipitation events	442	491	429	548	611	649	611	784	657	1381	673
RMSE-total >=5-day precipitation events	165	175	168	171	174	185	166	1085	177	509	189

Best performing model is highlighted in grey. Models M1 to M11 can be referred from Table 4.6.

($RMSE_{avg.} = 0.9$ mm) than the snow-free months ($RMSE_{avg.} = 1$ mm) and better in the TR test ($bias_{avg.} = 0.8$ mm) than the SR test ($bias_{avg.} = 1$ mm). Again the performance of SDSM.sig and GLM.sig models toward simulating mean wet-day precipitation intensity was found to be inferior to SDSM and GLM models. This suggests that the performance of SDSM and GLM models (formulated by considering all atmospheric variables) deteriorates by the selection of most relevant set of predictors for prediction.

The occurrences of 1-day and 2–4-day precipitation events were best simulated by SP method–based models, whereas GLM model performs best in simulating 5 or more day precipitation events as evident in Table 4.8. SDSM model was found to underestimate the occurrence frequency of 1-day and 2–4 day precipitation events and overestimate the occurrence frequency of 5 or more day precipitation events. The performances of SP models were again found to be superior in the TR test ($RMSE_{avg.} = 154$, 420, 132, respectively, for 1-day, 2–4 days, and 5 or more days precipitation events) than in the SR test ($RMSE_{avg.} = 298$, 743, 223 respectively for 1-day, 2–4 days, and 5 or more days precipitation event). Further SP method–based models were found to perform better in snow-free months than in the snow-covered months for 1-day and 2–4-day precipitation events ($RMSE_{avg.}$ found lower by 32 and 178 for 1 day and 2–4 day precipitation events, respectively). However, they were found to perform better in the snow-covered months than snow-free months for more than 5-day precipitation events ($RMSE_{avg.}$ found lower by 58). The performance of SDSM.sig and GLM.sig models toward simulating 1-day and 2–4-day precipitation events was found to be inferior than the SDSM and GLM models. However, in the case of 5 or more day precipitation events, SDSM.sig model was found to perform better than SDSM model. GLM.sig model still performs inferiorly to GLM model in the case of 5 or more day precipitation events.

Overall SP method–based models were found to perform better than the SDSM and GLM-based models. This was also evident when index specific bias associated with each individual model was normalized and combined. Index specific RMSE values were normalized so that intermodel comparisons can be made taking into consideration all seven indices. Overall, based on average normalized RMSE ($RMSE_{ANB}$), models ranked as: (1) SP_LC_elev ($RMSE_{ANB} = 0.02$), (2) SP ($RMSE_{ANB} = 0.03$), (3) SP_LC ($RMSE_{ANB} = 0.04$), (4) SPS9x9 ($RMSE_{ANB} = 0.16$), (5) SPS3x3 ($RMSE_{ANB} = 0.19$), (6) SPS5x5 ($RMSE_{ANB} = 0.22$), (7) GLM

($RMSE_{ANB} = 0.22$), (8) SPS7x7 ($RMSE_{ANB} = 0.28$), and (9) SDSM ($RMSE_{ANB} = 0.91$) in terms of model performance across indices. It was noted that SP method–based models, SP, SP_LC and SP_LC_elev, perform significantly better than all other models considered for evaluation. This suggested that the inclusion of neighborhood land-cover and elevation configuration didn't considerably improve SP model performance in the study region.

5.6 Conclusions From SP and SPS Model Validation

Results obtained from extensive validation of SP and SPS models have been described in this section. Downscaled NARR climate data obtained from SP and SPS models are found to be associated with acceptable amount of error. Additionally, the models are found to perform better than other state-of-the-art statistical downscaling models: SDSM and GLM. These findings establish the models as reliable downscaling tools for the southern Saskatchewan region. The unique advantage of these models is highlighted in the following section where they are used to identify land-cover change–induced hydroclimatic changes in future. Such an analysis is not possible with other state-of-the-art statistical downscaling models for the reasons discussed in Section 4.

6. CASE STUDY: USE OF SP AND SPS METHODS TO IDENTIFY FUTURE LAND-COVER CHANGE–INDUCED CHANGES IN HYDROCLIMATIC VARIABLES

In this section we present a case study where SP and SPS methods were used to identify hydroclimatic changes projected as a result of future land-cover change in four catchments located in the southern Saskatchewan region of Canada (Gaur and Simonovic, 2018). The locations of the selected catchments are shown by black crosses in Fig. 4.1. HYDAT is a flow database maintained by the Water Survey of Canada, an organization that is responsible for collecting and disseminating flow data in Canada. In Gaur and Simonovic (2018), catchments with HYDAT IDs: 05EG006, 05EG008, 05MC004, and 11AF005 were chosen for analysis. Hydrometric details of the selected catchments are summarized in Table 4.9. It can be noted that all four catchments were small (with catchment area less than 1000 km^2) and hence can be modeled as lumped hydrologic systems. Additionally, all are nonregulated catchments which mean that changes in outflow from these

Table 4.9 List of Catchments Analyzed in Gaur and Simonovic (2018)

HYDAT ID	Name	Drainage Area (Km^2)	Elevation (Masl)	Data Length
05EG006	Birling creek near Paynton	426	593	1970–92
05EG008	Page creek near Iffley	921	673	1971–87
05MC004	Conjuring creek near Preeceville	255	594	1965–92
11AF005	Beaver creek at international boundary	387	773	1977–95

catchments can be attributed to changes in external climate forcing and changes in the physical characteristics of the catchments.

6.1 Data Used

NASA satellite elevation and land-cover products were used to satisfactorily represent catchment physical characteristics. To this end, Shuttle Radar Topographic Mission (SRTM) elevation product (Jarvis et al., 2008) and MODIS land-cover data, MCQ12Q1 in the University of Maryland (UMD) classification scheme, were used. The elevation and land-cover data have a spatial resolution of 90 and 500 m respectively. The land-cover data has an annual temporal frequency and is considered for the period 2006–12 in this study.

ANUSPLIN is a 10-km gridded precipitation, maximum temperature, and minimum temperature data developed by applying thin plate spline smoothing algorithms on gauged climate records across Canada (Hopkinson et al., 2011 and Hutchinson et al., 2009). ANUSPLIN data grids falling within the selected catchments were considered for analysis. Daily air temperature and precipitation at 947 climate gauging stations located in Alberta, Saskatchewan, and Manitoba provinces of Canada were acquired from Environment Canada. The gridded and gauged climate data were considered for the periods 1961–2012 and 2006–12, respectively, in this study. It was found that together these gauging stations characterize all of University of Maryland (UMD) land-cover classes (listed in Table 4.2) and hence were considered appropriate for calibrating the SP and SPS down-scaling models.

Daily flow data for the selected catchments were collected from the HYDAT database. Flow data available between the period 1961 and 2012 was collected. A summary of the flow data length available at each flow

Table 4.10 List of GCMs Considered for Analysis in Gaur and Simonovic (2018)

GCM	Model	Resolution	Source
1	IAP-FGOALS	1.66° × 2.81°	Institute of Atmospheric Physics, Chinese Academy of Sciences, China
2	MRI-CGCM3	1.08° × 2.16°	Meteorological Research Institute, Japan
3	NorESM1-M	2° × 2°	Norwegian Climate Centre, Norway

gauging station, drainage area, mean elevation, and HYDAT ID is provided in Table 4.9.

Daily reanalysis precipitation rate and near-surface air temperature were acquired for the period 2006–12 from North American Regional Reanalysis (NARR) data repository (Mesinger et al., 2006). The data are spatially interpolated across all gauging stations. The interpolated reanalysis climate data are used to calibrate downscaling models considered in this study.

Future land-cover projections for southern Saskatchewan region corresponding to three climate models listed in Table 4.10 and emission scenarios RCP 2.6 and RCP 8.5 were prepared in Gaur and Simonovic (2016a) by downscaling and reclassifying future harmonized land-use projections from two integrated assessment models: IMAGE and MESSAGE (Hurtt et al., 2011). In Gaur and Simonovic (2018) annual land-cover projection at the selected catchments for the periods 2041–60 and 2081–2100 were acquired from the land-cover data produced in Gaur and Simonovic (2016a). In Gaur and Simonovic (2016a) variations in land-cover for different GCMs were found negligible as compared to variations with different emission scenarios, a single land-cover projection was considered for all three GCMs in this study.

An analysis of future land-cover trajectories over the 21st century across the selected catchments for RCP2.6 and RCP8.5 in Gaur and Simonovic (2018) suggested that parts of catchments 05EG006 and 05EG008 will transition from forested land to BSV land-cover class under both RCPs; however, the transition is expected to be more drastic in case of RCP 8.5 as compared to RCP 2.6. In case of catchment 05EG004 under RCP 2.6, a section of the forested land was projected to transition to grasslands, while under RCP 8.5 a transition from forested land to BSV was expected. Lastly, in the case of catchment 11AF005 under RCP 2.6 a transition from croplands to grasslands was projected while in the case of RCP 8.5 a conversion of grasslands to BSV was projected.

GCM-based daily air temperature and precipitation projections for the period 2006–2100 were collected from the coupled model inter-comparison project-phase 5 (CMIP5) data repository (Taylor et al., 2012). Data corresponding to climate models listed in Table 4.3 and for two Representative Concentration Pathways (RCPs)—RCP 2.6 and RCP 8.5 was acquired. The choice of climate models was made based on the availability of future land-cover data as developed in Gaur and Simonovic (2016a).

6.2 Sacramento Hydrologic Model

Hydrologic models are used to estimate catchment flow response provided by a set of catchment geophysical and climatic characteristics. In Gaur and Simonovic (2018), Sacramento hydrologic model is used to estimate catchment outflow under future projected climatic and land-cover trajectories. The Sacramento model (Burnash, 1995) flow is generated by distributing precipitation falling at a location to overland flow, interflow, and baseflow components accounting for losses due to evapotranspiration and interception. Groundwater movement is modeled by considering upper zone and lower zone storages. Runoff is contributed by five different processes: (1) direct runoff from permanent and temporary impervious areas, (2) surface runoff due to precipitation occurring at a rate faster than percolation and interflow that take place when both upper zone storages are full, (3) interflow resulting from the lateral drainage of a temporary free water storage, (4) supplemental base flow, and (5) primary base flow. The model has 13 free parameters which are optimized in the model.

Routing in this model is performed using exponential form of unit hydrograph with explicit slow and quick flow components. The routing scheme involves three free parameters which are optimized. Snow-melt is modeled offline using a temperature index modeling approach discussed in Walter et al. (2005). Model is calibrated to optimize an objective function which is a weighted sum of coefficient of determination (R^2) and relative bias. The shuffled complex evolution global optimization method (Duan et al., 1993) followed by a local optimization method (Nelder and Mead, 1965) with multistart options is used to calibrate the model. A lumped version of this hydrologic model is considered sufficient for this study because the catchments considered are small (drainage area $<1000\ km^2$).

6.3 Analysis and Results

In Gaur and Simonovic (2018), SP and SPS downscaling models were calibrated over the period 2006–12 using gauged climate data as predictant variable and model specific predictor variables as described before. Precipitation and temperature downscaling models were calibrated separately for each month. The calibrated temperature downscaling models were used to downscale future GCM maximum and minimum temperature, and calibrated precipitation downscaling models were used to downscale GCM precipitation projections. Downscaling was performed for historical (2006–16), 2050s (2041–60), and 2090s (2081–2100) timelines with all MODIS grid cells falling within the selected catchments.

Future projected changes in precipitation, maximum and minimum temperature, and flow variables were analyzed between (1) historical and 2050s; and (2) historical and 2090s, for the selected catchments. An analysis of the projected changes suggested that both minimum and maximum temperatures can be expected to increase across the four catchments in future. Under RCP 2.6, minimum temperatures were projected to increase by 4.1 (2.3) K, 4.1 (2.4) K, 3.7 (2.1) K, and 4.1 (2.5) K by 2050s (2090s) for catchments 05EG006, 05EG008, 05MC004, and 11AF005, respectively. Under RCP 8.5 minimum temperatures were projected to increase by 4.1 (7.9) K, 4 (7.9) K, 4.3 (8.1) K, and 4.2 (8.6) K by 2050s (2090s) for catchments 05EG006, 05EG008, 05MC004, and 11AF005, respectively. On the other hand, under RCP 2.6 maximum temperatures were projected to increase by 3.9 (2.5) K, 3.7 (2.4) K, 3.9 (2.9) K, and 4.4 (3.4) K by 2050s (2090s) for catchments 05EG006, 05EG008, 05MC004, and 11AF005, respectively. Under RCP 8.5 maximum temperatures were projected to increase by 2.4 (6) K, 2.8 (6.6) K, 3.8 (7.6) K, and 3.3 (6.9) K by 2050s (2090s) for catchments 05EG006, 05EG008, 05MC004, and 11AF005, respectively. Overall larger changes are projected for minimum temperature as compared to maximum temperature in all four catchments.

Changes in precipitation were investigated by accessing projected changes in mean precipitation, dry days (days with precipitation less than 0.1 mm), frequency of small (lasting 1 day), moderate (lasting 2–4 consecutive days), and heavy (lasting more than 4 consecutive days) precipitation events between historical and future timelines. Mean precipitation was projected to increase slightly across the catchments. Under RCP 2.6 mean precipitation was projected to increase by 0.03 (0.03), 0.02 (0.03), 0.05 (0.06), and 0.01 (0.04) mm/day by 2050s (2090s) for catchments 05EG006, 05EG008, 05MC004, and 11AF005, respectively. Under RCP 8.5 mean precipitation was projected to increase by 0.07 (0.13), 0.06 (0.12), 0.05 (0.1), and 0.06 (0.08) mm/day by 2050s (2090s) for catchments 05EG006, 05EG008, 05MC004, and 11AF005, respectively. In accordance with these projected changes in mean precipitation, a decrease in the number of dry days and increase in the frequency of small, moderate, and heavy precipitation events were also projected across the selected catchments.

Next, Sacramento hydrologic model was calibrated for each catchment using ANUSPLIN-based gridded climate data and discharge data available at each discharge gauging station. ANUSPLIN gridded precipitation, maximum temperature, and minimum temperature gridded data were used to calibrate the hydrologic model because only a few or none of the climate gauging stations were found to be located within the selected catchments. The available daily discharge data length at each catchment is provided in Table 4.9. The entire flow data length was considered while calibrating the model to obtain robust hydrologic parameters which can then be used to predict future flows in the selected catchments. Nash-Sutcliffe efficiency and correlation between observed and modeled streamflow were analyzed for all catchments and satisfactory calibration results were obtained.

The calibrated Sacramento model was thereafter used to predict catchment outflow using downscaled climate projections. Above mentioned changes in climate variables were also reflected in the projected changes in catchment outflow which was largely projected to increase across the four catchments. Under RCP 2.6 flow rate was projected to change by −0.01 (−0.01), 0.03 (0.02), 0.05 (0.01), and 0.04 (0.20) m^3/s by 2050s (2090s) for catchments 05EG006, 05EG008, 05MC004, and 11AF005, respectively. On the other hand, under RCP 8.5 flow rate was projected to change by 0.02 (−0.01), 0.08 (0.04), 0.1 (0.1), and 0.14 (0.16) m^3/s by 2050s (2090s) for catchments 05EG006, 05EG008, 05MC004, and 11AF005, respectively. In summary, all three hydroclimatic variables of interest, that is, precipitation, temperature, and flow were projected to increase across the selected catchments in future.

To further investigate land-cover driven hydroclimatic changes in the catchments, quantitative and statistical differences between the projected hydroclimatic changes with and without considering land-cover change were analyzed. Here former (i.e., climatic changes with land-cover change) are presented as mean hydroclimatic change obtained from SP, SPS3x3, SPS5x5, SPS7x7, SPS9x9 downscaling models while latter (i.e., climatic changes without land-cover change) were presented as the mean hydro-climatic change obtained from SP_LC downscaling model. In Gaur and Simonovic (2018) SP_LC model was referred to a version of SP model where land-cover is omitted from the list of predictors used for downscaling in the SP model. In other words only large-scale climate and elevation are used as predictors for downscaling GCM projections in the SP_LC model.

Statistical significance of the influence of land-cover change was explored by performing Wilcoxon signed rank test (Wilcoxon, 1945) on projected temperature, precipitation, and flow time-series obtained from SP_LC and other downscaling models for timelines 2050s and 2090s. Test results identified statistically significant differences in the projected time-series for all three hydroclimatic variables, and for both future timelines.

The magnitudes of land-cover driven climatic changes and its flow-related implications were explored. In the case of mean temperature (average of minimum and maximum temperature), a land-cover change driven warming of −0.01 (−0.01), −0.01 (−0.01), −0.01 (−0.02), and 0.16 (0.19) K, averaged across both emission scenarios was obtained by 2050s (2090s) for catchments 05EG006, 05EG008, 05MC004, and 11AF005, respectively. In the case of mean precipitation, a land-cover change driven increase of −0.02 (0.05), 0.01 (0.03), 0.02 (0.01), and −0.06 (−0.11) mm/day, averaged across GCMs and both emission scenarios were obtained by 2050s (2090s) for catchments 05EG006, 05EG008, 05MC004, and 11AF005, respectively. Finally in the case of flow, a land-cover change driven increase of −0.014 (0.001), −0.022 (−0.002), −0.004 (−0.004), and 0.002 (0.002) m^3/s averaged across GCMs and both emission scenarios was obtained by 2050s (2090s) for catchments 05EG006, 05EG008, 05MC004, and 11AF005, respectively.

Finally, the distribution of land-cover—driven changes across different quantiles was also evaluated. It was found that for all catchments, higher quantiles were associated with higher changes. In other words hydro-climatic extremes were found to be more influenced from the land-cover change than the lower climate and flow distribution quantiles.

6.4 Conclusions From SP Model Case Study

In Gaur and Simonovic (2018) climatic effects of land-cover change and their implications on flow were explored. Four small catchments located in the southern Saskatchewan region were selected for investigation. Future climate and hydrologic projections are made by downscaling future precipitation and temperature projections made by three GCMs and two emission scenarios. By analyzing changes projected between baseline (2006—16) and future timelines, 2050s and 2090s, it can be concluded that minimum temperature, maximum temperature, precipitation magnitude and intensity, and flow rate are set to increase in future over the selected catchments.

Climatic projections with future land-cover change are obtained from SP model versions: SP, SPS3x3, SPS5x5, SPS7x7, SPS9x9 and projections without considering future land-cover change are obtained using SP model version: SP_LC. The downscaled climate projections are used to derive future streamflow response from all four catchments using Sacramento hydrologic model. A comparison of these two sets of hydroclimatic projections for 2050s and 2090s showed statistically significant differences between them. These results are in line with the findings from other dynamic downscaling-based studies which have found notable land-cover change—induced climatic changes at local to regional scales (for example, Malyshev et al., 2015). An analysis of the variation of slope of the projected change with quantiles indicates an increasing trend with the quantile value for all three variables analyzed. This suggests higher land-cover—induced changes in hydroclimatic extremes as compared to the means.

Most importantly, this case study clearly demonstrated the usefulness of SP and SPS models in incorporating regional geophysical characteristics in model formulation, as well as their ability to model climatic influences of geophysical changes within a statistical downscaling framework.

7. WORKING EXAMPLE OF SP AND SPS METHODS IN R PROGRAMMING LANGUAGE

In this section, we demonstrate how SP and SPS methods can be applied to downscale GCM projections in the R programming language. First we provide a gentle introduction to the R programming language, and then present a step-by-step demonstration of SP and SPS model downscaling using it.

7.1 Basics of R Programming Language

R is a programming language and software environment for statistical analysis, graphics representation, and reporting. R was created by Ross Ihaka and Robert Gentleman at the University of Auckland, New Zealand, and is currently developed by the R Development Core Team (R Development Core Team, 2008). R implements many common statistical procedures, as well as provides excellent graphics functionalities through its libraries and packages. It is an open source language which means that the users are able to acquire it free of cost, as well as can contribute towards its research and development. In this subsection, basic features of R programming language have been reviewed with an intention that this can help the readers to get started with R.

7.1.1 Downloading and Installing R

R programming language for windows operating system can be downloaded from https://cran.r-project.org/bin/windows/base/. R for Linux and Mac operating systems can also be downloaded from the R project for statistical computing website: https://www.r-project.org/. It is also common to use R Studio to run and edit R codes as this software provides enhanced code editing, debugging, and visualizing capabilities to the R users. R Studio for Windows, Linux, and Mac operating systems can be downloaded from https://www.rstudio.com/products/rstudio/download/. R and R Studio softwares can then be installed on the computer by running the executables obtained from above sources.

7.1.2 Common Data-Types in R Programming

Some of the most common data-types used in R programming language are vector, data-frame, and list. A vector can be a sequence of numbers, logical values, or character strings. A vector with three numeric values can be defined as follows in the R Studio command-line. Below the user commands are provided in lines commencing with the symbol: ">" while output from the R (if any) is provided in lines commencing with the symbol: "[1]".

```
C1 > c(3, 4, 7)
[1] 3 4 7
```

A vector with three logical values can be defined as follows.

```
C2 > c(TRUE, FALSE, FALSE)
[1] TRUE FALSE FALSE
```

A vector with three character values can be defined as follows. The resulting vector has also been saved into a variable named: vec.char.

```
C3 > vec.char = c("a", "b", "c")
```

A vector element can be extracted using the "[]" brackets with the index of the element to be extracted as shown below.

```
C4 > vec.char[1]
[1] a
```

A data-frame is used to store vectors of equal length in the form of tables. Below is the command to create a data-frame "df" with three columns and four rows representing the marks obtained by four students named: "A," "B," "C," "D" in two subjects: maths and physics.

```
C5>df=data.frame(name=c("A","B","C","D"),maths=c(85,90,97,76),
        physics=c(88,66,76,98))
```

Data-frame element(s) can be extracted using the "[]" brackets with the row number and column number indices of the element to be extracted. The entire row or column can be extracted by specifying the same in the command. An entire column can also be extracted by specifying the column name together with the "$" operator as shown below.

```
C6 > df[1,2]
[1] 85
C7>df[1,]
[1] A 85 88
C8 > df[,1]
[1] A B C D
C9>df$name
[1] A B C D
```

A list can be used to store vectors of equal or unequal lengths. Below is an R command to create a list with 3 numeric, 4 logical, and 5 character elements stored in the first, second, and third elements. The list is stored as a variable named "lst."

```
C10 > lst = list (c(1, 5, 7), c(TRUE, TRUE, FALSE, FALSE), c("a", "b",
      "c", "d", "e"))
```

List elements can be extracted by using the "[[]]" brackets along with the list element number that needs to be extracted. Further, subelements within a list element can also be accessed using "[]" brackets with the index of the subelement number as demonstrated below.

```
C11 > lst[[1]]
[1] 1 5 7
C12 > lst[[1]][3]
[1] 7
```

7.1.3 Relevant R Packages

R packages are a collection of R codes, functions, data, and compiled code in a well-defined format. R comes with a standard set of packages. Other packages can be downloaded and installed separately by the users based on their needs. Here we provide a brief introduction of a few R packages (apart from the standard packages) that are very useful in performing downscaling of GCM data by SP and SPS methods in R. These packages are:

- MODIS and MODISTools: The intended purpose of these packages is to facilitate acquisition and processing of MODIS data-products. MODIS package contains functions to gain automated access to the global online data archives and processing capabilities such as file conversion, mosaicking, subsetting, and time-series filtering (Mattiuzzi, 2016). The package can be downloaded from Comprehensive R Archive Network (CRAN) by running the following command on the command line: install.packages("MODIS", repos="http://R-Forge.R-project.org"). MODISTools package also allows users to extract MODIS data time-series at one or more than one locations without downloading the image rasters (Tuck and Phillips, 2016). The package can be downloaded from Comprehensive R Archive Network (CRAN) by running the following command on the command line: install.packages("MODISTools", repos="http://R-Forge.R-project.org").
- ncdf4: This package is designed to work with NetCDF libraries version 4 in R, which is the most commonly used NetCDF version currently. Another package "ncdf" can be used to access NetCDF version 3 libraries. In "ncdf4" package, utilities like chunking and compression have also been included. The package can be downloaded from CRAN by running the following command on the command line: install.packages("ncdf4", repos="http://R-Forge.R-project.org").
- raster: This package is intended to facilitate raster processing in R. Among other functions, the package contains functions that can read and write rasters, perform raster operations such as reprojection, resampling, filtering, merging, etc.; perform raster calculations; and visualize raster data (Hijmans, 2015). The package can be downloaded from CRAN by running the following command on the command line: install.packages("raster", repos"http://R-Forge.R-project.org").
- lubridate: The lubridate package is intended to facilitate easy handling of date-time data in R. Among others, it contains functions that can be used to extract components of a date-time such as year, month, day, hour, minute, and seconds, and perform algebraic manipulation on the date-time objects (Grolemund and Wickham, 2011). The package can be downloaded from CRAN by running the following command on the command line: install.packages("lubridate", repos =" http://R-Forge.R-project.org").

- reshape2: This R package is extremely useful to transform data between wide and long formats. A wide format has a column for each variable while a long format has a column for possible variable types and another column for the values of these variables (Wickham, 2007). The package can be downloaded from CRAN by running the following command on the command line: install.packages("reshape2", repos="http://R-Forge.R-project.org").
- ggplot2: This package is meant for "declaratively" creating graphic by telling ggplot2 how to map variables to aesthetics and what graphical primitives to use. It produces plots following the grammar of graphics (Wilkinson, 2005) where essential building blocks of a graph, that is, data, aesthetic mapping, geometric object, statistical transformations, scales, coordinate system, position adjustments, and faceting are specified by the user (Wickham, 2009). The package can be downloaded from CRAN by running the following command on the command line: install.packages("ggplot2", repos="http://R-Forge.R-project.org").
- mgcv: This package is very helpful in performing GAM regression. It includes several methods for estimating regression parameters, smoothing functions, and link functions in GAMs in computationally efficient manner (Wood, 2011). The package can be downloaded from CRAN by running the following command on the command line: install.packages("ggplot2", repos="http://R-Forge.R-project.org").

7.1.4 Other Relevant R Functions

Apart from the functions available in aforementioned packages, plenty of useful functions are available in R base library. They have been used extensively while downscaling GCM projections using SP and SPS models. A description of some of those functions that have been most extensively used is provided below. A more detailed description of any of these functions (or any other function) can be obtained by running "?"function-name command in R.

- which(): This function is used to know the position of elements of a logical vector that are TRUE.
- subset(): This function is used to extract section of a data-frame with rows that meet a particular criteria. For instance in a data-frame that stores monthly discharge data time-series, this function can be used to select part of the time-series that corresponds to January or have month values equal to 1.

- sapply/lapply: The sapply() and other similar functions are an alternate for looping in R. Their usage is recommended as their usage can make the R codes run much faster than when for instance for() loops are used. The "apply" family of functions have many variants like sapply() which stands for simplify and apply, lapply() which stands for list and apply, vapply(), tapply(), etc.
- rbind/cbind(): Both rbind() and cbind() are used to combine two data-frames or matrices. It is essential that data-frames (or matrices) have the same number of columns when rbind() is used to combine them. Similarly, data-frames (or matrices) should have equal number of rows when cbind() is used.
- do.call(): The do.call() function executes a function call over all list elements passed to it. This function is commonly used in conjunction with lapply() function where "list" output obtained from lapply() are further analyzed or manipulated using do.call() function.

7.2 Application of SP and SPS Model Downscaling in R

In this section a demonstration of how R programming language can be used to downscale climate model projections using SP and SPS methods is presented. A discussion on how to extract, organize, and prepare remotely sensed and climate model datasets in R is provided first. This is followed by a working example where SP and SPS models are applied to downscale a sample future temperature projection dataset.

Climate model datasets from GCMs or reanalysis products are generally available in Network Common Data From (NetCDF) format. NetCDF is a self-describing, machine independent data format that supports the creation, access, and sharing of array oriented data. By self-describing, it means that information about specifications of the file, the data it stores, and its layout that is stored within the file. In R NetCDF format files can be accessed using "ncdf4" package. All packages need to be downloaded and loaded in the R session prior to their usage. Packages can be loaded by using command "library" (package-name) in the R session. Following commands can be used to access and extract data from NetCDF files in R:

```
C13 > file.read = nc_open(file.name)
```

Above function nc_open reads a NetCDF file. The location of.nc file should be specified in the argument file.name of the function. The details of the.nc file are stored in the variable file.read.

```
C14 > file.read$dim$names
```

The variable file.read can be used to extract import file characteristics such as file dimensions, which conveys the layout of data stored in the.nc file. Above command prints out detailed description of all dimensions associated with the file.read variable.

```
C15 > file.read$var$names
```

Variables convey information about the data stored in the.nc files. Above command prints out detailed description of all dimensions associated with the file.read variable.

```
C16 > ncvar_get(file.read, var.name)
C17 > ncvar_get(file.read, varid=var.name, start=c(1,1,1),
      count=c(dim1,dim2,1))
C18 > sapply(1:10, function(x) ncvar_get(file.read, varid=var.name,
      start=c(1,1,x), count=c(dim1,dim2,1)))
```

The function "ncvar_get" is used to extract data stored in a particular variable in the.nc file. Command C16 will extract the values of a variable with name "var.name" stored in the.nc file. Sometimes due to data volume it is not possible or desirable to import and save all of the data in one variable in one go. In those cases, "ncvar_get" command can also be specified other attributes so that the data is read in manageable chunks. For instance, command C17 provided above reads only the first array element of the variable "var.name". Variables "dim1" and "dim2" denote the x and y dimensions of the data array. Command C18 performs the same operation iteratively for array elements 1 to 10 using the "sapply" function. The indices for which the data needs to be extracted is ascertained by examining the time-indices for which data is provided in the.nc file and then finding indices that contain data for the user-defined time-period of interest.

Above discussion provides a brief introduction on how R programming language can be used to access GCM and reanalysis climate data that is typically available in the.nc format using the ncdf4 package.

Next, we discuss how MODIS data can be extracted using R programming language. MODIS-based climatic and land-cover data are very useful for performing SP and SPS model—based downscaling, and they can be easily extracted and managed using R programming language. We demonstrate the use of two packages: "MODISTools" and "MODIS" towards downloading and analyzing MODIS data in R. The first package

MODISTools can be used to download spatiotemporal MODIS data using "MODISSubsets" function provided below:

```
C19 > MODISSubsets(LoadDat, Products, Bands, Size)
```

In "MODISSubsets" function, argument "LoadDat" reads in a data-frame with details about coordinates and IDs of all locations where data needs to be extracted, as well as the start and end dates of the data to be downloaded. The argument "Products" reads in the product code, which can be obtained from the function GetProducts(). The argument "Bands" is supplied the band names to be downloaded. For a particular product, a list of bands can be obtained using the Getbands() command. The argument "Size" is supplied with the spatial scale at which the data should be extracted. A value of c(0,0) is supplied if only data at the location of interest needs to be extracted. Other values such as c(1,1) provide values spatially averaged over an area of 2 km^2 from the location of interest. Following command C20 will extract the day-time and night-time surface temperature data sensed by Terra satellite over the London (Ontario) city (lon = −81.25, lat = 42.98) for the year 2012.

```
C20>MODISSubsets(LoadDat=data.frame(lat=42.98,long=-81.25,
            start.date=2012,end.date=2012,id=1),
            Products = "MOD11A1"
            Bands = c("LST_Day_1km", "LST_Night_1km"),
            Size = c(0,0))
```

Downloading MODIS data using "MODISTools" is most advantageous when the data needs to be extracted at a limited number of locations. However, if data needs to be extracted over a region or a country, it is beneficial to extract a raster image encompassing the region of interest for analysis. The runGdal function provided in the MODIS package is extremely useful in such cases. Following is a description of the function:

```
C21 > runGdal(product, begin, end, extent)
```

In the "runGdal" function, the argument "product" needs to be supplied with the MODIS product ID, beginning and end dates for data extraction are supplied in the arguments: "begin" and "end" respectively. The area for which the data needs to be extracted is supplied through the "extent" argument. The area to be supplied can be selected interactively, by supplying a shapefile or a raster file, by specifying the country name or an extent object. Other important arguments such as "outProj" which supplies the output raster projection "pixelSize" which can be supplied with the output data spatial resolution, and "dataFormat" which can be used to specify the data-format of output raster image. The following command can be used to extract surface temperature data sensed by Terra satellite over Canada for the year 2012.

```
C22 > runGdal(product="MOD11A1",begin="2,012,001",
      end = "2,012,366",extent="canada")
```

So far the discussion provided in this section dealt with the preparation of geophysical and climatic data required for the application of SP and SPS method downscaling models. Once the climatic and geophysical data are prepared, the downscaling models can be used to downscale GCM projections. We demonstrate this by downscaling future temperature and precipitation projections obtained from a GCM: MRI-CGCM3 under RCP2.6 using SP and SPS models. The data for downscaling model calibration and for making future projections can be downloaded from https://drive.google.com/drive/folders/0B86JMqp3jy6CcmYzT2dQNXlMQW8?usp=sharing. The datasets are provided in .rds format which is a format to save and load R objects as files in the system. File named "Model calibration data.rds" contains data needed to calibrate SP and SPS models and file named "Future prediction data.rds" contains data needed for making downscaled future climatic projections.

Once the folder is placed in a directory (termed as "Fakepath" below), the data can be imported into R using following commands:

```
C23 > cal.data=readRDS("Fakepath/Model calibration data.rds")
C24 > pred.data=readRDS("Fakepath/Future prediction data.rds")
```

The function "readRDS" used above reads in any R object with .rds extension. Here model calibration data are imported into an R data-frame object "cal.data" and prediction data are imported into an R object "pred.data." The datasets can be examined using two very useful R functions: "summary" and "head"/"tail" as shown below.

```
C25 > summary(cal.data)
C26 > head(cal.data, 50)
```

The former function provides a summary of the data stored in each column of the data-frame. The latter functions "head" or "tail" shows user requested number (which in above case is 50) of first or last few lines in a data-frame, respectively. Table 4.11 provides a description of different columns in data-frames: "cal.data" and "val.data."

Table 4.11 Descriptions of Columns Present in the Calibration and Prediction Data

Column Name	Description
elev	Elevation of the gauging stations
P.obs	Precipitation recorded at the gauging stations
T.obs	Temperature recorded at the gauging stations
C.3x3/C.5x5/C.7x7/C.9x9	Cropland fraction in 3x3/5x5/7x7/9x9 neighborhood scale
G.3x3/G.5x5/G.7x7/G.9x9	Grassland fraction in 3x3/5x5/7x7/9x9 neighborhood scale
BSV.3x3/BSV.5x5/BSV.7x7/BSV.9x9	Barren or Sparsely Vegetated fraction in 3x3/5x5/7x7/9x9 neighborhood scale
OS.3x3/OS.5x5/OS.7x7/OS.9x9	Open Shrublands in 3x3/5x5/7x7/9x9 neighborhood scale
W.3x3/W.5x5/W.7x7/W.9x9	Water fraction in 3x3/5x5/7x7/9x9 neighborhood scale
ENF.3x3/ENF.5x5/ENF.7x7/ENF.9x9	Evergreen Needleleaf Forest fraction in 3x3/5x5/7x7/9x9 neighborhood scale
UB.3x3/UB.5x5/UB.7x7/UB.9x9	Urban fraction in 3x3/5x5/7x7/9x9 neighborhood scale
DBF.3x3/DBF.5x5/DBF.7x7/DBF.9x9	Deciduous Broadleaf Forest fraction in 3x3/5x5/7x7/9x9 neighborhood scale
DNF.3x3/DNF.5x5/DNF.7x7/DNF.9x9	Deciduous Needleleaf Forest fraction in 3x3/5x5/7x7/9x9 neighborhood scale
MF.3x3/MF.5x5/MF.7x7/MF.9x9	Mixed Forest fraction in 3x3/5x5/7x7/9x9 neighborhood scale
S.3x3/S.5x5/S.7x7/S.9x9	Savannas fraction in 3x3/5x5/7x7/9x9 neighborhood scale
WS.3x3/WS.5x5/WS.7x7/WS.9x9	Woody Savannas fraction in 3x3/5x5/7x7/9x9 neighborhood scale
CS.3x3/CS.5x5/CS.7x7/CS.9x9	Closed Shrublands fraction in 3x3/5x5/7x7/9x9 neighborhood scale
EBF.3x3/EBF.5x5/EBF.7x7/EBF.9x9	Evergreen Broadleaf Forest fraction in 3x3/5x5/7x7/9x9 neighborhood scale
ef.elev.3x3/ef.elev.5x5/ef.elev.7x7/ef.elev.9x9	Relative elevation fraction in 3x3/5x5/7x7/9x9 neighborhood scale
T.NARR	NARR temperature interpolated at the gauging station location
P.NARR	NARR precipitation interpolated at the gauging station location
LC	Land-cover at the gauging station
GCM.ppt	GCM precipitation interpolated at the gauging station location
GCM.tas	GCM temperature interpolated at the gauging station location
date	Date associated with the data

SP method calibration can be performed using "gam" function available in the package "mgcv." This can be performed with either of the following commands:

```
C28 > SP.mod.T = gam(T.obs ~ s(T.NARR) + LC+ s(elev), data = cal.data)
C29 > SP.mod.T.jan = gam(T.obs ~ s(T.NARR) + LC+ s(elev),
      data = subset(cal.data, month(date) %in% 1))
```

The command C28 fits a GAM function on the observations provided in the "cal.data" object and the resulting model is stored in another R object: "SP.mod." If the calibration needs to be performed using data belonging to all days in the month of January, this can be done using command C29. The code subsets only the data belonging to January month by using "subset" function and checking which months corresponding to the "dates" column equal to 1. The model can be calibrated for other months in a similar fashion.

SPS model for downscaling temperature can be calibrated in a similar way. Command C30 calibrates SPS3x3 model using entire "cal.data" series and stores the calibrated model in an R object named: "SPS3x3.mod.T." Appropriate modifications in predictor variables can be made when calibrating SPS models for other neighborhood scales.

```
C30 > SPS3x3.mod.T = gam(T.obs ~ s(T.NARR) + s(LC) + s(elev) +
                         s(C.3x3) + s(G.3x3) + s(BSV.3x3) + s(OS.3x3) +
                         s(W.3x3) + s(ENF.3x3) + s(UB.3x3) + s(DBF.3x3) +
                         s(DNF.3x3) + s(MF.3x3) + s(S.3x3) + s(WS.3x3) +
                         s(CS.3x3) + s(EBF.3x3) + s(ef.elev.3x3),
                         data = cal.data)
```

Performing downscaling using the calibrated models is also very straightforward and intuitive. For instance, following commands can be used to predict downscaled temperature data using above calibrated SP and SPS3x3 models:

```
C31 > colnames(pred.data)[which(colnames(pred.data) %in% "GCM.tas")]
      ="T.NARR"
C32 > pred.data$SP.pred.T = predict(SP.mod.T, newdata = pred.data)
C33 > pred.data$SPS3x3.pred.T = predict(SPS3x3.mod.T, newdata
      = pred.data)
```

For making predictions using the calibrated GAM model, the predictor variables present in the calibration data should also be present in the validation data. Therefore the column "GCM.tas" which contains interpolated GCM data is renamed as "T.NARR" which contained NARR-derived model data in the calibration dataset. This is performed by command

C31 presented above. Following this, the calibrated models "SP.mod" and "SPS3x3.mod" are used to downscale prediction dataset "pred.data" using commands provided in the commands C32 and C33 above.

Precipitation downscaling by SP and SPS models involves two steps as discussed in Section 3. First, dry and wet-day sequences are predicted using a logistic regression model and then, wet-day precipitation intensity is predicted using a GAM model. The calibration of logistic regression model and GAM models can be performed in R using following set of commands:

```
C34 > cal.data$r.switch=0
C35 > cal.data$r.switch[which(cal.data$P.obs>0.1)]=1
C36 > occ.mod.SP.P=glm(r.switch~P.NARR+elev+LC,data=cal.data,
      family="binomial")
C37 > int.mod.SP.P=gam(P.obs~s(P.NARR)+s(elev)+s(LC),data=
      subset(cal.data,r.switch==1))
C38 > occ.mod.SPS3x3.P = glm(r.switch ~ P.NARR + elev + LC + C.3x3 +
      G.3x3 + BSV.3x3 + OS.3x3 + W.3x3 + ENF.3x3 + UB.3x3 + DBF.3x3 +
      DNF.3x3 + MF.3x3 + S.3x3 + WS.3x3 + CS.3x3 + EBF.3x3 +
      ef.elev.3x3, data=cal.data, family="binomial")
C39 > int.mod.SPS3x3.P = gam(P.obs ~ s(P.NARR) + s(elev) +
      s(LC) + s(C.3x3) + s(G.3x3) + s(BSV.3x3) + s(OS.3x3) +
      s(W.3x3) + s(ENF.3x3) + s(UB.3x3) + s(DBF.3x3) +
      s(DNF.3x3) + s(MF.3x3) + s(S.3x3) + s(WS.3x3) +
      s(CS.3x3) + s(EBF.3x3) + s(ef.elev.3x3), data=subset(cal.data,
      r.switch==1))
```

In command C34 and C35 a new column with a rainfall switch defining rainfall (1) or no-rainfall (0) state is added to the calibration data-frame: "cal.data". All days with daily precipitation magnitude greater than 0.1 mm is allotted a value of 1 in the column "r.switch" whereas all days with precipitation magnitudes less than or equal to 0.1 mm are allotted a value of 0. Command C36 calibrates the logistic regression model for SP model using "r.switch" as the predictant variable and columns "P.NARR", "elev," and "LC" as predictor variables. In R, logistic regression can be performed within the generalized linear modelling (glm) framework. A glm framework in R can model regression functions of many families including gaussian, binomial, poisson, gamma, inverse gaussian, and quasi. For defining a logistic regression, we specify the family of the regression function as "binomial" and link function as "logit" (not specified above as it is the default link function for binomial family in R). Command C37 calibrates a GAM model on rainfall intensities only using data for days when the value of "r.switch" is equal to 1 (in other words only using data for wet days). Commands C38 and C39 perform similar calibration of logistic regression and GAM models for the SPS3x3 downscaling model. The only difference is that in this case all neighborhood predictors corresponding to the scale 3x3 are also used in defining the regression models.

Prediction of downscaled precipitation using SP and SPS models involves first predicting dry and wet-day sequences and secondly predicting wet precipitation magnitudes. This can be performed in R using passing following set of commands:

```
C40 > colnames(pred.data)[which(colnames(pred.data) %in%
      "GCM.ppt")]="P.NARR"
C41 > pred.data$r.switch=as.numeric(predict(occ.mod.SP.P,
      newdata=pred.data, type="response"))
C42 > pred.data$r.switch[which(pred.data$r.switch>=0.5)]=1
C43 > pred.data$r.switch[which(pred.data$r.switch<0.5)]=0
C44 > pred.data$SP.pred.P=0
C45 > pred.data$SP.pred.P[which(pred.data$r.switch>=0.5)] =
      as.numeric(predict(int.mod.SP.P, newdata = subset(pred.data,
      r.switch> = 0.5)))
C46 > pred.data$r.switch=as.numeric(predict(occ.mod.SPS3x3.P,
      newdata=pred.data, type="response"))
C47 > pred.data$r.switch[which(pred.data$r.switch>=0.5)]=1
C48 > pred.data$r.switch[which(pred.data$r.switch<0.5)]=0
C49 > pred.data$SPS3x3.pred.P=0
C50 > pred.data$SPS3x3.pred.P[which(pred.data$r.switch>=0.5)] =
      as.numeric(predict(int.mod.SPS3x3.P, newdata = sub
      set(pred.data, r.switch> = 0.5)))
```

The command C40 again renames the column "GCM.ppt" to "P.NARR" for prediction to ensure that the predictors chosen to calibrate the occurrence and precipitation intensity models are also present in the prediction dataset "pred.data." The command C41 predicts the probabilities of a particular day to be rainy given the values of predictors "P.NARR", "LC", and "elev" for each day. Next, in lines C42 and C43 we choose a probability threshold that can be used to decide on the predictant state given the predicted probabilities. In Gaur and Simonovic (2018) this has been calibrated for different models. For this demonstration a value of 0.5 is chosen as the probability threshold value above which the predictant, that is, "r.switch" is considered as having a rainy (1) state or else it is considered to have a nonrainy (0) state. In lines C44 and C45 the values of precipitation for days with "r.switch" values equals to zero are taken to be zero. For days with "r.switch" equals to 1 (or for rainy or wet days) the precipitation magnitude is predicted using the model "int.mod.SP.P."

Commands C46 to C50 perform precipitation downscaling using SPS3x3 model using a similar process but using the logistic regression model "occ.mod.SPS3x3.P" and wet-day precipitation magnitude prediction GAM model "int.mod.SPS3x3.P."

Finally the results can be visualized effectively using several useful functions available in the "ggplot2" package. The function "melt" in package "reshape2" is also extremely useful for preparing data to be used in "ggplot2" package. A small example is presented here where yearly maximums of GCM and downscaled precipitation data (from models SP and SPS3x3) stored in the "pred.data" data-frame are plotted using "ggplot2" package.

```
C51 > ymax.P.GCM = sapply(2014:2100, function(x) max(subset
      (pred.data, year(date) %in% x)$P.NARR, na.rm=T))
C52 > ymax.P.SP = sapply(2014:2100, function(x) max(subset
      (pred.data, year(date) %in% x)$SP.pred.P, na.rm=T))
C53 > ymax.P.SPS3x3 = sapply(2014:2100, function(x) max(subset
      (pred.data, year(date) %in% x)$SPS3x3.pred.P, na.rm=T))
C54 > ymax.P=data.frame(year=2014:2100, GCM=ymax.P.GCM,
      SP=ymax.P.SP, SPS3x3=ymax.P.SPS3x3)
C55 > data.plot = melt(ymax.P, id="year")
C56 > ggplot()+geom_line(data=data.plot,aes(x=year,y=value,
      group=variable),size=1)+
              geom_point(data=data.plot,aes(x=year,y=value),
              size=3)+
              facet_wrap(~variable, scales="free")+theme_bw()+
              xlab("Year")+ylab("Precipitation(mm)")+
              theme(legend.title = element_text(size=22,
              face = "italic"),legend.position="bottom")+
              theme(axis.title.y=element_text(face="bold",
              size=22),
                    axis.title.x=element_text(face="bold",
                    size=22),
                    text=element_text(size=22))+
                         theme(axis.text.x=element_text(size=18),
              axis.text.y=element_text(size=18))+theme(legend.
              key.width = unit(3,"cm"))
C57 > ggsave("Fakepath/Sample_plot.png",width=14,height=10)
```

The extraction of yearly maximum precipitation as simulated by GCMs and downscaling models SP and SPS3x3 is performed in commands C51, C52, and C53, respectively. Next the yearly maximum results are aggregated into one data-frame "ymax.P". Thereafter the wide-format data-frame with four columns is converted into a long-format data-frame "ymax.P" with three columns using a function "melt" from "reshape2" package.

The long-format data-frame "data.plot" has three columns named "year", "variable", and "value." These three columns store the year values, that is, 2014 to 2100, variable values, that is, GCM, SP, and SPS3x3, and value of precipitation maximums (in mm) corresponding to each combination of year and variable name. The function ggplot() which we are using to plot the graphs needs an input data-frame in the long-format in order to do the plotting.

The data-frame "data.plot" is used to plot yearly maximum precipitation from GCMs, and SP, SPS3x3 downscaling models using the ggplot() function. As explained before, ggplot2 package is built on the grammar of graphics. It can be noted from command C56 that and data, aesthetic mapping, geometric object, statistical transformations, scales, coordinate system, position adjustments, and faceting arguments are passed along with the ggplot() function. The output generated from this command is shown in Fig. 4.4. In the plot, the three panels show annual maximum precipitation values for GCM, SP, and SPS3x3-based precipitation projections. The generated plot can be saved by using the ggsave() function in C57, where among other arguments, the location where file needs to be saved, file-type, plot dimensions, etc., are specified to save the plot in the system.

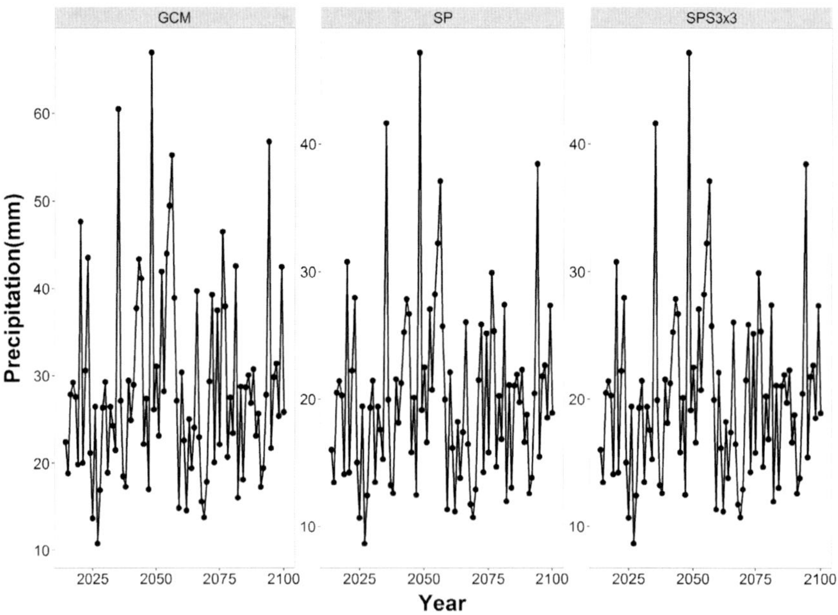

Figure 4.4 Annual precipitation maximum magnitudes (in mm) as plotted by the ggplot() function in R.

8. CONCLUSIONS AND FUTURE WORK

This discussion provided in this chapter is centered on a novel statistical downscaling model: physical scaling (SP) and its extension: SPS model. The motivation behind the development of the model and the theoretical research gap that this model fills is discussed first. The proposed model aims to bridge the gap between dynamic and statistical downscaling by building on the positives of both approaches. The state-of-the-art statistical downscaling models do not account for regional geophysical characteristics when performing downscaling. For this reason, climatic effects of geophysical characteristics as well as of changes in those characteristics can't be quantified currently within a statistical downscaling framework.

This limitation is overcome in SP and SPS models by considering covariates which represent regional geophysical characteristics like elevation and land-cover and their distribution in the model definition. The predictor and predictant variables are linked using a GAM function which is fully nonparametric in nature and can be used to link a range of predictor and predictant variables. Temperature downscaling by SP model is performed in one step by linking locally observed temperature with model-based temperature, land-cover, and elevation of the location of interest. In case of precipitation, downscaling is performed by formulating a logistic regression function which models wet-dry day precipitation sequences, as well as a GAM regression function that models precipitation magnitudes on wet days using model-based precipitation, land-cover, and elevation of the location of interest as predictors. In SPS models, land-cover and elevation characteristics of the location of interest, as well as its neighborhoods, are incorporated into the model formulation. This is done at a chosen neighborhood scale which has been taken as 3x3, 5x5, 7x7, and 9x9 in previous studies.

Both SP and SPS models have been extensively validated for their skill to downscale NARR temperature and precipitation gridded data in previous studies (Gaur and Simonovic, 2016a,b, 2018). This has been done both by quantifying model error over the historical time-period, as well as by comparing its performance with other state-of-the-art downscaling models like SDSM and GLM. The results from validation studies have been summarized in this chapter. They clearly show that the developed SP and SPS models are reliable downscaling tools that can be used to downscale GCM projections in southern Saskatchewan region of Canada.

Apart from being a fast, reliable, and accurate downscaling model, the major advantage of SP and SPS models is that they can be used to identify climatic impacts of future geophysical changes. In this chapter results from a case-study which quantified hydroclimatic impacts of future land-cover change at four catchments in southern Saskatchewan region of Canada are also presented and discussed. The case-study clearly demonstrates a sample analysis that can be performed using SP and SPS models that is not possible using other traditional statistical downscaling models.

Finally, a detailed presentation of how SP and SPS models can be formulated and used using an open source R programming language is made. To make the tutorial accessible for early users of R, an overview of R and its key components has also been provided. The tutorial then presents commands to read climate model—based atmospheric and climatic data from .nc files, and MODIS remotely sensed data directly from R. The tutorial then presents commands and discussion to calibrate SP and SPS models in R, and to use them for downscaling future precipitation and temperature projections from a representative GCM. Finally R is used to plot and compare downscaled future precipitation results with the raw GCM projections. From the tutorial it can be seen that SP and SPS models are very easily implementable in an open source platform and therefore they can be easily used by the scientific community across the globe without any additional costs.

The presented research can be extended in many possible directions in future. SP method can be modified to account for snow-cover during the winter months. Snow-cover is a very important physical parameter which affects the climatology and hydrology of any region and is especially relevant in the Canadian context. An appropriate spatial scale for the calibration of SP method needs to be ascertained. The spatial scale chosen should be a compromise between accuracy and robustness of the downscaled results. The applicability of SP method can be evaluated in other regions of Canada and the globe. It will be interesting to see how the model performs in regions that have more complex physiography than the region considered in this study. Further, case studies can be performed on catchments located in other regions of the globe to better understand the underlying dynamics of the land-cover change driven climatic changes. It will be interesting to compare land-cover change driven hydrologic changes with land-cover change driven climatic changes which in turn affects catchment hydrology. The study can be performed at catchments located in different climatic regions and biomes to compare and contrast the results.

REFERENCES

Abebe, M.T., Megento, T.L., 2016. The City Of Addis Ababa From 'Forest City' To 'Urban Heat Island'. Assessment Of Urban Green Space Dynamics, J. Urban Environ. Eng. 10 (2), 254–262.

Adachi, S.A., Kimura, F., Kusaka, H., Inoue, T., Ueda, H., 2012. Comparison of the impact of global climate changes and urbanization on summertime future climate in the Tokyo Metropolitan Area. J. Appl. Meteorol. Climatol. 51, 1441–1454.

Anandhi, A., Frei, A., Pierson, D.C., Schneiderman, E.M., Zion, M.S., Lounsbury, D., Matonse, A.H., 2011. Examination of change factor methodologies for climate change impact assessment. Water Resour. Res. 47 (3), W03501.

Apipattanavis, S., Podesta, G., Rajagopalan, B., Katz, R.W., 2007. A semiparametric multivariate and multisite weather generator. Water Resour. Res. 43 (11), W11401.

Argueso, D., Evans, J.P., Fita, L., Bormann, K.J., 2014. Temperature response to future urbanization and climate change. Clim. Dyn. 42 (7-8), 2183–2199.

Barradas, V.L., 1991. Air temperature and humidity and human comfort index of some city parks of Mexico City. Int. J. Biometeorol. 35 (1), 24–28.

Beecham, S., Rashid, M., Chowdhury, R.K., 2014. Statistical downscaling of multi-site daily rainfall in a South Australian catchment using a Generalized Linear Model. Int. J. Climatol. 34, 3654–3670.

Bellone, E., Hughes, J.P., Guttorp, P., 2000. A hidden Markov model for downscaling synoptic atmospheric patterns to precipitation amounts. Clim. Res. 15, 1–12.

Benestad, R., Hanssen-Bauer, I., Chen, D., 2008. Empirical-Statistical Downscaling. World Scientific Publishing Company, Singapore, p. 228.

Bergant, K., Kajfez-Bogataj, L., 2005. N-PLS regression as empirical downscaling tool in climate change studies. Theor. Appl. Climatol. 81, 11–23.

Best, M.J., Grimmond, C.S.B., Villani, M.G., 2006. Evaluation of the urban tile in MOSES using surface energy balance observations. Bound. Lay. Meteorol. 118, 503–525.

Boucher, O., Myhre, G., Myhre, A., 2004. Direct human influence of irrigation on atmospheric water vapor and climate. Clim. Dyn. 22, 597–603.

Bounoua, L., Zhang, P., Mostovoy, G., Thome, K., Masek, J., Imhoff, M., Shepherd, M., Quattrochi, D., Santanello, J., Silva, J., 2015. Impact of urbanization on US surface climate. Environ. Res. Lett. 10, 101001.

Brunsdon, C., McClatchey, J., Unwin, D.J., 2001. Spatial variations in the average rainfall-altitude relationship in Great Britain: an approach using geographically weighted regression. Int. J. Climatol. 21, 455–466.

Burnash, R.J.C., 1995. The NWS river Forecast system – catchment modeling. In: Singh, V.P. (Ed.), Computer Models of Watershed Hydrology, Revised edition. Water Resources Publications, Highlands Ranch, Colo. c1995. http://www.wrpllc.com/books/cmwh.html

Campra, P., Carcia, M., Canton, Y., Palacios-Orueta, A., 2008. Surface greenhouse cooling trends and negative radiative forcing due to land use change toward greenhouse farming in southeastern Spain. J. Geophys. Res. 113, D18109.

Cao, Q., Yu, D., Georgescu, M., Wu, J., 2016. Impacts of urbanization on summer climate in China: an assessment with coupled land-atmospheric modeling. J. Geophys. Res. Atmos. 121, 10505–10521.

Cavazos, T., Hewitson, B.C., 2005. Performance of NCEP-NCAR reanalysis variables in statistical downscaling of daily precipitation. Clim. Res. 28, 95–107.

Charles, S.P., Bari, M.A., Kitsios, A., Bates, B.C., 2007. Effect of GCM bias on downscaled precipitation and runoff projections for the Serpentine catchment, western Australia. Int. J. Climatol. 27, 1673–1690.

Chang, C.R., Li, M.H., Chang, S.D., 2007. A preliminary study on the local cool island intensity of Taipei city parks. Landsc. Urban Plan. 80 (4), 386–395.
Chen, L., Frauenfeld, O.W., 2016. Impacts of urbanization on future climate in China. Clim. Dyn. 47 (1-2), 345–357.
Chen, Y., Wong, N.H., 2006. Thermal benefits of city parks. Energy Build. 38, 105–120.
Chiew, F.H.S., Kirono, D.G.C., Kent, D.M., Frost, A.J., Charles, S.P., Timbal, B., Nguyen, K.C., Fu, G., 2010. Comparison of runoff modelled using rainfall from different downscaling methods for historical and future climates. J. Hydrol. 387 (1–2), 10–23.
Clarke, L.E., Edmonds, J.A., Jacoby, H.D., Pitcher, H., Reilly, J.M., Richels, R., 2007. Scenarios of greenhouse gas emissions and atmospheric concentrations. Sub-report 2.1a of Synthesis and Assessment Product 2.1. In: Climate Change Science Program and the Subcommittee on Global Change Research, Washington DC.
Collins, M., Knutti, R., Arblaster, J., Dufresne, J.-L., Fichefet, T., Friedlingstein, P., Gao, X., Gutowski, W.J., Johns, T., Krinner, G., Shongwe, M., Tebaldi, C., Weaver, A.J., Wehner, M., 2013. Long-term climate change: projections, commitments and irreversibility. In: Stocker, T.F., Qin, D., Plattner, G.-K., Tignor, M., Allen, S.K., Boschung, J., Nauels, A., Xia, Y., Bex, V., Midgley, P.M. (Eds.), Climate Change 2013: The Physical Science Basis. Contribution of Working Group I to the Fifth Assessment Report of the Intergovernmental Panel on Climate Change. Cambridge University Press, Cambridge, United Kingdom and New York, NY, USA.
Coulibaly, P., 2004. Downscaling daily extreme temperatures with genetic programming. Geophys. Res. Lett. 31, L16203.
Craigmile, P.F., Guttorp, P., 2011. Space-time modelling of trends in temperature series. J. Time Anal. 32, 378–395.
De, U.S., Rao, G.S.P., 2004. Urban climate trends - the Indian scenario. J. Ind. Geophys. Union 8, 199–203.
Dibike, Y.B., Coulibaly, P., 2005. Hydrologic impact of climate change in the Saugenay Watershed: comparison of downscaling methods and hydrologic models. J. Hydrol. 307, 144–163.
Diem, J.E., Mote, T.L., 2005. Interepochal changes in summer precipitation in the southeastern United States: evidence of possible urban effects near Atlanta, Georgia. J. Appl. Meteorol. 44, 717–730.
Dixon, P.G., Mote, T.L., 2003. Patterns and causes of Atlanta's urban heat island-initiated precipitation. J. Appl. Meteorol. 42, 1273–1284.
Duan, Q., Gupta, V.K., Sorooshian, S., 1993. A shuffled complex evolution approach for effective and efficient optimization. J. Optim. Theory Appl. 76 (3), 501–521.
Emmanuel, R., Johansson, E., 2006. Influence of urban morphology and sea breeze on hot humid microclimate: the case of Colombo, Sri Lanka. Clim. Res. 30, 189–200.
Eum, H.-I., Simonovic, S.P., 2012. Assessment on variability of extreme climatic events for the upper Thames river basin in Canada. Hydrol. Process. 26 (4), 485–499.
Fahey, B., Jackson, R.B., 1997. Hydrological impacts of converting native forests and grasslands to pine plantations, South Island, New Zealand. Agric. Forest Meteorol. 84, 69–82.
Fall, S., Niyogi, D., Pielke, R., Gluhovsky, A., Kalnay, E., Rochon, G., 2010a. Impacts of land use land cover on temperature trends over the continental United States: assessment using the North American Regional Reanalysis. Int. J. Climatol. 30, 1980–1993.
Fall, S., Diffenbaugh, N., Niyogi, D., Peilke Sr, R., Rochon, G., 2010b. Temperature and equivalent temperature over the United States (1979–2005). Int. J. Climatol. 30, 2045–2054.
Fast, J.D., Torcolini, J.C., Redman, R., 2005. Pseudovertical temperature profiles and the urban heat island measured by a temperature datalogger network in Phoenix, Arizona. J. Appl. Meteorol. 44, 3–13.

Fealy, R., Sweeney, J., 2007. Statistical downscaling of precipitation for a selection of sites in Ireland employing a generalized linear modeling approach. Int. J. Climatol. 27, 2083–2094.
Fowler, H.J., Blenkinsop, S., Tebaldi, C., 2007. Review: linking climate change modeling to impacts studies: recent advances in downscaling techniques for hydrological modeling. Int. J. Climatol. 27, 147–178.
Friederichs, P., Hense, A., 2007. Statistical downscaling of extreme precipitation events using censored quantile regression. Mon. Weather Rev. 135, 2365–2378.
Friedl, M.A., Sulla-Menashe, D., Tan, B., Schneider, A., Ramankutty, N., Sibley, A., Huang, X., 2010. MODIS Collection 5 global land cover: algorithm refinements and characterization of new datasets. Remote Sens. Environ. 114, 168–182.
Frost, A.J., Charles, S.P., Timbal, B., Chiew, F.H.S., Mehrotra, R., Nguyen, K.C., Chandler, R.E., McGregor, J.L., Fu, G., Kirono, D.G.C., Fernandez, E., Kent, D.M., 2011. A comparison of multi-site daily rainfall downscaling techniques under Australian conditions. J. Hydrol. 408 (1–2), 1–18.
Fujibe, F., Togawa, H., Sakata, M., 2009. Long-term change and spatial anomaly of warm season afternoon precipitation in Tokyo. Sci. Online Lett. Atmos. 5, 17–20.
Fujino, J., Nair, R., Kainuma, M., Masui, T., Matsuoka, Y., 2006. Multigas mitigation analysis on stabilization scenarios using aim global model. Energy J. 343–354. Special issue #3.
Gaur, A., Simonovic, S.P., 2015. Towards reducing climate change impact assessment process uncertainty. Environ. Process. 2 (2), 275–290.
Gaur, A., Simonovic, S.P., 2016a. Accessing vulnerability of land-cover types to climate change using physical scaling downscaling model. Int. J. Climatol. https://doi.org/10.1002/joc.4887.
Gaur, A., Simonovic, S.P., 2016b. Extension of Physical Scaling method and its application towards downscaling climate model based near surface air temperature. Int. J. Climatol. https://doi.org/10.1002/joc.4921.
Gaur, A., Simonovic, S.P., 2018. Application of physical scaling towards downscaling climate model precipitation data. Theor. Appl. Climatol. 132 (1-2), 287–300.
Ge, J., 2010. MODIS observed impacts of intensive agriculture on surface temperature in the southern Great Plains. Int. J. Climatol. 30, 1994–2003.
Georgescu, M., et al., 2013. Summer-time climate impacts of projected megapolitan expansion in Arizona. Nat. Clim. Change 3, 37–41.
Ghosh, S., Mujumdar, P.P., 2008. Statistical downscaling of GCM simulations to streamflow using relevance vector machine. Adv. Water Resour. 31, 132–146.
Giorgi, F., Hurrell, J.W., Marinucci, M.R., Beniston, M., 1996. Elevation dependency of the surface climate change signal: a model study. J. Clim. 10, 288–296.
Giorgi, F., Hewitson, B.C., Christensen, J., Hulme, M., von Storch, H., Whetton, P., Jones, R., Mearns, L., Fu, C., 2001. Regional climate information – evaluation and projections. In: Houghton, J.T., Ding, Y., Griggs, D.J., Noguer, M., van der Linden, P.J. (Eds.), Climate Change 2001: The Scientific Basis. Contribution of Working Group I to the Third Assessment Report of the Intergovernmental Panel on Climate Change. Cambridge University Press, Cambridge, New York, USA, p. 881.
Grolemund, G., Wickham, H., 2011. Dates and times made easy with lubridate. J. Stat. Softw. 40 (3), 1–25.
Hale, R.C., Gallo, K.P., Owen, T.W., Loveland, T.R., 2006. Land use/land cover change effects on temperature trends at U.S. Climate Normals stations. Geophys. Res. Lett. 33, L11703.
Hale, R.C., Gallo, K.P., Loveland, T.R., 2008. Influences of specific land use/land cover conversions on climatological normals of near-surface temperature. J. Geophys. Res. 113, D14113.

Halfon, N., Levin, Z., Alpert, P., 2009. Temporal rainfall fluctuations in Israel and their possible link to urban and air pollution effects. Environ. Res. Lett. 4, 025001.

Hamdi, R., Van de Vyver, H., De Troch, R., Termonia, P., 2014. Assessment of three dynamical urban climate downscaling methods: Brussels's future urban heat island under an A1B emission scenario. Int. J. Climatol. 34, 978–999.

Han, J.Y., Baik, J.J., Lee, H., 2014. Urban impacts on precipitation. Asia-Pac. J. Atmos. Sci. 50 (1), 17–30.

Hashmi, M.Z., Shamseldin, A.Y., Melville, B.W., 2011. Comparison of LARS-WG and SDSM for simulation and downscaling of extreme precipitation events in a Watershed. Stoch. Environ. Res. Risk Assess. 25, 475–484.

Hertig, E., Jacobeit, J., 2008. Downscaling future climate change: temperature scenarios for the Mediterranean area. Global Planet. Change 63, 127–131.

Hewitson, B.C., Crane, R.G., 1996. Climate downscaling: techniques and application. Clim. Res. 7, 85–95.

Hijioka, Y., Matsuoka, Y., Nishimoto, H., Masui, T., Kainuma, M., 2008. Global GHG emission scenarios under GHG concentration stabilization targets. J. Global Environ. Eng. 13, 97–108.

Hijmans, R.J., 2015. Raster: Geographic Data Analysis and Modeling. R package version 2.5-2. https://CRAN.R-project.org/package=raster.

Hope, P., Timbal, B., Fawcett, R., 2010. Associations between rainfall variability in the southwest and southeast of Australia and their evolution through time. Int. J. Climatol. 30, 1360–1371.

Hopkinson, R.F., McKenney, D.W., Milewska, E.J., Hutchinson, M.F., Papadopol, P., Vincent, L.A., 2011. Impact of aligning climatological day on gridding daily maximum-minimum temperature and precipitation over Canada. J. Appl. Meteorol. Climatol. 50, 1654–1665.

Houghton, R.A., 1999. The annual net flux of carbon to the atmosphere from changes in land use 1850–1990. Tellus 51B, 298–313.

Hua, W., Chen, H., Sun, S., Zhou, L., 2015. Assessing climatic impacts of future land use and land cover change projected with the CanESM2 model. Int. J. Climatol. 35, 3661–3675.

Hughes, J.P., Guttorp, P., 1994. A class of stochastic models for relating synoptic atmospheric patterns to regional hydrologic phenomena. Water Resour. Res. 30, 1535–1546.

Hughes, J., Guttorp, P., Charles, S.P., 1999. A non-homogeneous hidden Markov model for precipitation occurrence. Appl. Stat. 48, 15–30.

Hurtt, G.C., Frolking, S., Fearon, M.G., Moore, B., Shevliakova, E., Malyshev, S., Pacala, S.W., Houghton, R.A., 2006. The underpinnings of land-use history: three centuries of global gridded land-use transitions, wood harvest activity, and resulting secondary lands. Global Change Biol. 12, 1208–1229.

Hurtt, G.C., et al., 2011. Harmonization of land-use scenarios for the period 1500–2100: 600 years of global gridded annual land-use transitions, wood harvest, and resulting secondary lands. Clim. Change 109 (1–2), 117–161.

Hutchinson, M.F., McKenney, D.W., Lawrence, K., Pedlar, J.H., Hopkinson, R.F., Milewska, E., Papadopol, P., 2009. Development and testing of Canada-wide interpolated spatial models of daily minimum-maximum temperature and precipitation for 1961–2003. J. Appl. Meteorol. Climatol. 48, 725–741.

Huth, R., 2002. Statistical downscaling of daily temperature in central Europe. J. Clim. 15, 1731–1742.

Inoue, T., Kimura, F., 2004. Urban effects on low-level clouds around the Tokyo metropolitan area on clear summer days. Geophys. Res. Lett. 31, L05103.

Jackson, R.B., Canadell, J., Ehleringer, J.R., Mooney, H.A., Sala, O.E., et al., 1996. A global analysis of root distributions for terrestrial biomes. Oecologia 108, 389–411.
Jansson, C., Jansson, P.E., Gustafsson, D., 2007. Near surface climate in an urban vegetated park and its surroundings. Theor. Appl. Climatol. 89 (3–4), 185–193.
Jarvis A., Reuter H. I., Nelson A., Guevara E., Hole-filled SRTM for the globe Version 4, available from the CGIAR-CSI SRTM 90m Database (http://srtm.csi.cgiar.org), 2008.
Jauregui, E., 1991. Influence of a large urban park on temperature and convective precipitation in a tropical city. Energy Build. 15 (3), 457–463.
Jonsson, P., 2004. Vegetation as an urban climate control in the subtropical city of Gaborone, Botswana. Int. J. Climatol. 24 (10), 1307–1322.
Kigobe, M., McIntyre, N., Wheater, H., Chandler, R., 2011. Multi-site stochastic modelling of daily rainfall in Uganda. Hydrol. Sci. J. 56 (1), 17–33.
Kikon, N., Singh, P., Singh, S.K., Vyas, A., 2016. Assessment of urban heat islands (UHI) of Noida City, India using multi-temporal satellite data. Sustain. Cities Soc. 22, 19–28.
King, L.M., Irwin, S., Sarwar, R., McLeod, A.I., Simonovic, S.P., 2012. The effects of climate change on extreme precipitation events in the upper Thames river basin: a comparison of downscaling approaches. Can. Water Resour. J. 37 (3), 253–274.
King, L.M., McLeod, I., Simonovic, S.P., 2016. Improved weather generator algorithm for multisite simulation of precipitation and temperature. J. Am. Water Resour. Assoc. 51 (5), 1305–1320.
Kishtawal, C.M., Niyogi, D., Tewari, M., Pielke Sr., R.A., Shepherd, J.M., 2010. Urbanization signature in the observed heavy rainfall climatology over India. Int. J. Climatol. 30, 1908–1916.
Kruger, E., Givoni, B., 2007. Outdoor measurements and temperature comparisons of seven monitoring stations: preliminary studies in Curitiba, Brazil. Build. Environ. 42 (4), 1685–1698.
Kuchar, L., 2004. Using WGENK to generate synthetic daily weather data for modelling of agricultural processes. Math. Comput. Simulat. 65, 69–75.
Kug, J.-S., Ahn, M.S., 2013. Impact of urbanization on recent temperature and precipitation trends in the Korean peninsula. Asia-Pac. J. Atmos. Sci. 49, 151–159.
Kumar, S., Dirmeyer, P.A., Merwade, V., DelSole, T., Adams, J.M., Niyogi, D., 2013. Land use/cover change impacts in CMIP5 climate simulations: a new methodology and 21st century challenges. J. Geophys. Res. Atmos. 118, 6337–6353.
Kusaka, H., Hara, M., Takane, Y., 2012. Urban Climate Projection by the WRF Model at 3-km Horizontal Grid Increment: Dynamical Downscaling and Predicting Heat Stress in the 2070's August for Tokyo, Osaka, and Nagoya Metropolises. J. Meteorol. Soc. Jpn. 90B, 47–63.
Kwon, H.H., Sivakumar, B., Moon, Y.I., Kim, B.S., 2011. Assessment of change in design flood frequency under climate change using a multivariate downscaling model and a precipitation-runoff model. Stoch. Environ. Res. Risk Assess. 25, 567–581.
Lawrence, P.J., Chase, T.N., 2010. Investigating the climate impacts of global land cover change in the Community Climate System Model (CCSM). Int. J. Climatol. 30, 2066–2087.
Lee, E., Chase, T.N., Rajagopalan, B., Barry, R.G., Biggs, T.W., Lawrence, P.J., 2009. Effects of irrigation and vegetation activity on early Indian summer monsoon variability. Int. J. Climatol. 29, 573–581.
Lee, T-c., Chan, K-y, Chan, H-s, Kok, M-h, 2011. Projections of extreme rainfall in Hong Kong in the 21st century. Acta Meteorol. Sin. 25 (6), 691–709.
Lee, T., Ouarda, T.B.M.J., Jeong, C., 2012. Nonparametric multivariate weather generator and an extreme value theory for bandwidth selection. J. Hydrol. 452–453, 161–171.
Li, Y., Smith, I., 2009. A statistical downscaling model for southern Australia winter rainfall. J. Clim. 22, 1142–1158.

Li, Z., He, Y., Theakstone, W.H., Wang, X., Zhang, W., Cao, W., Du, J., Xin, H., Chang, L., 2012. Altitude dependency of trends of daily climate extremes in southwestern China, 1961–2008. J. Geogr. Sci. 22, 416.

Li, X., Chen, H., Liao, H., Hua, W., Sun, S., Ma, H., Li, X., Gao, C., Zhu, S., 2017. Potential effects of land cover change on temperature extremes over Eurasia: current versus historical experiments. Int. J. Climatol. https://doi.org/10.1002/joc.4976.

Li, X., Jiang, F., Li, L., Wang, G., 2011. Spatial and temporal variability of precipitation concentration index, concentration degree and concentration period in Xinjiang, China. Int. J. Climatol. 31, 1679–1693.

Liu, X., Cheng, Z., Yan, L., Yin, Z.-Y., 2009. Elevation dependency of recent and future minimum surface air temperature trends in the Tibetan Plateau and its surroundings. Global Planet. Change 68 (3), 164–174.

Liu, Z., Xu, Z., Charles, S.P., Fu, G., Liu, L., 2011. Evaluation of two statistical downscaling models for daily precipitation over an arid basin in China. Int. J. Climatol. 31 (13), 2006–2020.

Lowry, W.P., 1977. Empirical estimation of the urban effects on climate: a problem analysis. J. Appl. Meteorol. 16, 129–135.

Lydolph, P.E., 1985. The Climate of the Earth. Rowman and Littlefield Publishers Inc.

Mahmood, R., Foster, S.A., Keeling, T., Hubbard, K.B., Carlson, C., Leeper, R., 2006. Impacts of irrigation on 20th century temperature in the northern Great Plains. Global Planet. Change 54, 1–18.

Malyshev, S., Shevliakova, E., Stouffer, R.J., Pacala, S.W., 2015. Contrasting local versus regional effects of land-use-change-induced heterogeneity on historical climate: analysis with the GFDL earth system model. J. Clim. 28, 5448–5469.

Manzanas, R., Brands, S., San-Martin, D., Lucero, A., Limbo, C., Gutierrez, J.M., 2015. Statistical downscaling in the tropics can be sensitive to reanalysis choice: a case study for precipitation in the Philippines. J. Clim. 28, 4171–4184.

Masson, D., Knutti, R., 2011. Spatial-scale dependence of climate model performance in the CMIP3 ensemble. J. Clim. 24, 2680–2692.

Mattiuzzi, M., 2016. MODIS: MODIS Acquisition and Processing. R package version 0.10-34/r510. https://R-Forge.R-project.org/projects/modis/.

McCarthy, M.P., Harpham, C., Goodess, C.M., Jones, P.D., 2012. Simulating climate change in UK cities using a regional climate model, HadRM3. Int. J. Climatol. 32 (12), 1875–1888.

McPherson, R.A., Stensrud, D.J., Crawford, K.C., 2004. The impact of Oklahoma's winter wheat belt on the mesoscale environment. Mon. Weather Rev. 132, 405–422.

Mesinger, F., et al., 2006. North American regional reanalysis. Bull. Am. Meteorol. Soc. 87, 343–360.

Mote, T.L., Lacke, M.C., Shepherd, J.M., 2007. Radar signatures of the urban effect on precipitation distribution: a case study for Atlanta, Georgia. Geophys. Res. Lett. 34, L20710.

Neale R. B., et al., Description of the NCAR Community Atmosphere Model CAM 5.0, NCAR Tech. Note NCAR/TN-486+STR, 2012, 289pp., Natl. Cent. For Atmos. Res., Boulder, Colo.

Nelder, J.A., Mead, R., 1965. A simplex method for function minimization. Comput. J. 7, 308–313.

Neset, T.-S., Opach, T., Lion, P., Lilja, A., Johansson, J., 2016. Map-based web tools supporting climate change adaptation. Prof. Geogr. 68 (1), 103–114.

Niyogi, D., Kishtawal, C., Tripathi, S., Govindaraju, R.S., 2010. Observational evidence that agricultural intensification and land use change may be reducing the Indian summer monsoon rainfall. Water Resour. Res. 46, W03533.

Nosetto, M.D., Jobbagy, E.G., Paruelo, J.M., 2005. Land-use change and water losses: the case of grassland afforestation across a soil textural gradient in central Argentina. Global Change Biol. 11, 1101–1117.

Offerle, B., et al., 2005. Urban modification of the surface energy balance in the West African Sahel: Ouagadougou, Burkina Faso. J. Clim. 18, 3983–3995.

Oke, T.R., 1982. The energetic basis of the urban heat island. Quart. J. R. Meteorol. Soc. 108, 1–24.

Oke, T.R., Crowther, J.M., McNaughton, K.G., Monteith, J.L., Gardiner, B., 1989. The micrometeorology of the urban forest [and discussion]. Philos. Trans. R. Soc. Lond. B 324 (1223), 335–349.

Pan, Y., Jackson, R.T., 2008. Ethnic difference in the relationship between acute inflammation and serum ferritin in US adult males. Epidemiol. Infect. 136, 421–431.

Pepin, N., Bradley, R.S., Dias, H.F., et al., 2015. Elevation-dependent warming in mountain regions of the world. Nat. Clim. Change 5, 424–430.

Pielke, R.A., Adegoke, J., Beltran-Przekurat, A., Hiemstra, C.A., Lin, J., Nair, U.S., Niyogi, D., Nobis, T.E., 2006. An overview of regional land-use and land-cover impacts on rainfall. Tellus 59B, 587–601.

Pitman, A.J., 2003. The evolution of, and revolution in, land surface schemes designed for climate models. Int. J. Climatol. 23, 479–510.

Pitman, A.J., Narisma, G.T., 2005. The role of land surface processes in regional climate change: a case study of future land cover change over south western Australia. Meteorol. Atmos. Phys. 89, 235–249.

Puvaneswaran, K.M., Smithson, P.A., 1991. Precipitation — elevation relationships over Sri Lanka. Theor. Appl. Climatol. 43, 113.

Qiao, Z., Guangjin, T., Lixiao, Z., Xu, X., 2014. Influences of urban expansion on urban heat island in Beijing during 1989–2010. Adv. Meteorol. 11, 187169.

Quesada, B., Devaraju, N., de Noblet-Ducoudré, N., Arneth, A., 2017. Reduction of monsoon rainfall in response to past and future land use and land cover changes. Geophys. Res. Lett. 44 https://doi.org/10.1002/2016GL070663.

Rao, V.B., Fernandez, J.P.R., Franchito, S.H., 2004. Quasi-stationary waves in the Southern Hemisphere during the El Niño and La Niña events. Ann. Geophys. 22, 789–806.

R Development Core Team, 2008. R: A Language and Environment for Statistical Computing. R Foundation for Statistical Computing, Vienna, Austria, ISBN 3-900051-07-0. http://www.R-project.org.

Räisänen, J., Ylhäisi, J.S., 2011. How much should climate model output be smoothed in space? J. Clim. 24, 867–880.

Riahi, K., Krey, V., Rao, S., Chirkov, V., Fischer, G., Kolp, P., Kindermann, G., Nakicenovic, N., Rafai, P., 2007. Exploring the consequence of high emission trajectories. Climatic Change 109, 133.

Richardson, C.W., 1981. Stochastic simulation of daily precipitation, temperature, and solar radiation. Water Resour. Res. 17, 182–190.

Robertson, A.W., Kirshner, S., Smyth, P., 2004. Downscaling of daily rainfall occurrence over northeast Brazil using a hidden Markov model. J. Clim. 17, 4407–4424.

Robertson, A.W., Ines, A.V.M., Hansen, J.W., 2007. Downscaling of seasonal precipitation for crop simulation. J. Appl. Meteorol. Climatol. 46, 677–693.

Rogerson, P.A., 2001. Statistical Methods for Geography. Sage, London.

Roy, S.S., Mahmood, R., Niyogi, D., Lei, M., Foster, S.A., Hubbard, K.G., Douglas, E., Pielke Sr., R., 2007. Impacts of the agricultural Green Revolution–induced land use changes on air temperatures in India. J. Geophys. Res. 112, D21108.

Salathé, E.P., 2003. Comparison of various precipitation downscaling methods for the simulation of streamflow in a rainshadow river basin. Int. J. Climatol. 23, 887–901.

Salathé, E.P., 2005. Downscaling simulations of future global climate with application to hydrologic modeling. Int. J. Climatol. 25, 419–436.

Samuels, R., Rimmer, A., Alpert, P., 2009. Effect of extreme rainfall events on the water resources of the Jordan River. J. Hydrol. 375 (3–4), 513–523.

Savic, D.A., Morley, M.S., Khoury, M., 2016. Serious gaming for water systems planning and management. Water 8 (10), 456.

Schlünzen, K.H., Hoffmann, P., Rosenhagen, G., Riecke, W., 2010. Long-term changes and regional differences in temperature and precipitation in the metropolitan area of Hamburg. Int. J. Climatol. 30, 1121–1136.

Schoof, J.T., Pryor, S.C., Robeson, S.M., 2007. Downscaling daily maximum and minimum air temperatures in the midwestern USA: a hybrid empirical approach. Int. J. Climatol. 27, 439–454.

Schoof, J.T., Pryor, S.C., Surprenant, J., 2010. Development of daily precipitation scenarios for the United States based on probabilistic downscaling. J. Geophys. Res. 115, D13106.

Schoof, J.T., 2012. Scale issues in the development of future precipitation scenarios. J. Contemp. Water Res. Educ. 147, 8–16.

Schoof, J.T., 2013. Statistical downscaling in climatology. Geogr. Compass 7, 249–265.

Segal, M., Pan, Z., Turner, R.W., Takle, E.S., 1998. On the potential impact of irrigation areas in North America on summer rainfall caused by large-scale systems. J. Appl. Meteorol. 37, 325–331.

Sellers, P.J., 1992. Biophysical models of land surface processes. In: Trenberth, K.E. (Ed.), Climate System Modeling. Cambridge University Press, pp. 451–490.

Semenov, M.A., Barrow, E.M., 1997. Use of a stochastic weather generator in the development of climate change scenarios. Climatic Change 35, 397–414.

Sharif, M., Burn, D.H., 2006. Simulating climate change scenarios using an improved K-Nearest neighbor model. J. Hydrol. 325, 179–196.

Sharif, M., Burn, D.H., 2007. Improved K-nearest neighbour weather generating model. J. Hydrol. Eng. 12, 42–51.

Sharma, D., Das Gupta, A., Babel, M.S., 2007. Spatial disaggregation of bias-corrected GCM precipitation for improved hydrologic simulation: Ping River Basin, Thailand. Hydrol. Earth Syst. Sci. 11, 1373–1390.

Shepherd, J.M., Pierce, H., Negri, A.J., 2002. Rainfall modification by major urban areas: observations from spaceborne rain radar on the TRMM satellite. J. Appl. Meteorol. 41, 689–701.

Smith, S.J., Wigley, T.M.L., 2006. MultiGas forcing stabilization with minicam. Energy J. 373–392. Special issue #3.

Soltani, A., Hoogenboom, G., 2003. Minimum data requirements for parameter estimation of stochastic weather generators. Clim. Res. 25, 109–119.

Spronken-Smith, R.A., Oke, T.R., 1998. The thermal regime of urban parks in two cities with different summer climates. Int. J. Remote Sens. 19 (11), 2085–2104.

Srivastav, R.K., Simonovic, S.P., 2015. Multi-site, multivariate weather generator using maximum entropy bootstrap. Clim. Dyn. 44, 3431–3448.

Stocker, T.F., et al., 2013. Summary for policymakers. In: Stocker, T.F., Qin, D., Plattner, G.K., Tignor, M., Allen, S.K., Boschung, J., Nauels, A., Xia, Y., Bex, V., Midgley, P.M. (Eds.), Climate Change 2013: The Physical Science Basis. Contribution of Working Group I to the Fifth Assessment Report of the Intergovernmental Panel on Climate Change. Cambridge University Press, Cambridge, United Kingdom and New York, NY, USA.

Stone, M., 1974. Cross-validatory choice and assessment of statistical predictions (with discussion). J. R. Stat. Soc. B 36, 111–147.

Sun, C.Y., 2011. A street thermal environment study in summer by the mobile transect technique. Theor. Appl. Climatol. 106 (3–4), 433–442.

Taylor, K.E., Stouffer, R.J., Meehl, G.A., 2012. An Overview of CMIP5 and the experiment design. Bull. Am. Meteorol. Soc. 93, 485–498.

Teutschbein, C., Seibert, J., 2012. Bias correction of regional climate model simulations for hydrological climate-change impact studies: review and evaluation of different methods. J. Hydrol. 456–457, 12–29.

Timbal, B., Hope, P., Charles, S., 2008. Evaluating the consistency between statistically downscaled and global dynamical model climate change projections. J. Clim. 20, 6052–6059.

Timbal, B., Jones, D.A., 2008. Future projections of winter rainfall in southeast Australia using a statistical downscaling technique. Climatic Change 86, 165–187.

Tripathi, S., Srinivas, V.V., Nanjundiah, R.S., 2006. Downscaling of precipitation for climate change scenarios: a support vector machine approach. J. Hydrol. 330, 621–640.

Tuck, S., Phillips, H., 2016. MODISTools: MODIS Subsetting Tools. R package version 0.95.0. http://cran.r-project.org/package=MODISTools.

Turner II, B.L., Clark, W.C., Kates, R.M., Richards, J.F., Mathews, J.T., Meyer, W.B. (Eds.), 1990. The Earth as Transformed by Human Action: Global and Regional Changes in the Biosphere over the Past 300 Years. Cambridge Univ Press, Cambridge, p. 713.

Van Vuuren, D.P., Eickhout, B., Lucas, P.L., den Elzen, M.G.J., 2006. Long-term multi-gas scenarios to stabilise radiative forcing—exploring costs and benefits within an integrated assessment framework. Energy J. 27, 201–233.

Van Vuuren, D.P., Den Elzen, M.G.J., Lucas, P.L., Eickhout, B., Strengers, B.J., Van Ruijven, B., Wonink, S., Van Houdt, R., 2007. Stabilizing greenhouse gas concentrations at low levels: an assessment of reduction strategies and costs. Climatic Change 81, 119–159.

Van Vuuren, D.P., Edmonds, J., Kainuma, M., et al., 2011. The representative concentration pathways: an overview. Climatic Change 109, 5–31.

Velazquez-Lozada, A., Gonzalez, J.E., Winter, A., 2006. Urban heat island effect analysis for San Juan, Puerto Rico. Atmos. Environ. 40, 1731–1741.

Voogt, J.A., Oke, T.R., 2003. Thermal remote sensing of urban climate. Remote Sens. Environ. 86, 370–384.

Walter, M.T., Brooks, E.S., McCool, D.K., King, L.G., Molnau, M., Boll, J., 2005. Process-based snowmelt modeling: does it require more input data than temperature-index modeling? J. Hydrol. 300 (1–4), 65–75.

Wan, Z., 2006. Collection-5 MODIS Land Surface Temperature Products Users Guide.

Wang, Y., Berardi, U., Akbari, H., 2015. The urban heat island effect in the city of Toronto. Procedia Eng. 118, 137–144.

Wickham, H., 2007. Reshaping data with the reshape package. J. Stat. Softw. 21 (12), 1–20.

Wickham, H., 2009. ggplot2: Elegant Graphics for Data Analysis. Springer-Verlag, New York.

Widmann, M., Bretherton, C.S., Salathé, E.P., 2003. Statistical precipitation downscaling over the Northwestern United States using numerically simulated precipitation as a predictor. J. Clim. 16, 799–816.

Wilby, R.L., Wigley, T.M.L., 1997. Downscaling general circulation model output: a review of methods and limitations. Prog. Phys. Geogr. 21, 530–548.

Wilby, R.L., Dawson, C.W., Barrow, E.M., 2002. SDSM – a decision support tool for the assessment of regional climate change impacts. Environ. Model. Softw. 17 (2), 145–157.

Wilby, R.L., Charles, S.P., Zorita, E., Timbal, B., Whetton, P., Mearns, L.O., 2004. Guidelines for Use of Climate Scenarios Developed from Statistical Downscaling Methods, Supporting Material of the Intergovernmental Panel on Climate Change. Available from: the DDC of IPCC TGCIA, 27.

Wilcoxon, F., 1945. Individual comparisons by ranking methods. Biometrics 1, 80–83.

Wilkinson, L., 2005. The grammar of graphics. In: Statistics and Computing, second ed. Springer, New York.

Wilks, D.S., 1999. Multisite downscaling of precipitation with a stochastic weather generator. Clim. Res. 11, 125–136.

Wilks, D.S., 2010. Use of stochastic weather generators for precipitation downscaling. WIREs Clim. Change 1 (6), 898–907.

Wilks, D.S., 2012. Stochastic weather generators for climate-change downscaling, part II: multivariable and spatially coherent multisite downscaling. WIREs Clim. Change 3, 267–278.

Wise, M., Calvin, K., Thomson, A., Clarke, L., Bond-Lamberty, B., Sands, R., Smith, S.J., Janetos, A., Edmonds, J., 2009. Implications of limiting CO2 concentrations for land use and energy. Science 324, 1183–1186.

Wood, S.N., 2000. Modelling and smoothing parameter estimation with multiple quadratic penalties. J. R. Stat. Soc. B 62 (2), 413–428.

Wood, A.W., Leung, L.R., Sridhar, V., Lettenmaier, D.P., 2004. Hydrologic implications of dynamical and statistical approaches to downscaling climate model outputs. Climatic Change 62, 189–216.

Wood, S.N., 2011. Fast stable restricted maximum likelihood and marginal likelihood estimation of semiparametric generalized linear models. J. R. Stat. Soc. 73 (1), 3–36.

Wong, N.H., Yu, C., 2005. Study of green areas and urban heat island in a tropical city. Habitat Int. 29 (3), 547–558.

Xue, Y., Shukla, J., 1996. The influence of land surface properties on Sahel climate. Part II: Afforestation. J. Clim. 9, 3260–3275.

Yan, H., Fan, S., Guo, C., Hu, J., Dong, L., 2014. Quantifying the impact of land cover composition on intra-urban air temperature variations at a mid-latitude city. PLoS One 9 (7), e102124.

Yan, L., Liu, X., 2014. Has climatic warming over the Tibetan Plateau paused or continued in recent years? J. Earth Ocean Atmos. Sci. 1, 13–28.

Yokobori, T., Ohta, S., 2009. Effect of land cover on air temperatures involved in the development of an intra-urban heat island. Clim. Res. 39 (1), 61–73.

Zhang, P., Bounoua, L., Thome, K., Wolfe, R., 2016. Modeling impact of urbanization in us cities using simple biosphere model SiB2. In: Geoscience and Remote Sensing Symposium (IGARSS), 2016 IEEE International. https://doi.org/10.1109/IGARSS.2016.7730764.

Zhao, C., Jiang, Q., Sun, Z., Zhong, H., Lu, S., 2013. Projected urbanization impacts on surface climate and energy budgets in the Pearl River Delta of China. Adv. Meteorol. 542086.

Zong-Ci, Z., Yong, L., Jian-Bin, H., 2013. Are there impacts of urban heat island on future climate change? Adv. Clim. Change Res. 4 (2), 133–136.

Zorita, E., von Storch, H., 1999. The analog method as a simple statistical downscaling technique: comparison with more complicated methods. J. Clim. 12, 2474–2489.

FURTHER READING

Cai, W., Santoso, A., Wang, G., Yeh, S.-W., An, S., Cobb, K.M., Collins, M., Guilyardi, E., Jin, F., Kug, J.-S., Lengaigne, M., McPhaden, M.J., Takahashi, K., Timmermann, A., Vecchi, G., Watanabe, M., Wu, L., 2015. ENSO and greenhouse warming. Nat. Clim. Change 5, 849–859.

Rizwan, A.M., Dennis, L.Y.C., Liu, C., 2008. A review on the generation, determination and mitigation of Urban Heat Island. J. Environ. Sci. 20 (1), 120–128.

Sitch, S., Brovkin, V., von Bloh, W., van Vuuren, D., Eickhout, B., Ganopolski, A., 2005. Impacts of future land cover changes on atmospheric CO2 and climate. Global Biogeochem. Cycles 19, GB2013.

CHAPTER 5

Trends and Changes in Streamflow With Climate

Ashish Sharma, Conrad Wasko
School of Civil and Environmental Engineering, University of New South Wales, Sydney, NSW, Australia

Contents

1. INTRODUCTION

With the advent of anthropogenic climate change, the long held notion that the hydrologic cycle is stationary is no longer tenable (Milly et al., 2008). Widespread changes in streamflow (runoff) are predicted with global warming (Collins et al., 2013). The primary drivers of changes in streamflow are changes in circulation patterns leading to differences in rainfall, and an increase in temperature resulting in greater evaporation and a drier

Trends and Changes in Hydroclimatic Variables
ISBN 978-0-12-810985-4
https://doi.org/10.1016/B978-0-12-810985-4.00005-0

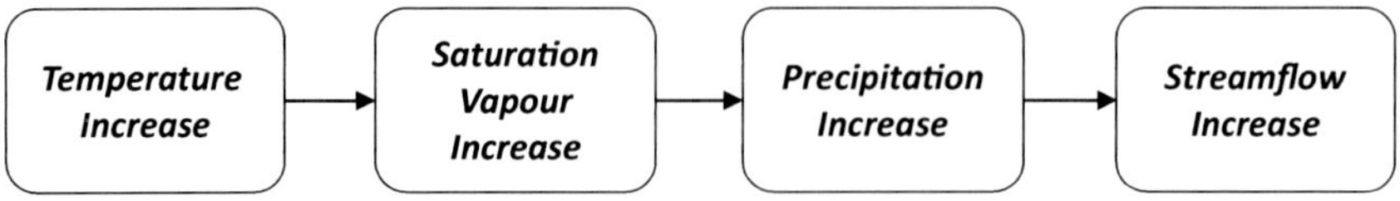

Figure 5.1 Simple conceptual physical basis for increases in streamflow resulting from higher temperature.

earth surface. Along with these, changes are also expected in precipitation and streamflow extremes, a key input in the design of most buildings, roads, drainage, and other engineering infrastructure that gives society a low-risk platform to exist and thrive in. This chapter aims to assess the extent of these changes, both from a water supply (volume) perspective, as well as from a design (extremes) perspective, assessing evidence using both observations recorded over time, as well as simulations of the future climate under assumptions of likely future greenhouse gas concentrations.

The simplest conceptualization of the impact of temperature increase on hydrology is best made while ignoring changes to atmospheric circulation. This conceptualization, presented in Fig. 5.1, shows that as temperature increases the saturation vapor pressure increases, which in turn results in greater moisture in the atmosphere. This, in general, results in greater precipitation, which results in more runoff volume being produced, having implications both from a water supply as well as a design point of view.

This physical reasoning, in the absence of other changes and a constant water balance, implies that evapotranspiration should also increase, resulting in what is commonly termed an intensification of the water cycle (Huntington, 2006). However, changes are not that straightforward. Firstly, this simple conceptualization ignores large-scale atmospheric circulation changes, as well as changes in major climatic anomalies such as the El Niño-Southern Oscillation. Secondly, there are many catchment-specific factors which are likely to impact streamflow in the future, such as anthropogenic changes to water courses or changes to land coverings needed to supply water to our ever increasing population. Thirdly, it ignores the changes in vegetation resulting from greater temperature, moisture, and rainfall, as well as their distribution in time and space, creating circumstances for changes in vegetation attributes, forcing changes in transpiration, as well as surface roughness. Fourthly, it ignores increase in drying resulting from an increased evapotranspiration, which in turn results in a decrease in overland flow volumes, as well as extremes. Finally, there are also changes in the mechanisms generating streamflow as higher temperatures result in less precipitation falling as snow, as well as changed snowmelt timing. The result is that on the global scale, unifying explanations of changes to streamflow are unlikely, and on a local scale, the myriad of factors affecting streamflow will inevitably result in variability which masks climate change signals.

The impact of changed streamflow is far-reaching. Reductions in streamflow, and in particular mean or low-flow events, puts pressure on freshwater resources and environments which depend on such water arriving with consistency year after year. This includes implications for biodiversity, as well as for water supply to existing and future communities. At the other end of the spectrum, high-flow events, typically termed floods, have a large damage cost and risk to life associated with them. As a result, understanding streamflow changes is critical for societies to both secure water availability as well as adequately design infrastructure to reduce the risk to life.

2. TRENDS IN MEAN PRECIPITATION AND STREAMFLOW VOLUME

The Intergovernmental Panel on Climate Change (IPCC) Fifth Assessment Report (AR5) was published in 2013 and forms the most comprehensive compendium of the scientific understanding of climate change impacts on the environment we live in. AR5 concluded that there is a low confidence for increasing trends in global river discharges (Hartmann et al., 2013). However, very large variations were noted spatially, as well as due to the methods applied in assessing trends. Given the uncertainty and variability in this assessment, it is necessary to look at the causative variables in order to decrease the uncertainty, and also to improve our ability to forecast future changes. One common issue that arises in assessing trends in streamflow is human intervention in the catchments or watersheds which transport the runoff. Development of artificial storages and diversions will reduce streamflow, while urbanization will reduce infiltration and increase discharge. Although these changes form part of the anthropogenic climate change which results in streamflow changes, from a climatic perspective it is necessary to identify discharges that are unaffected by catchment changes in order to capture any trends from climate nonstationarity. Generally, increases of precipitation with increases in temperature are expected to result in increases in runoff. The physical basis is first discussed before examining observed changes in precipitation as a precursor to changes in streamflow.

2.1 Physical Basis

An understanding of streamflow changes requires an understanding of the basis of the changes that occur. There are many local mechanisms responsible for generating streamflow in a catchment and there is no unifying theory that provides an explanation for the streamflow generation process (Brutsaert, 2005) and the relative contributions of each mechanism. The conceptualization discussed here, in its simplest form, assumes that the resultant streamflow depends on the intensity of the rainfall event and the overall water balance before the rainfall event takes place. This type of rainfall-runoff process is generally modeled through a lumped conceptual system (Boughton and Droop, 2003; Moore, 2007) where the streamflow is a function of precipitation and evaporation and a state variable representing the storage in the subsurface.

Precipitation falls on a catchment, conceptualized as a store of a particular shape, in this case triangular (Fig. 5.2). The storage fills as a function of the precipitation minus the evaporation. The evaporation is in turn a function of the moisture storage as the more water in storage, the greater the evaporation. As the storage fills and reaches capacity, the surface of the storage, any additional precipitation (minus evaporation) becomes streamflow. This streamflow is the runoff and the overland response to a precipitation event. Streamflow is additionally fed through a proportion of the water in the moisture storage termed as the base flow. The base flow is generally conceptualized as the streamflow in a waterway that is not directly related to a rainfall event. In this way, streamflow is positively related to precipitation, negatively related to evaporation, and in addition, related to

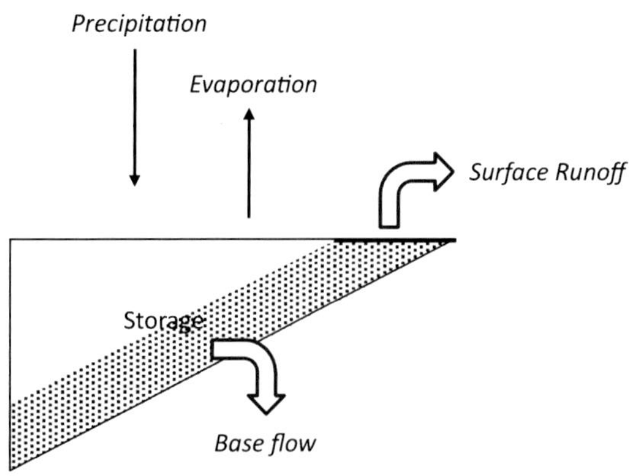

Figure 5.2 Conceptual diagram of streamflow generation.

the moisture in storage prior to the precipitation event. For example, if two identical precipitation events were to coincide closely in time, it might be assumed that the storage would be full from the first event resulting in greater surface runoff for the second event, as precipitation is no longer going to fill the moisture store. This conceptual model forms the basis of our discussions throughout this chapter. As the state variable storage is largely a function of the precipitation, the focus of the causative factors in this chapter remains on precipitation.

2.2 Precipitation Observations

Historical estimates of changes in mean precipitation present approximately a 2.4mm increase in annual rainfall per decade, averaged over the planet (Dai et al., 1997), though estimates vary depending on the period analyzed and dataset used (Table 5.1). A remarkable amount of variability exists globally (Allan et al., 2010; Zhang et al., 2007). Most of the increases appear to have taken place in the mid-latitudes of the northern hemisphere (Hartmann et al., 2013). For example, total rainfall has increased 7% over the last century over the United States (Groisman et al., 2004), which is in excess of the global average. It could be expected that with higher temperatures, snow would preferentially fall as precipitation (Trenberth, 2011) and this has been confirmed by many studies (Knowles et al., 2006; Serquet et al., 2011; Takeuchi et al., 2008) giving confidence to the fact that there is decreased snowfall in regions where increased temperatures are observed (Hartmann et al., 2013).

2.3 Streamflow Observations

There is much conflicting research on the historical changes in streamflow. Using the largest 925 rivers in the world, accounting for 73% of the global runoff, sites showing statistically significant decreases (45) outnumbered those showing increases (19) suggesting an overall drying trend worldwide (Dai et al., 2009). This was generally consistent with results from

Table 5.1 Global Precipitation Trends and 90% Confidence Intervals (Hartmann et al., 2013)

Dataset	1901–2008 (Trend in mm per Decade)
CRU TS3.10.01	2.77 ± 1.46
GHCN V2	2.08 ± 1.66
GPCC V6	1.48 ± 1.65

Milliman et al. (2008) where no historical trend in global discharge was found from 137 representative rivers representing nearly half the global discharge. These results are however in contrast to previous studies, which found overall increasing trends globally (Labat et al., 2004; Probst and Tardy, 1987). Despite global trends not displaying a consensus, consensus appears on where increases and decreases have occurred. Increases in streamflow have been observed at the highest northern latitudes and in many equatorial regions with decreases elsewhere (Huntington, 2006). Broadly, these streamflow trends follow historical rainfall patterns (see Fig. 5.3) but exhibit huge spatial variability (Table 5.2).

Many studies have confirmed that trends in streamflow follow trends in precipitation (Dai et al., 2009; Espinoza Villar et al., 2009; Groisman et al., 2001; McCabe and Wolock, 2002). However, exceptions remain, for example, the Arctic (Adam and Lettenmaier, 2008). Reasons for increases not being linked to precipitation include increased permafrost thawing (Adam and Lettenmaier, 2008; St. Jacques and Sauchyn, 2009) or increased groundwater discharge (Smith et al., 2007). It is expected that streamflow trends will be less than precipitation due to increased evapotranspiration (Adam and Lettenmaier, 2008) but global evapotranspiration appears to have decreased (Hartmann et al., 2013) as has soil moisture (Jung et al., 2010). A lack of a unifying global picture on streamflow changes necessitates a focus on local studies and these are summarized in Table 5.2.

The United States has shown generally increasing trends in mean flows, in contrast to Canada which has shown decreases, though not all the studies agree. Decreases in total streamflow are also exhibited in Australia, China, the Amazon, and southeast Europe. Northern Europe however, exhibits increases. For those studies that examined a range of flows (that is, low flows and high flows) it appears that when a trend is found, all flows indices exhibit the same positive or negative trend.

2.4 Predictions

Climate models predict increasing annual mean precipitation in most areas of the deep tropics and mid-latitudes, with decreases in large areas of the subtropics. This pattern is generally consistent with the expectation of an intensification of the hydrological cycle (Huntington, 2006) as the wet areas of the world get wetter and dry areas of the world get drier (Held and Soden, 2006). On a global scale, general circulation models predict

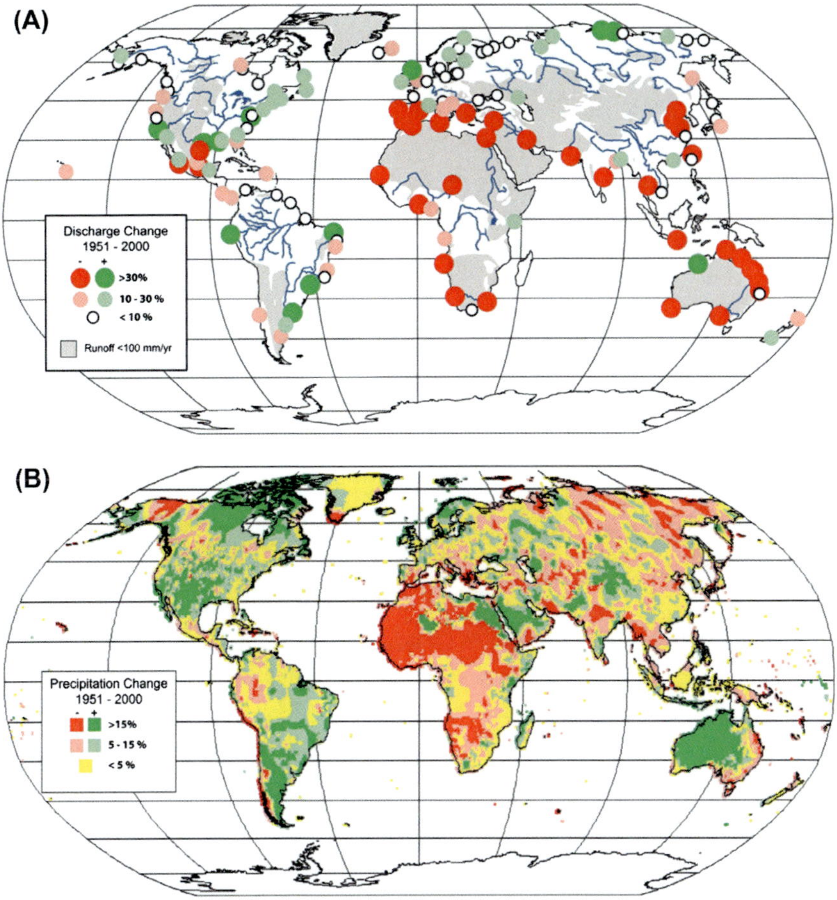

Figure 5.3 (A) 1951–2000 discharge trends for 137 global rivers, 109 of which are based on ≥45 years of data. With few exceptions (Kolyma, Lena, Fraser), <10% changes are not statistically significant. Shaded regions represent water-scarce areas, mean annual runoff <100 mm/year. (B) 1951–2000 global precipitation trends based on 0.5-degree grid cells; data from Climatic Research Unit (CRU), University of East Anglia. *(Reproduced from Fig. 2 (Milliman, J.D., Farnsworth, K.L., Jones, P.D., Xu, K.H., Smith, L.C., 2008. Climatic and anthropogenic factors affecting river discharge to the global ocean, 1951–2000.* Glob. Planet. Chang. *62, 187–194. https://doi.org/10.1016/j.gloplacha.2008.03.001).)*

approximately a 0.05 mm/day increase per degree change (Collins et al., 2013) or approximately 2% per degree Celsius (Allan et al., 2014; Kharin et al., 2013). But the fact that some areas are decreasing and some are increasing means that those areas that are increasing are exhibiting sensitivities much greater than 2% per degree.

Table 5.2 Summary of Selected Studies Examining Regional Streamflow Trends

References	Region	Trend
Lins and Slack (1999)	U.S.A	General increasing trend in low and mean flows with decreases only in isolated regions (Pacific Northwest and Southeast).
Groisman et al. (2001, 2004)	U.S.A	Increases in streamflow.
McCabe and Wolock (2002)	U.S.A	Increase in annual minimum and median daily streamflows.
Zhang et al. (2001)	Canada	Mean annual streamflow has generally decreased.
St. Jacques and Sauchyn (2009)	Canada	Using 23 stream gauges, with more than 30 years of record, in the Canadian northwest territories upward trend in winter baseflow mean annual flow.
Monk et al. (2011)	Canada	Decreasing trend in the annual low-flow indices.
Keim et al. (1995)	Louisiana	Increase in total runoff following precipitation trends.
Zhang et al. (2016)	Australia	Decreases in total streamflow in the south and increases in the north. This trend is generally replicated for low and high flows.
Peterson et al. (2002)	Arctic	Six largest Eurasian rivers show a total increase of 7% in discharge.
Aizen et al. (1997)	Central Asia (Tien Shan)	Decline in annual runoff.
Milly and Dunne (2001)	Mississippi	Increase in total runoff.
Piao et al. (2010) and Zhai et al. (2010)	China	Decreases in streamflow in most major basins with the exception of parts of the Yangtze River in the north.
Espinoza Villar et al. (2009)	Amazon	Discharge in 18 basins showed a decrease in low stage and mean runoff, but mixed results in the high stage runoff.
Giuntoli et al. (2013)	France	Decreases in low-flow indices present increasing drought conditions.
Smith et al. (2007)	Eurasia	Low flows have increases in 111 rivers and are more numerous and at a faster rate than corresponding increases in mean flow.
Stahl et al. (2010, 2012)	Europe	Negative annual streamflow trends in south and east, and positive elsewhere.

The changes predicted in precipitation follow General Circulation Model (GCM) model simulations for runoff to an extent, with decreases in streamflow predicted through large parts of Europe, Central U.S., Amazon, and southern Africa with increases elsewhere, particularly in the high northern latitudes and parts of the tropics (Collins et al., 2013; Milly et al., 2005). As evaporation is predicted to increase globally, it is expected that streamflow will not increase in proportion to precipitation. Predictions of evaporation minus precipitation exhibit a strong correlation with the predictions in runoff (Fig. 5.4). Despite great variability, these changes still translate to an overall increase in streamflow globally, if, for example, carbon dioxide concentrations are doubled (Cao et al., 2010).

3. TRENDS IN PRECIPITATION EXTREMES

Precipitation extremes form the basis of designing most civil engineering structures that are needed to withstand flooding and its associated impacts. Hence, any changes to these extremes have major implications on the infrastructure that is needed to contain resulting flood damage. Despite likely increases in extreme precipitation (Hartmann et al., 2013), the IPCC AR5 concludes there is little confidence in changes in sign, magnitude, or frequency of flooding. There is a general expectation, as in many climates precipitation is the primary driver of extreme flooding, that increases in floods will follow increases in extreme rainfalls (Seneviratne et al., 2012). However, despite a strong physical basis for extreme precipitation changes, there exists a significant problem in identifying trends in both precipitation (Kendon et al., 2008) and flooding (Hall et al., 2014) due to large variability, multiple factors influencing change (such as deforestation and urbanization), and the relatively short records that exist. Trend analysis is additionally hindered by short, poor quality, and spare gauge networks, both for precipitation and flows (Groisman et al., 2005). Hence an understanding of the potential changes in extreme floods is often largely based on the physical reasons for changes in precipitation. The broad definition of what is extreme varies, but from an engineering point of view, these are events that are large enough to be assigned a return interval, usually greater than or equal to 1 year.

3.1 Physical Basis

The fundamental principle that precipitation extremes, and hence floods, must change is steeped in thermodynamics. The Clausius-Clapeyron

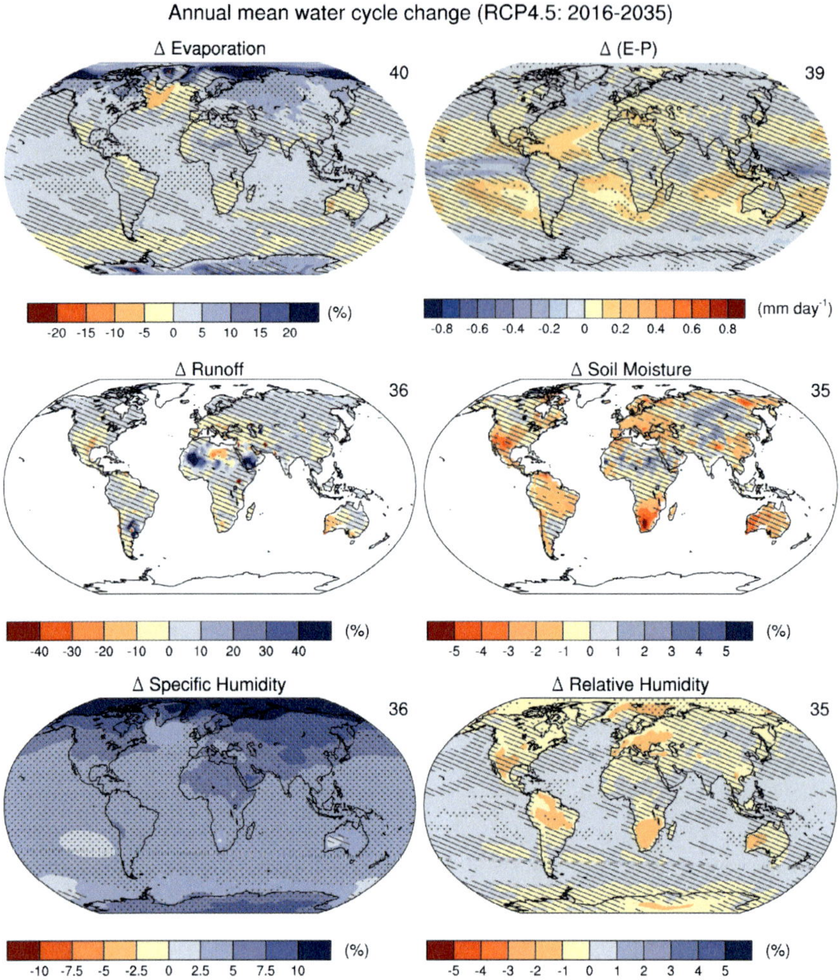

Figure 5.4 CMIP5 multimodel annual mean projected changes for the period 2016–35 relative to 1986–2005 under RCP4.5 for: (A) evaporation (%), (B) evaporation minus precipitation (E - P, mm/day), (C) total runoff (%), (D) soil moisture in the top 10 cm (%), (E) relative change in specific humidity (%), and (F) absolute change in relative humidity (%). The number of CMIP5 models used is indicated in the upper right corner of each panel. Hatching in left-hand panels indicates areas where projected changes are small compared to the internal variability (i.e., smaller than one standard deviation of estimated internal variability), and stippling indicates regions where the multimodel mean projections deviate significantly from the simulated 1986–2005 period (by at least two standard deviations of internal variability) and where at least 90% of the models agree on the sign of change. The number of models considered in the analysis is listed in the top-right portion of the panels; from each model one ensemble member is used. *(Reproduced from Fig. 11.14 (Kirtman, B., Power, S., Adedoyin, J., Boer, G., Bojariu, R., Camilloni, I., Doblas-Reyes, F., Fiore, A., Kimoto, M., Meehl, G., Prather, M., Sarr, A., Schär, C., Sutton, R., van Oldenborgh, G., Vecchi, G., Wang, H., 2013. Near-term climate change: projections and predictability. In: Stocker, T., Plattner, G-K., Tignor, M., Allen, S., Boschung, J., Nauels, A., Xia, Y., Bex, V., Midgley, P. (Eds.), Climate Change 2013: The Physical Science Basis. Contribution of Working Group I to the Fifth Assessment Report of the Intergovernmental Panel on Climate Change. Cambridge University Press, Cambridge, United Kingdom and New York, NY, USA, 953–1028).)*

relation can be used to describe the equilibrium pressure between water vapor and liquid water as a function of temperature. As temperature increases, so does the equilibrium pressure resulting in a greater fraction of water (in a closed system) existing in the vapor phase. Practically, the result is, as the earth's temperature increases there is a potential for more water to exist in the atmosphere in the vapor phase. The rate of this increase depends on the temperature of the system, but approximates to between 6% and 7% per degree Celsius between 0 and 24 degree Celsius (Westra et al., 2014). Hence, it is postulated that if an extreme precipitation event occurs due to a saturated atmosphere, assuming no changes in relative humidity, extreme precipitation will increase at a similar rate to the Clausius-Clapeyron rate with global warming (Trenberth, 2011; Trenberth et al., 2003).

3.2 Historical Trends

Global studies of extreme precipitation have overwhelmingly shown increases on a global scale (Table 5.3) with these increases at times faster than the rate of increase for average precipitation (Alexander et al., 2006). Using a number of high-quality daily precipitation observations, a global gridded dataset of extreme precipitation and temperature indices (HadEX) was produced for the years 1951–2003. The extreme precipitation indices

Table 5.3 Summary of Selected Manuscripts Presenting Global Extreme Precipitation Trends

References	Data	Remarks
Alexander et al. (2006)	Daily extreme indices (HadEX)	Tendency to wetter conditions
Donat et al. (2013)	Daily extreme indices (HadEX2)	Tendency to wetter conditions
Westra et al. (2013)	Daily annual maxima	Two-thirds of the stations show increase and these increases are shown to be associated with temperature
Groisman et al. (2005)	Daily 0.3% tail of regional averages across the world	Increases in extreme precipitation universally found
Barbero et al. (2017)	Subdaily U.S. rainfall	Association with temperature is less for subdaily extremes than daily extremes
Wentz et al. (2007)	Global satellite observations	Increases similar to Clausius-Clapeyron

considered showed an overall trend to wetter conditions. For example, of all the grid cells analyzed 7% showed a statistically significant increase in the annual maximum 1-day precipitation while 2.7% showed a decrease (Alexander et al., 2006). This analysis was extended to 1901–2010 (HadEX2), and despite great spatial variability, similar results were found (Donat et al., 2013). A similar study using annual maxima from worldwide gauge data presented 10.0% of stations globally had statistically significant positive associations with the annual global mean near-surface temperature and 2.2% had significant negative associations. The rate of change was found to be between 5.9% and 7.7% per degree Celsius (Westra et al., 2013).

There have been limited studies looking at subdaily rainfall on a large scale, primarily due to a lack of available records. However, the most recent findings, using hourly maxima from 733 conterminous United States, found statistically significant increases at approximately 5% of the stations (Barbero et al., 2017). However, their sensitivity to temperature was recorded at ~4% per degree Celsius (Barbero et al., 2017) less than what many studies report for daily precipitation maxima (Westra et al., 2013), which is contrary to expectation (Wasko et al., 2015). A confounding factor is possibly the observation of the increasing frequency of short duration events (Molnar et al., 2015). There is, however, evidence to suggest that subdaily increase in precipitation is likely to be of a greater magnitude as compared to that at a daily scale (Hardwick Jones et al., 2010; Westra and Sisson, 2011).

3.3 Precipitation–Temperature Sensitivities

Compared to the trend estimation studies above, a clearer argument can be made when assessing the relationship between temperature increase and precipitation extremes. The physical basis for changes in precipitation and the relation to temperature has led to a number of studies examining the historical relationship between precipitation and temperature (Berg et al., 2013; Lenderink and van Meijgaard, 2008; Wasko and Sharma, 2014). These studies plot the observed rainfall against the observed temperature on a daily basis and then form relationships between the precipitation extremes and their associated temperature (Wasko and Sharma, 2014). Such studies have a large advantage over historical trend studies, in that, by using all available historical data, they overcome the signal to noise ratio problem and circumvent the problem of climate models that are unable to capture the correct rainfall distribution (Stephens et al., 2010). However, there is no guarantee that these relationships may hold in a future climate despite evidence suggesting historical precipitation extremes follow historical temperature anomalies (Lenderink and Attema, 2015; Wasko and Sharma, 2017). However, if we put this controversy to one side, examining the historical sensitivity of extreme precipitation to temperature, and departures in this sensitivity from the postulated 6%–7% per degree Celsius, is informative to debating the mechanisms controlling rainfall changes in a future climate.

Fig. 5.5 presents the sensitivity of hourly precipitation extremes resulting from a degree rise in temperature, underlain by the sensitivity of the absolute humidity to temperature, using long-term high-quality rainfall and temperature records across all climate zones in Australia. This assessment indicates that there are increases in precipitation intensity related to the increase in moisture-holding capacity on warmer days. This argument is also consistent with the broad conceptualization presented in Fig. 5.1, which outlined that a warmer atmosphere will lead to more moisture being

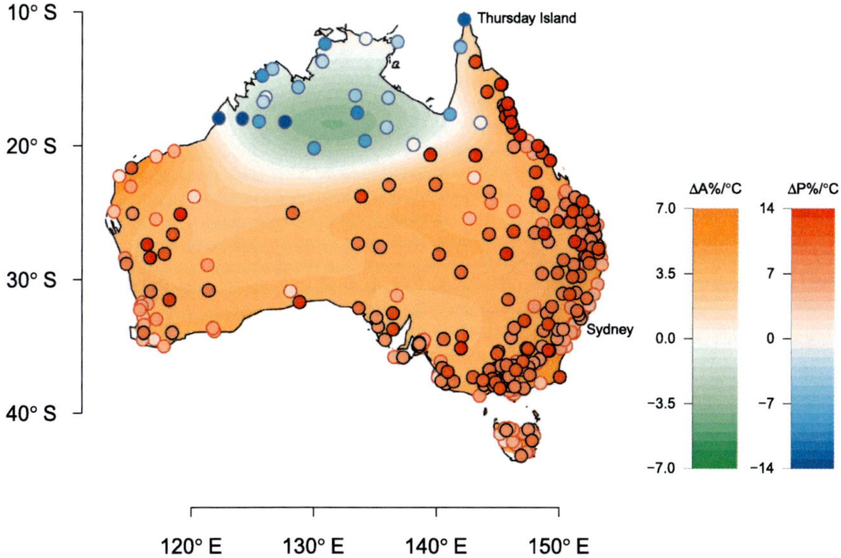

Figure 5.5 Scaling of 99th percentile hourly precipitation burst intensity (P) with temperature underlain by the median scaling of the absolute humidity (A) with temperature using quantile regression. Dots with black outline signify a precipitation burst intensity scaling in excess of 7% per degree Celsius. *(Reproduced from Wasko, C., Sharma, A., Johnson, F., 2015. Does storm duration modulate the extreme precipitation-temperature scaling relationship?* Geophys. Res. Lett. *42, 8783–8790. https://doi.org/10.1002/2015GL066274.)*

present, which in turn (in its extreme saturated state) will lead to more intense rainfall events and possibly larger flood peaks. However, this conceptualization requires the assumption that the moisture being supplied to this warmer atmosphere is locally created, and not coming from an external source as is the case with events such as fronts or cyclones. As this assumption can be argued to hold with greater confidence when considering short-duration rainfall events, such increases in rainfall are associated more with the shorter duration extremes that are responsible for localized and often urban flooding, than with longer duration events that could be of importance in regional scale flooding in larger river systems. Furthermore, even at the short durations of interest in Fig. 5.5, there are locations where the increase in intensity is not present. While reasons for this are many (external moisture sources for one and the difficulty in saturating a very large atmospheric moisture holding capacity in warmer regions another), this creates questions on the universality of the argument that precipitation extremes will increase in a warmer climate. Negative sensitivity leading to suggestion that atmospheric moisture may be limited at higher temperatures (Hardwick Jones et al., 2010; Prein et al., 2016; Wasko et al., 2015), particularly in the tropics (Utsumi et al., 2011; Wasko and Sharma, 2017), does not necessarily imply a cause-and-effect relationship. Rather than higher temperatures resulting in lower rainfall, it has been observed in dry regions, less rainfall is correlated with higher temperatures due to the surface heating up due to the generally dry conditions (Trenberth and Shea, 2005). Recent model studies suggest that the precipitation—temperature sensitivity curve simply shifts with increasing temperature (Drobinski et al., 2016; Prein et al., 2016) in line with or greater than the Clausius-Clapeyron relationship (Bao et al., 2017; Wang et al., 2017).

From an engineering point of view, an increase in precipitation extremes is only one of the many factors that can increase the resulting flood in a catchment. The other key factor that is used in engineering design is the pattern of the rainfall within the duration it is occurring in. This pattern is referred to as the design temporal pattern, and is used along with a design precipitation intensity to estimate the design flood magnitude. In a recent study, Wasko and Sharma (2015) assessed changes to the temporal patterns for varying storm durations using a long record of high-quality gauges representing all climate zones within Australia. Their findings, reproduced in Fig. 5.6, found that these temporal patterns were intensifying, or, the storms were getting "peakier" within the duration of interest, which in turn was having a significant impact on the resulting flood

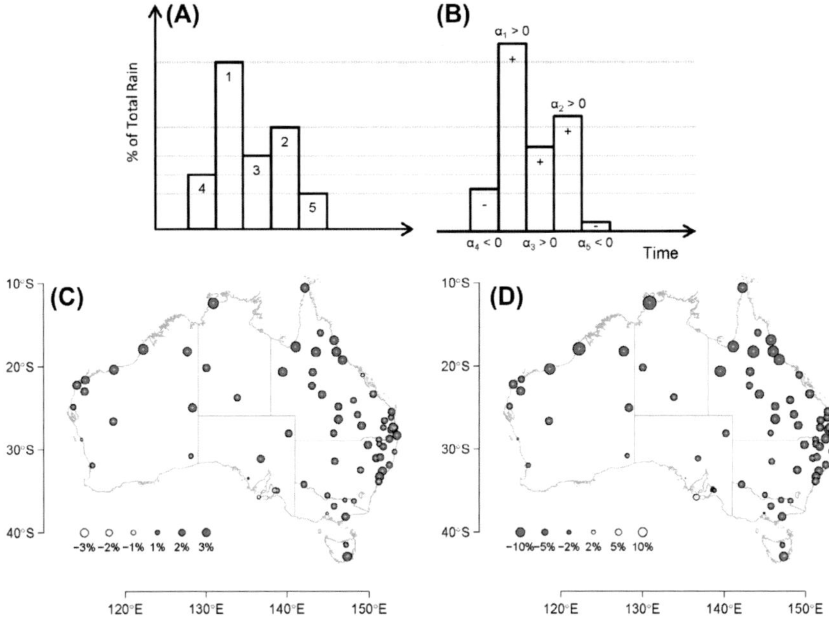

Figure 5.6 Changes in temporal patterns as a result of temperature increase (A) Typical hourly storm burst split into five 12-min segments (B) Hypothesis that the segments with the most rain will decrease while those with the least will decrease (C) Percentage change of the first segment with degree Celsius increase in temperature (D) Percentage change of the last segment with degree Celsius increase in temperature (Wasko and Sharma, 2015).

estimate. What made this finding special was that (1) there was a universal impact of rising temperatures across all gauges analyzed, and (2) this was the clearest impact of a warming climate that had been observed on rainfall extremes till date, with past studies (such as the results in Fig. 5.5) focusing on intensity and resulting in both increases and decreases depending on the location and durations being analyzed.

In a follow-up study, Wasko et al. (2016) hypothesized that the intensification of extreme storms is also occurring in space, with the spatial extent of a warmer storm being smaller than would be the case under colder conditions. This, in turn, is expected to further increase the resulting flood, and will result in even greater magnitudes once all three changes (intensity, temporal, and spatial) are factored in. While the above changes are clearest for shorter-duration rainfall events where the moisture can be hypothesized to be resulting from the warmer atmosphere creating the event, similar arguments are applicable even when moisture sources are not local.

3.4 Predictions

Prediction of precipitation is split between three main approaches: (1) extrapolating historical trends; (2) extrapolating historical precipitation-temperature sensitivities based on model temperature predictions; or (3) using GCM results directly, subject to bias correction or downscaling. As the previous two sections addressed the first two possibilities, this section briefly summarizes GCM predictions.

The general consensus of model studies is that precipitation extremes increase with a sensitivity close to the Clausius-Clapeyron scaling rate (see, for example, Lau et al., 2013), despite predicted decreases in the relative humidity (O'Gorman and Muller, 2010). There is agreement between models, on a global level, that precipitation extremes increase and the increase is exacerbated as greenhouse gas emissions increase (Sillmann et al., 2013). But these results vary widely from region to region, with decreases, for example, predicted in Central America, South Africa, and Australia. These decreases are predicted alongside increases in consecutive dry days suggesting drying conditions rather than wetter conditions. The greatest wetting trends are observed in the northernmost latitudes (Sillmann et al., 2013) and tropics (Lau et al., 2013), which are consistent with the trends and predictions in mean precipitation and streamflow presented earlier (Fig. 5.4).

Examining more extreme events, the return period of the 20-year Annual Recurrence Interval (ARI) daily precipitation is almost consistently predicted to decrease globally, to 12.3 years on average by about the mid-century (Kharin et al., 2013) and to 10 years or even more frequently by 2100 (Collins et al., 2013). Contrary to the spatially variable results suggested in the above paragraph, these results are consistent globally suggesting that the more extreme the rainfall the more consistent the changes become. Indeed, increases in the 20-year rainfall are also predicted at an average rate of approximately 6% per degree increase in warming (Kharin et al., 2013).

4. STREAMFLOW EXTREMES

The main cause of changes to flooding is either increases in extreme precipitation (Smith and Ward, 1998) or increases in snowmelt (Trenberth, 2011) as a result of increased temperatures. As stated previously, the IPCC AR5 concluded there is little confidence in changes in the sign, magnitude, or frequency of flooding. This is despite clear evidence of increasing precipitation extremes, a strong physical basis for their increase, and predictions of widespread increase in precipitation extremes. Because flood hazard is influenced by the catchment-specific conditions on which the precipitation happens to fall and runoff, changes in extreme precipitation may not translate to changes in flooding. Johnson et al. (2016) suggest the resultant flood is influenced by the:

- Location of the precipitation
- Pattern of precipitation
- Duration of precipitation
- Wetness state of catchment prior to the event
- Hydraulic characteristics of the catchment

And hence changes in all of these will affect the resultant flood. Floods are consistently the most hazardous extreme event the world faces. Hence a knowledge of how flooding will change with climate change is of upmost importance; however, there remains a paucity of large-scale studies investigating trends in flooding.

4.1 Physical Basis

As in many climates precipitation is the primary driver of extreme flooding, it could be expected that trends in historical precipitation extremes should follow trends in streamflow. However, this assumes that there are no changes in all the other factors possibly affecting the flood event; in particular, nonstationarity in the catchment conditions. To consider all possible interactions between meteorological and catchment conditions is very challenging. To focus on climatic nonstationarity, for the time being, we ignore catchment nonstationarity induced by manmade catchment intervention. If we return to the lumped conceptual model approach presented in Fig. 5.2, then the flood is a function of the precipitation, evaporation, and soil moisture stores. As extremes are being considered, it is generally assumed that the evaporation and soil moisture store play a secondary role to precipitation, and hence flood trends are generally expected to follow precipitation trends even more strongly than mean changes.

4.2 Historical Trends

Historical trend studies suffer from the already discussed issues of paucity in records and large natural variability in the data. Moreover, there is an inherent difficulty in identifying catchments that are free from human development as any human development confounds any climate signal.

In addition, most historical trend studies focus on annual maxima series, when it is easier to identify nonstationarity using flood peaks rather than block maxima (Villarini et al., 2009a). Using 774 stream gauge stations across the United States (158) 20% of these stations showed statistically significant changes in annual maxima with the majority (101 out of 158) showing increases. This is in comparison to 34% presenting increasing flood frequency and 9% decrease in frequency when a peak over threshold approach is used (Mallakpour and Villarini, 2015). Despite this recent evidence of increases in flooding, it is not consistent with previous studies, which presented more evidence for a decrease in extremes than an increase (Lins and Slack, 1999). Groisman et al. (2001) also failed to find increases in peak streamflow in the United States and McCabe and Wolock (2002) found mixed evidence for increases in annual maxima. A global study of flood trends also found more evidence for decreases in flooding compared to increases (Do et al., 2017). This conflicting evidence for data-rich parts of the world shows the difficulty in finding trends in flooding and how generalization of results is not possible (Archfield et al., 2016). It is worth commenting that an increasing frequency of flood events would be equivalent to an increase in flood peak from a design point of view, for both a reduced Average Recurrence Interval (ARI) of a particular event, or, complementarily increase the magnitude of a certain ARI. However, from an ecological perspective the frequency of flows may be more important.

A summary of selected localized studies is presented in Table 5.4. There is little uniform trend across these studies. Recent reviews suggest increasing flood trends throughout the UK and northern Europe, but decreases in the south (Hall et al., 2014) and increases in flooding in the United States, but this is conflicting (Archfield et al., 2016). In other regions of the world, such as Australia, decreases are more likely. There does however appear a consistent shift to earlier flooding in spring due to increased temperatures (Trenberth, 2011). Despite difficulty in identifying a unifying pattern in flooding, increasing flood risk has been attributed to climatic change (Pall et al., 2011) as have more intense precipitation extremes (Min et al., 2011).

4.3 Predicted Changes in Flooding

As flooding is a local-scale phenomena resulting from precipitation, this necessitates the use of a hydrologic model to predict the impact of changes in precipitation on discharge from bias corrected GCM precipitation. The general expectation is floods will increase with climate change

Table 5.4 Selected Regional Manuscripts Investigating Trends in Annual Maxima

References	Region	Remarks
Villarini et al. (2009a)	USA	No statistically significant trend.
Mallakpour and Villarini (2015)	USA	Increasing frequency of floods rather than changing peaks.
Cunderlik and Ouarda (2009)	Canada	Shift to earlier spring snowmelt floods. Decreased magnitude of snowmelt floods.
Shiklomanov et al. (2007)	Russia/Artic	Equal numbers of significant positive and negative trends. Shift to earlier spring discharge.
Lindström and Bergström (2004)	Sweden	Increasing (nonsignificant) trend.
Petrow and Merz (2009)	Germany	Many more positive than negative trends.
Renard et al. (2008)	France	Increasing flood peaks in north, decreasing in the Pyrenees. Earlier snowmelt flooding.
Mediero et al. (2014)	Spain	Decreasing trends in magnitude and frequency.
Hannaford and Marsh (2008)	UK	Significant upward trend.
Bhutiyani et al. (2008)	Himalayas	Significant upward trend.
Ishak et al. (2013) and Zhang et al. (2016)	Australia	More gauges show statistically significant decreasing trends than increasing trends.

(Milly et al., 2002). Though catchment scale studies exist (Cameron, 2006; Kay et al., 2009; Prudhomme and Davies, 2009), here we focus on global trends. One particular scenario, using a global hydrologic model on a 0.5×0.5 degree grid resolution, predicted that by 2050 the current 100-year flood could occur at least twice as frequently across 40% of the globe (Arnell and Gosling, 2016). A similar study, using nine global hydrological models, found that the 30-year return level flood could be occurring more than once in 5 years across 5%–30% of the global land mass (Dankers et al., 2014). However, there is large disagreement between models and large variability in the results, with localized decreases predicted across Australia, Brazil, USA, and Europe (Arnell and Gosling, 2016; Dankers et al., 2014). Other studies suggest increases worldwide with the exception of much of Europe and North America (Hirabayashi et al., 2013). In general, it appears that a greater proportion of global flood-prone population is exposed to an increase in flood frequency than a decrease.

The hydrologic modeling suggests that higher temperatures lead to greater evaporation, offsetting increases in precipitation and also change in the proportion of precipitation falling as snow, thus reducing the flood as a result of snowmelt which also occurs earlier (Arnell and Gosling, 2016; Dankers et al., 2014). These studies suggest that changes in flooding are not able to be explained through changes in precipitation alone, with a more complex interaction of flood generating mechanisms resulting in change (Hirabayashi et al., 2008). It needs to be noted that the applicability of these results is limited. In smaller catchments, flooding occurs on shorter timescales, and where the longer term balance between precipitation and evaporation is not as relevant, flooding is expected to follow precipitation trends.

5. ADDITIONAL CONSIDERATIONS

Changes in precipitation are likely to be the largest causative variable affecting streamflow. However, trends in streamflow do not always follow precipitation trends, necessitating the investigation and consideration of additional factors. These other factors can be other climatic variables or catchment nonstationarity due to human intervention.

5.1 Treatment of Additional Climatic Variables

An analysis using 390 watersheds across the U.S.A found that the 99th percentile precipitation only results in 99th percentage discharge 36% of the time (Ivancic and Shaw, 2015). This implies that trends in precipitation will not always directly translate to flooding. The impact of antecedent conditions on flooding is undisputed with a modeling study using 45 different catchments of ranging size in southeast Australia demonstrating consistent flood underestimation by up to 30% when only 1 or 2 days of antecedent rainfall were considered (Pathiraja et al., 2012). It has been shown that the precipitation leading up to a peak precipitation event does not have to follow the same trend historically as the peak precipitation (Table 5.5) and indeed decreases in flooding have been related to increases in temperature suggesting a decrease in moisture stores prior to the precipitation event (Wasko and Sharma, 2017). This implies that changes in future flooding will not always simply follow changes in predicted peak precipitation (Wasko and Sharma, 2017) and changes to the overall rainfall (which would largely be similar to preextreme rainfall in Table 5.5) need to be factored in. Changes to the overall rainfall are more likely a result of

Table 5.5 Annual Maximum Rainfall Trends (Woldemeskel and Sharma, 2016)

Region	Extreme Rainfall Trend	Preextreme Antecedent Precipitation Trend (API)
North America	Positive	No Trend
South America	Positive	No Trend
Africa	Positive	Positive
Eurasia	Positive	Negative
Australia	Negative	Positive

changed circulation patterns, with regions that are wet getting wetter, and those dry getting drier, on average.

We discussed initially how changes in evaporation and the moisture storage will impact streamflow and flooding. Broadly, there is correlation between trends in precipitation and streamflow, but not always, suggesting something else must be changing also. However, there is much conflicting information about changes in evaporation and soil storage. Decreases in global evapotranspiration have been attributed to decreases in soil moisture (Albergel et al., 2013; Jung et al., 2010), as well as changes to wind speeds. But even then a small wetting trend in the global soil moisture and decreasing drought conditions have been observed (Sheffield and Wood, 2008), which is consistent with increases in mean precipitation.

5.2 Urbanization

There is a relative paucity of streamflow gauging information in urbanized areas, but it is clear from the limited studies and physical intuition that urbanization will increase runoffs (Villarini et al., 2009b). Urbanization is likely to not only increase pervious surface areas increasing the volume of runoff, but the construction of integrated storm water systems increases the rate at which the catchment is able to transmit runoff reducing the travel time of the catchment (Shuster et al., 2005). This double impact increases runoff peaks and volumes, in particular, creating a possibility of increases in flash flooding. In rural catchments, deforestation is also shown to increase runoff (Pathiraja et al., 2016; Peña-Arancibia et al., 2012).

Reservoir construction and increased consumptive use, for example, for irrigation can have the opposite effect and reduce streamflow (Vörösmarty and Sahagian, 2000). Finally, construction of urban cities has the effect of creating "heat islands," which increase the temperature of the air mass and, among other factors, cause precipitation intensification (Huff and Changnon, 1973).

6. CONCLUSIONS

This chapter sought to provide evidence to the changes that are occurring in the hydrological cycle as a result of climate change and a shift toward increasing urbanization. These changes were separated into changes to rainfall and streamflow volumes, as well as changes to rainfall and streamflow extremes. The assessment presented results for studies that have quantified such changes using observed rainfall and flow records, as well as projected their analysis into the future using a range of climate change model simulations.

The best evidence to date suggests that average precipitation will increase and average streamflow may respond accordingly. Precipitation extremes will likely increase at a rate close to the Clausius-Clapeyron relationship with possible increases above this depending on the location and the duration of the storm event of interest. Resultantly, flood magnitudes may also increase, although evidence of such changes in historical records is limited. There is some consensus however that streamflow due to snowmelt is starting earlier in the season due to higher temperatures and decreasing the subsequent streamflow magnitude due to lower snowfalls.

Given that streamflow and flooding are incredibly scale-dependent, and vary depending on local catchment characteristics, it is certain that hydrologic modeling will have to play a role in explaining how flooding has changed historically, and how it will change in a future climate. Further, as measurements of streamflow are relatively sparse in comparison to measurements of precipitation—and this becomes even worse when one considers streamflow measurement in urban catchments—numerical modeling to translate precipitation changes to streamflow will remain a cornerstone of research in this field.

Given the cost related to flood hazard, understanding changes to flood hazard is of utmost importance. Despite decades of research many open questions still remain. In particular, a holistic understanding of how changes in precipitation, evaporation, and soil moisture storage will all change together, contributing to changed streamflow and flooding, is still lacking, particularly at the catchment scale.

REFERENCES

Adam, J.C., Lettenmaier, D.P., 2008. Application of new precipitation and reconstructed streamflow products to streamflow trend attribution in northern Eurasia. J. Clim. 21, 1807–1828. https://doi.org/10.1175/2007JCLI1535.1.

Aizen, V.B., Aizen, E.M., Melack, J.M., Dozier, J., 1997. Climatic and hydrologic changes in the Tien Shan, Central Asia. J. Clim. 10, 1393–1404. https://doi.org/10.1175/1520-0442(1997)010<1393:CAHCIT>2.0.CO;2.

Albergel, C., Dorigo, W., Reichle, R.H., Balsamo, G., de Rosnay, P., Muñoz-Sabater, J., Isaksen, L., de Jeu, R., Wagner, W., 2013. Skill and global trend analysis of soil moisture from reanalyses and microwave remote sensing. J. Hydrometeorol. 14, 1259–1277. https://doi.org/10.1175/JHM-D-12-0161.1.

Alexander, L.V., Zhang, X., Peterson, T.C., Caesar, J., Gleason, B., Klein Tank, A.M.G., Haylock, M., Collins, D., Trewin, B., Rahimzadeh, F., Tagipour, A., Rupa Kumar, K., Revadekar, J., Griffiths, G., Vincent, L., Stephenson, D.B., Burn, J., Aguilar, E., Brunet, M., Taylor, M., New, M., Zhai, P., Rusticucci, M., Vazquez-Aguirre, J.L., 2006. Global observed changes in daily climate extremes of temperature and precipitation. J. Geophys. Res. 111, D05109. https://doi.org/10.1029/2005JD006290.

Allan, R.P., Liu, C., Zahn, M., Lavers, D.A., Koukouvagias, E., Bodas-Salcedo, A., 2014. Physically consistent responses of the global atmospheric hydrological cycle in models and observations. Surv. Geophys. 35, 533–552. https://doi.org/10.1007/s10712-012-9213-z.

Allan, R.P., Soden, B.J., John, V.O., Ingram, W., Good, P., 2010. Current changes in tropical precipitation. Environ. Res. Lett. 5, 25205. https://doi.org/10.1088/1748-9326/5/2/025205.

Archfield, S.A., Hirsch, R.M., Viglione, A., Blöschl, G., 2016. Fragmented patterns of flood change across the United States. Geophys. Res. Lett. 43, 10232–10239. https://doi.org/10.1002/2016GL070590.

Arnell, N.W., Gosling, S.N., 2016. The impacts of climate change on river flood risk at the global scale. Clim. Chang. 134, 387–401. https://doi.org/10.1007/s10584-014-1084-5.

Bao, J., Sherwood, S.C., Alexander, L.V., Evans, J.P., 2017. Future increases in extreme precipitation exceed observed scaling rates. Nat. Clim. Chang. 7, 128–132. https://doi.org/10.1038/nclimate3201.

Barbero, R., Fowler, H.J., Lenderink, G., Blenkinsop, S., 2017. Is the intensification of precipitation extremes with global warming better detected at hourly than daily resolutions? Geophys. Res. Lett. 1–10. https://doi.org/10.1002/2016GL071917.

Berg, P., Moseley, C., Haerter, J.O., 2013. Strong increase in convective precipitation in response to higher temperatures. Nat. Geosci. 6, 181–185. https://doi.org/10.1038/ngeo1731.

Bhutiyani, M.R., Kale, V.S., Pawar, N.J., 2008. Changing streamflow patterns in the rivers of northwestern Himalaya: implications of global warming in the 20th century. Curr. Sci. 95, 618–626.

Boughton, W., Droop, O., 2003. Continuous simulation for design flood estimation—a review. Environ. Model. Softw. 18, 309–318. https://doi.org/10.1016/S1364-8152(03)00004-5.

Brutsaert, W., 2005. Hydrology - An Introduction. Cambridge University Press, New York.

Cameron, D., 2006. An application of the UKCIP02 climate change scenarios to flood estimation by continuous simulation for a gauged catchment in the northeast of Scotland, UK (with uncertainty). J. Hydrol. 328, 212–226. https://doi.org/10.1016/j.jhydrol.2005.12.024.

Cao, L., Bala, G., Caldeira, K., Nemani, R., Ban-Weiss, G., 2010. Importance of carbon dioxide physiological forcing to future climate change. Proc. Natl. Acad. Sci. U.S.A. 107, 9513–9518. https://doi.org/10.1073/pnas.0913000107.

Collins, M., Knutti, R., Arblaster, J., Dufresne, J.-L., Fichefet, T., Friedlingstein, P., Gao, X., Gutowski, W.J., Johns, T., Krinner, G., Shongwe, M., Tebaldi, C., Weaver, A.J., Wehner, M., 2013. Long-term climate change: projections, commitments and irreversibility. In: Stocker, T., Qin, D., Plattner, G.-K., Tignor, M., Allen, S.K.,

Boschung, J., Nauels, A., Xia, Y., Bex, V., Midgley, P.M. (Eds.), Climate Change 2013: The Physical Science Basis. Contribution of Working Group I to the Fifth Assessment Report of the Intergovernmental Panel on Climate Change. Cambridge University Press, Cambridge, UK and New York, NY, USA, pp. 1029–1136.

Cunderlik, J.M., Ouarda, T.B.M.J., 2009. Trends in the timing and magnitude of floods in Canada. J. Hydrol. 375, 471–480. https://doi.org/10.1016/j.jhydrol.2009.06.050.

Dai, A., Fung, I.Y., Del Genio, A.D., 1997. Surface observed global land precipitation variations during 1900–88. J. Clim. 10, 2943–2962. https://doi.org/10.1175/1520-0442(1997)010<2943:SOGLPV>2.0.CO;2.

Dai, A., Qian, T., Trenberth, K.E., Milliman, J.D., 2009. Changes in continental freshwater discharge from 1948 to 2004. J. Clim. 22, 2773–2792. https://doi.org/10.1175/2008JCLI2592.1.

Dankers, R., Arnell, N.W., Clark, D.B., Falloon, P.D., Fekete, B.M., Gosling, S.N., Heinke, J., Kim, H., Masaki, Y., Satoh, Y., Stacke, T., Wada, Y., Wisser, D., 2014. First look at changes in flood hazard in the inter-sectoral impact model intercomparison project ensemble. Proc. Natl. Acad. Sci. U.S.A. 111, 3257–3261. https://doi.org/10.1073/pnas.1302078110.

Donat, M.G., Alexander, L.V., Yang, H., Durre, I., Vose, R., Dunn, R.J.H., Willett, K.M., Aguilar, E., Brunet, M., Caesar, J., Hewitson, B., Jack, C., Klein Tank, A.M.G., Kruger, A.C., Marengo, J., Peterson, T.C., Renom, M., Oria Rojas, C., Rusticucci, M., Salinger, J., Elrayah, A.S., Sekele, S.S., Srivastava, A.K., Trewin, B., Villarroel, C., Vincent, L.A., Zhai, P., Zhang, X., Kitching, S., 2013. Updated analyses of temperature and precipitation extreme indices since the beginning of the twentieth century: the HadEX2 dataset. J. Geophys. Res. Atmos. 118, 2098–2118. https://doi.org/10.1002/jgrd.50150.

Do, H.X., Westra, S., Leonard, M., 2017. A global-scale investigation of trends in annual maximum streamflow. J. Hydrol. 552, 28–43. Online: http://linkinghub.elsevier.com/retrieve/pii/S0022169417304171.

Drobinski, P., Silva, N., Da, Panthou, G., Bastin, S., Muller, C., Ahrens, B., Borga, M., Conte, D., Fosser, G., Giorgi, F., Güttler, I., Kotroni, V., Li, L., Morin, E., Önol, B., Quintana-Segui, P., Romera, R., Torma, C.Z., 2016. Scaling precipitation extremes with temperature in the Mediterranean: past climate assessment and projection in anthropogenic scenarios. Clim. Dyn. 1–21. https://doi.org/10.1007/s00382-016-3083-x.

Espinoza Villar, J.C., Guyot, J.L., Ronchail, J., Cochonneau, G., Filizola, N., Fraizy, P., Labat, D., de Oliveira, E., Ordoñez, J.J., Vauchel, P., 2009. Contrasting regional discharge evolutions in the Amazon basin (1974–2004). J. Hydrol. 375, 297–311. https://doi.org/10.1016/j.jhydrol.2009.03.004.

Giuntoli, I., Renard, B., Vidal, J.-P., Bard, A., 2013. Low flows in France and their relationship to large-scale climate indices. J. Hydrol. 482, 105–118. https://doi.org/10.1016/j.jhydrol.2012.12.038.

Groisman, P.Y., Knight, R.W., Easterling, D.R., Karl, T.R., Hegerl, G.C., Razuvaev, V.N., 2005. Trends in intense precipitation in the climate record. J. Clim. 18, 1326–1350. https://doi.org/10.1175/JCLI3339.1.

Groisman, P.Y., Knight, R.W., Karl, T.R., 2001. Heavy precipitation and high streamflow in the contiguous United States: trends in the twentieth century. Bull. Am. Meteorol. Soc. 82, 219–246. https://doi.org/10.1175/1520-0477(2001)082<0219:HPAHSI>2.3.CO;2.

Groisman, P.Y., Knight, R.W., Karl, T.R., Easterling, D.R., Sun, B., Lawrimore, J.H., 2004. Contemporary changes of the hydrological cycle over the contiguous United States: trends derived from in situ observations. J. Hydrometeorol. 5, 64–85. https://doi.org/10.1175/1525-7541(2004)005<0064:CCOTHC>2.0.CO;2.

Hall, J., Arheimer, B., Borga, M., Brázdil, R., Claps, P., Kiss, A., Kjeldsen, T.R., Kriaučiūnienė, J., Kundzewicz, Z.W., Lang, M., Llasat, M.C., Macdonald, N., McIntyre, N., Mediero, L., Merz, B., Merz, R., Molnar, P., Montanari, A., Neuhold, C., Parajka, J., Perdigão, R.A.P., Plavcová, L., Rogger, M., Salinas, J.L., Sauquet, E., Schär, C., Szolgay, J., Viglione, A., Blöschl, G., 2014. Understanding flood regime changes in Europe: a state-of-the-art assessment. Hydrol. Earth Syst. Sci. 18, 2735–2772. https://doi.org/10.5194/hess-18-2735-2014.

Hannaford, J., Marsh, T.J., 2008. High-flow and flood trends in a network of undisturbed catchments in the UK. Int. J. Climatol. 28, 1325–1338. https://doi.org/10.1002/joc.1643.

Hardwick Jones, R., Westra, S., Sharma, A., 2010. Observed relationships between extreme sub-daily precipitation, surface temperature, and relative humidity. Geophys. Res. Lett. 37, L22805. https://doi.org/10.1029/2010GL045081.

Hartmann, D.L., Klein Tank, A.M.G., Rusticucci, M., Alexander, L.V., Broennimann, S., Charabi, Y., Dentener, F.J., Dlugokencky, E.J., Easterling, D.R., Kaplan, A., Soden, B.J., Thorne, P.W., Wild, M., Zhai, P.M., 2013. Observations: atmosphere and surface. In: Stocker, T., Qin, D., Plattner, G.-K., Tignor, M., Allen, S.K., Boschung, J., Nauels, A., Xia, Y., Bex, V., Midgley, P.M. (Eds.), Climate Change 2013: The Physical Science Basis. Contribution of Working Group I to the Fifth Assessment Report of the Intergovernmental Panel on Climate Change. Cambridge University Press, Cambridge, UK and New York, NY, USA, pp. 159–254.

Held, I.M., Soden, B.J., 2006. Robust responses of the hydrological cycle to global warming. J. Clim. 19, 5686–5699. https://doi.org/10.1175/JCLI3990.1.

Hirabayashi, Y., Kanae, S., Emori, S., Oki, T., Kimoto, M., 2008. Global projections of changing risks of floods and droughts in a changing climate. Hydrol. Sci. J. 53, 754–772. https://doi.org/10.1623/hysj.53.4.754.

Hirabayashi, Y., Mahendran, R., Koirala, S., Konoshima, L., Yamazaki, D., Watanabe, S., Kim, H., Kanae, S., 2013. Global flood risk under climate change. Nat. Clim. Chang. 3, 816–821. https://doi.org/10.1038/nclimate1911.

Huff, F.A., Changnon, S.A., 1973. Precipitation modification by major urban areas. Bull. Am. Meteorol. Soc. https://doi.org/10.1175/1520-0477(1973)054<1220:PMBMUA>2.0.CO;2.

Huntington, T.G., 2006. Evidence for intensification of the global water cycle: review and synthesis. J. Hydrol. 319, 83–95. https://doi.org/10.1016/j.jhydrol.2005.07.003.

Ishak, E.H., Rahman, A., Westra, S., Sharma, A., Kuczera, G., 2013. Evaluating the non-stationarity of Australian annual maximum flood. J. Hydrol. 494, 134–145. https://doi.org/10.1016/j.jhydrol.2013.04.021.

Ivancic, T.J., Shaw, S.B., 2015. Examining why trends in very heavy precipitation should not be mistaken for trends in very high river discharge. Clim. Chang. 133, 681–693. https://doi.org/10.1007/s10584-015-1476-1.

Johnson, F., White, C.J., van Dijk, A., Ekstrom, M., Evans, J.P., Jakob, D., Kiem, A.S., Leonard, M., Rouillard, A., Westra, S., 2016. Natural hazards in Australia: floods. Clim. Chang. 139, 21–35. https://doi.org/10.1007/s10584-016-1689-y.

Jung, M., Reichstein, M., Ciais, P., Seneviratne, S.I., Sheffield, J., Goulden, M.L., Bonan, G., Cescatti, A., Chen, J., de Jeu, R., Dolman, A.J., Eugster, W., Gerten, D., Gianelle, D., Gobron, N., Heinke, J., Kimball, J., Law, B.E., Montagnani, L., Mu, Q., Mueller, B., Oleson, K., Papale, D., Richardson, A.D., Roupsard, O., Running, S., Tomelleri, E., Viovy, N., Weber, U., Williams, C., Wood, E., Zaehle, S., Zhang, K., 2010. Recent decline in the global land evapotranspiration trend due to limited moisture supply. Nature 467, 951–954. https://doi.org/10.1038/nature09396.

Kay, A.L., Davies, H.N., Bell, V.A., Jones, R.G., 2009. Comparison of uncertainty sources for climate change impacts: flood frequency in England. Clim. Chang. 92, 41–63. https://doi.org/10.1007/s10584-008-9471-4.

Keim, B.D., Faiers, G.E., Muller, R.A., Grymes, J.M., Rohli, R.V., 1995. Long-term trends of precipitation and runoff in Louisiana, USA. Int. J. Climatol. 15, 531–541. https://doi.org/10.1002/joc.3370150505.

Kendon, E.J., Rowell, D.P., Jones, R.G., Buonomo, E., 2008. Robustness of future changes in local precipitation extremes. J. Clim. 21, 4280–4297. https://doi.org/10.1175/2008JCLI2082.1.

Kharin, V.V., Zwiers, F.W., Zhang, X., Wehner, M., 2013. Changes in temperature and precipitation extremes in the CMIP5 ensemble. Clim. Chang. 119, 345–357. https://doi.org/10.1007/s10584-013-0705-8.

Kirtman, B., Power, S., Adedoyin, J., Boer, G., Bojariu, R., Camilloni, I., Doblas-Reyes, F., Fiore, A., Kimoto, M., Meehl, G., Prather, M., Sarr, A., Schär, C., Sutton, R., van Oldenborgh, G., Vecchi, G., Wang, H., 2013. Near-term climate change: projections and predictability. In: Stocker, T., Plattner, G.-K., Tignor, M., Allen, S., Boschung, J., Nauels, A., Xia, Y., Bex, V., Midgley, P. (Eds.), Climate Change 2013: The Physical Science Basis. Contribution of Working Group I to the Fifth Assessment Report of the Intergovernmental Panel on Climate Change. Cambridge University Press, Cambridge, United Kingdom and New York, NY, USA, pp. 953–1028.

Knowles, N., Dettinger, M.D., Cayan, D.R., 2006. Trends in snowfall versus rainfall in the Western United States. J. Clim. 19, 4545–4559. https://doi.org/10.1175/JCLI3850.1.

Labat, D., Goddéris, Y., Probst, J.L., Guyot, J.L., 2004. Evidence for global runoff increase related to climate warming. Adv. Water Resour. 27, 631–642. https://doi.org/10.1016/j.advwatres.2004.02.020.

Lau, W.K.M., Wu, H.T., Kim, K.M., 2013. A canonical response of precipitation characteristics to global warming from CMIP5 models. Geophys. Res. Lett. 40, 3163–3169. https://doi.org/10.1002/grl.50420.

Lenderink, G., Attema, J., 2015. A simple scaling approach to produce climate scenarios of local precipitation extremes for The Netherlands. Environ. Res. Lett. 10, 85001. https://doi.org/10.1088/1748-9326/10/8/085001.

Lenderink, G., van Meijgaard, E., 2008. Increase in hourly precipitation extremes beyond expectations from temperature changes. Nat. Geosci. 1, 511–514. https://doi.org/10.1038/ngeo262.

Lindström, G., Bergström, S., 2004. Runoff trends in Sweden 1807–2002/Tendances de l'écoulement en Suède entre 1807 et 2002. Hydrol. Sci. J. 49, 69–83. https://doi.org/10.1623/hysj.49.1.69.54000.

Lins, H.F., Slack, J.R., 1999. Streamflow trends in the United States. Geophys. Res. Lett. 26, 227–230. https://doi.org/10.1029/1998GL900291.

Mallakpour, I., Villarini, G., 2015. The changing nature of flooding across the central United States. Nat. Clim. Chang. 5, 250–254. https://doi.org/10.1038/nclimate2516.

McCabe, G.J., Wolock, D.M., 2002. A step increase in streamflow in the conterminous United States. Geophys. Res. Lett. 29, 38-1–38-4. https://doi.org/10.1029/2002GL015999.

Mediero, L., Santillán, D., Garrote, L., Granados, A., 2014. Detection and attribution of trends in magnitude, frequency and timing of floods in Spain. J. Hydrol. 517, 1072–1088. https://doi.org/10.1016/j.jhydrol.2014.06.040.

Milliman, J.D., Farnsworth, K.L., Jones, P.D., Xu, K.H., Smith, L.C., 2008. Climatic and anthropogenic factors affecting river discharge to the global ocean, 1951–2000. Glob. Planet. Chang. 62, 187–194. https://doi.org/10.1016/j.gloplacha.2008.03.001.

Milly, P.C.D., Betancourt, J., Falkenmark, M., Hirsch, R.M., Kundzewicz, Z.W., Lettenmaier, D.P., Stouffer, R.J., 2008. Climate change: stationarity is dead: whither water management? Science (80-) 319, 573–574. https://doi.org/10.1126/science.1151915.

Milly, P.C.D., Dunne, K.A., 2001. Trends in evaporation and surface cooling in the Mississippi River Basin. Geophys. Res. Lett. 28, 1219–1222. https://doi.org/10.1029/2000GL012321.

Milly, P.C.D., Dunne, K. a, Vecchia, a V., 2005. Global pattern of trends in streamflow and water availability in a changing climate. Nature 438, 347–350. https://doi.org/10.1038/nature04312.

Milly, P.C.D., Wetherald, R.T., Dunne, K. a, Delworth, T.L., 2002. Increasing risk of great floods in a changing climate. Nature 415, 514–517. https://doi.org/10.1038/415514a.

Min, S.-K., Zhang, X., Zwiers, F.W., Hegerl, G.C., 2011. Human contribution to more-intense precipitation extremes. Nature 470, 378–381. https://doi.org/10.1038/nature09763.

Molnar, P., Fatichi, S., Gaál, L., Szolgay, J., Burlando, P., 2015. Storm type effects on super Clausius–Clapeyron scaling of intense rainstorm properties with air temperature. Hydrol. Earth Syst. Sci. 19, 1753–1766. https://doi.org/10.5194/hess-19-1753-2015.

Monk, W.A., Peters, D.L., Allen Curry, R., Baird, D.J., 2011. Quantifying trends in indicator hydroecological variables for regime-based groups of Canadian rivers. Hydrol. Process. 25 https://doi.org/10.1002/hyp.8137 n/a-n/a.

Moore, R.J., 2007. The PDM rainfall-runoff model. Hydrol. Earth Syst. Sci. 11, 483–499. https://doi.org/10.5194/hess-11-483-2007.

O'Gorman, P.A, Muller, C.J, 2010. How closely do changes in surface and column water vapor follow Clausius–Clapeyron scaling in climate change simulations? Environ. Res. Lett. 5, 025207. Online: http://stacks.iop.org/1748-9326/5/i=2/a=025207?key=crossref.e3222bcec4394fc1626baf06b07920c3.

Pall, P., Aina, T., Stone, D. a, Stott, P. a, Nozawa, T., Hilberts, A.G.J., Lohmann, D., Allen, M.R., 2011. Anthropogenic greenhouse gas contribution to flood risk in England and Wales in autumn 2000. Nature 470, 382–385. https://doi.org/10.1038/nature09762.

Pathiraja, S., Marshall, L., Sharma, A., Moradkhani, H., 2016. Detecting non-stationary hydrologic model parameters in a paired catchment system using data assimilation. Adv. Water Resour. 94, 103–119. https://doi.org/10.1016/j.advwatres.2016.04.021.

Pathiraja, S., Westra, S., Sharma, A., 2012. Why continuous simulation? The role of antecedent moisture in design flood estimation. Water Resour. Res. 48, W06534. https://doi.org/10.1029/2011WR010997.

Peña-Arancibia, J.L., van Dijk, A.I.J.M., Guerschman, J.P., Mulligan, M., (Sampurno) Bruijnzeel, L.A., McVicar, T.R., 2012. Detecting changes in streamflow after partial woodland clearing in two large catchments in the seasonal tropics. J. Hydrol. 416–417, 60–71. https://doi.org/10.1016/j.jhydrol.2011.11.036.

Peterson, B.J., Holmes, R.M., McClelland, J.W., Vörösmarty, C.J., Lammers, R.B., Shiklomanov, A.I., Shiklomanov, I.A., Rahmstorf, S., 2002. Increasing river discharge to the Arctic Ocean. Science (80-) 298, 2171–2173. https://doi.org/10.1126/science.1077445.

Petrow, T., Merz, B., 2009. Trends in flood magnitude, frequency and seasonality in Germany in the period 1951–2002. J. Hydrol. 371, 129–141. https://doi.org/10.1016/j.jhydrol.2009.03.024.

Piao, S., Ciais, P., Huang, Y., Shen, Z., Peng, S., Li, J., Zhou, L., Liu, H., Ma, Y., Ding, Y., Friedlingstein, P., Liu, C., Tan, K., Yu, Y., Zhang, T., Fang, J., 2010. The impacts of climate change on water resources and agriculture in China. Nature 467, 43–51. https://doi.org/10.1038/nature09364.

Prein, A.F., Rasmussen, R.M., Ikeda, K., Liu, C., Clark, M.P., Holland, G.J., 2016. The future intensification of hourly precipitation extremes. Nat. Clim. Chang. 7, 48–52. https://doi.org/10.1038/nclimate3168.

Probst, J.L., Tardy, Y., 1987. Long range streamflow and world continental runoff fluctuations since the beginning of this century. J. Hydrol. 94, 289–311. https://doi.org/10.1016/0022-1694(87)90057-6.

Prudhomme, C., Davies, H., 2009. Assessing uncertainties in climate change impact analyses on the river flow regimes in the UK. Part 2: future climate. Clim. Chang. 93, 197–222. https://doi.org/10.1007/s10584-008-9461-6.

Renard, B., Lang, M., Bois, P., Dupeyrat, A., Mestre, O., Niel, H., Sauquet, E., Prudhomme, C., Parey, S., Paquet, E., Neppel, L., Gailhard, J., 2008. Regional methods for trend detection: assessing field significance and regional consistency. Water Resour. Res. 44, W08419. https://doi.org/10.1029/2007WR006268.

Seneviratne, S., Nicholls, N., Easterling, D., Goodess, C., Kanae, S., Kossin, J., Luo, Y., Marengo, J., McInnes, K., Rahimi, M., Reichstein, M., Sorteberg, A., Vera, C., Zhang, X., 2012. Changes in climate extremes and their impacts on the natural physical environment. In: Field, C.B., Barros, V., Stocker, T.F., Qin, D., Dokken, D.J., Ebi, K.L., Mastrandrea, M.D., Mach, K.J., Plattner, G.-K., Allen, S.K., Tignor, M., Midgley, P.M. (Eds.), Managing the Risk of Extreme Events and Disasters to Advance Climate Change Adaptation. Cambridge University Press, Cambridge, UK, and New York, NY, USA, pp. 109–230. A Special Report of Working Groups I and II of the Intergovernmental Panel on Climate Change (IPCC).

Serquet, G., Marty, C., Dulex, J.-P., Rebetez, M., 2011. Seasonal trends and temperature dependence of the snowfall/precipitation-day ratio in Switzerland. Geophys. Res. Lett. 38 https://doi.org/10.1029/2011GL046976 n/a-n/a.

Sheffield, J., Wood, E.F., 2008. Global trends and variability in soil moisture and drought characteristics, 1950-2000, from observation-driven simulations of the terrestrial hydrologic cycle. J. Clim. 21, 432–458. https://doi.org/10.1175/2007JCLI1822.1.

Shiklomanov, A.I., Lammers, R.B., Rawlins, M.A., Smith, L.C., Pavelsky, T.M., 2007. Temporal and spatial variations in maximum river discharge from a new Russian data set. J. Geophys. Res. Biogeosci. 112 https://doi.org/10.1029/2006JG000352 n/a-n/a.

Shuster, W.D., Bonta, J., Thurston, H., Warnemuende, E., Smith, D.R., 2005. Impacts of impervious surface on watershed hydrology: a review. Urban Water J. 2, 263–275. https://doi.org/10.1080/15730620500386529.

Sillmann, J., Kharin, V.V., Zwiers, F.W., Zhang, X., Bronaugh, D., 2013. Climate extremes indices in the CMIP5 multimodel ensemble: Part 2. Future climate projections. J. Geophys. Res. Atmos. 118, 2473–2493. https://doi.org/10.1002/jgrd.50188.

Smith, K., Ward, R., 1998. Floods: Physical Processes and Human Impacts. John Wiley & Sons Ltd., Chichester, U.K.

Smith, L.C., Pavelsky, T.M., MacDonald, G.M., Shiklomanov, A.I., Lammers, R.B., 2007. Rising minimum daily flows in northern Eurasian rivers: a growing influence of groundwater in the high-latitude hydrologic cycle. J. Geophys. Res. Biogeosci. 112, G04S47. https://doi.org/10.1029/2006JG000327.

St. Jacques, J.-M., Sauchyn, D.J., 2009. Increasing winter baseflow and mean annual streamflow from possible permafrost thawing in the Northwest Territories, Canada. Geophys. Res. Lett. 36, L01401. https://doi.org/10.1029/2008GL035822.

Stahl, K., Hisdal, H., Hannaford, J., Tallaksen, L.M., van Lanen, H.A.J., Sauquet, E., Demuth, S., Fendekova, M., Jódar, J., 2010. Streamflow trends in Europe: evidence from a dataset of near-natural catchments. Hydrol. Earth Syst. Sci. 14, 2367–2382. https://doi.org/10.5194/hess-14-2367-2010.

Stahl, K., Tallaksen, L.M., Hannaford, J., Van Lanen, H.A.J., 2012. Filling the white space on maps of European runoff trends: estimates from a multi-model ensemble. Hydrol. Earth Syst. Sci. 16, 2035–2047. https://doi.org/10.5194/hess-16-2035-2012.

Stephens, G.L., L'Ecuyer, T., Forbes, R., Gettlemen, A., Golaz, J.-C., Bodas-Salcedo, A., Suzuki, K., Gabriel, P., Haynes, J., 2010. Dreary state of precipitation in global models. J. Geophys. Res. 115, D24211. https://doi.org/10.1029/2010JD014532.

Takeuchi, Y., Endo, Y., Murakami, S., 2008. High correlation between winter precipitation and air temperature in heavy-snowfall areas in Japan. Ann. Glaciol. 49, 7–10. https://doi.org/10.3189/172756408787814898.

Trenberth, K.E., 2011. Changes in precipitation with climate change. Clim. Res. 47, 123–138. https://doi.org/10.3354/cr00953.

Trenberth, K.E., Dai, A., Rasmussen, R.M., Parsons, D.B., 2003. The changing character of precipitation. Bull. Am. Meteorol. Soc. 84, 1205–1217. https://doi.org/10.1175/BAMS-84-9-1205.

Trenberth, K.E., Shea, D.J., 2005. Relationships between precipitation and surface temperature. Geophys. Res. Lett. 32, L14703. https://doi.org/10.1029/2005GL022760.

Utsumi, N., Seto, S., Kanae, S., Maeda, E.E., Oki, T., 2011. Does higher surface temperature intensify extreme precipitation? Geophys. Res. Lett. 38, L16708. https://doi.org/10.1029/2011GL048426.

Villarini, G., Serinaldi, F., Smith, J.A., Krajewski, W.F., 2009a. On the stationarity of annual flood peaks in the continental United States during the 20th century. Water Resour. Res. 45, W08417. https://doi.org/10.1029/2008WR007645.

Villarini, G., Smith, J.A., Serinaldi, F., Bales, J., Bates, P.D., Krajewski, W.F., 2009b. Flood frequency analysis for nonstationary annual peak records in an urban drainage basin. Adv. Water Resour. 32, 1255–1266. https://doi.org/10.1016/j.advwatres.2009.05.003.

Vörösmarty, C.J., Sahagian, D., 2000. Anthropogenic disturbance of the terrestrial water cycle. Bioscience 50, 753. https://doi.org/10.1641/0006-3568(2000)050[0753:ADOTTW]2.0.CO;2.

Wang, G., Wang, D., Trenberth, K.E., Erfanian, A., Yu, M., Bosilovich, M.G., Parr, D.T., 2017. The peak structure and future changes of the relationships between extreme precipitation and temperature. Nat. Clim. Chang. 1–8. https://doi.org/10.1038/nclimate3239.

Wasko, C., Sharma, A., 2017. Continuous rainfall generation for a warmer climate using observed temperature sensitivities. J. Hydrol. 544, 575–590. https://doi.org/10.1016/j.jhydrol.2016.12.002.

Wasko, C., Sharma, A., 2015. Steeper temporal distribution of rain intensity at higher temperatures within Australian storms. Nat. Geosci. 8, 527–529. https://doi.org/10.1038/ngeo2456.

Wasko, C., Sharma, A., 2014. Quantile regression for investigating scaling of extreme precipitation with temperature. Water Resour. Res. 50, 3608–3614. https://doi.org/10.1002/2013WR015194.

Wasko, C., Sharma, A., Johnson, F., 2015. Does storm duration modulate the extreme precipitation-temperature scaling relationship? Geophys. Res. Lett. 42, 8783–8790. https://doi.org/10.1002/2015GL066274.

Wasko, C., Sharma, A., Westra, S., 2016. Reduced spatial extent of extreme storms at higher temperatures. Geophys. Res. Lett. 43, 4026–4032. https://doi.org/10.1002/2016GL068509.

Wasko, C., Sharma, A., 2017. Global assessment of flood and storm extremes with increased temperatures. Sci. Rep. 7, 7945. Online: http://www.nature.com/articles/s41598-017-08481-1.

Wentz, F.J., Ricciardulli, L., Hilburn, K., Mears, C., 2007. How much more rain will global warming bring? Science (80-) 317, 233–235. https://doi.org/10.1126/science.1140746.

Westra, S., Alexander, L., Zwiers, F., 2013. Global increasing trends in annual maximum daily precipitation. J. Clim. 26, 3904–3918. https://doi.org/10.1175/JCLI-D-12-00502.1.

Westra, S., Fowler, H.J., Evans, J.P., Alexander, L., Berg, P., Johnson, F., Kendon, E.J., Lenderink, G., Roberts, N.M., 2014. Future changes to the intensity and frequency of short-duration extreme rainfall. Rev. Geophys. 52, 522–555. https://doi.org/10.1002/2014RG000464.

Westra, S., Sisson, S.A., 2011. Detection of non-stationarity in precipitation extremes using a max-stable process model. J. Hydrol. 406, 119–128. https://doi.org/10.1016/j.jhydrol.2011.06.014.

Woldemeskel, F., Sharma, A., 2016. Should flood regimes change in a warming climate? The role of antecedent moisture conditions. Geophys. Res. Lett. 43, 7556–7563. https://doi.org/10.1002/2016GL069448.

Zhai, J., Su, B., Krysanova, V., Vetter, T., Gao, C., Jiang, T., 2010. Spatial variation and trends in PDSI and SPI indices and their relation to streamflow in 10 large regions of China. J. Clim. 23, 649–663. https://doi.org/10.1175/2009JCLI2968.1.

Zhang, X., Harvey, K.D., Hogg, W.D., Yuzyk, T.R., 2001. Trends in Canadian streamflow. Water Resour. Res. 37, 987–998. https://doi.org/10.1029/2000WR900357.

Zhang, X., Zwiers, F.W., Hegerl, G.C., Lambert, F.H., Gillett, N.P., Solomon, S., Stott, P.A., Nozawa, T., 2007. Detection of human influence on twentieth-century precipitation trends. Nature 448, 461–465. https://doi.org/10.1038/nature06025.

Zhang, X.S., Amirthanathan, G.E., Bari, M.A., Laugesen, R.M., Shin, D., Kent, D.M., MacDonald, A.M., Turner, M.E., Tuteja, N.K., 2016. How streamflow has changed across Australia since the 1950s: evidence from the network of hydrologic reference stations. Hydrol. Earth Syst. Sci. 20, 3947–3965. https://doi.org/10.5194/hess-20-3947-2016.

CHAPTER 6

Detection of Temporal Changes in Droughts Over Indiana

Ganeshchandra Mallya[1], Shivam Tripathi[2], Rao S. Govindaraju[1]
[1]School of Civil Engineering, Purdue University, West Lafayette, IN, United States;
[2]Department of Civil Engineering, Indian Institute of Technology, Kanpur, India

Contents

Trends and Changes in Hydroclimatic Variables
ISBN 978-0-12-810985-4
https://doi.org/10.1016/B978-0-12-810985-4.00006-2

1. INTRODUCTION

1.1 Introduction to Droughts

Droughts are amongst the world's costliest disasters with an annual cost estimated in the range of \$6– \$8 billion (Federal Emergency Management Agency, 1995). Unlike other natural disasters such as floods and earthquakes, droughts manifest slowly and are already a serious threat before they are detected. Droughts have major impacts on agriculture, natural habitats and ecosystems, and economies of affected regions. Modeled precipitation and temperature results from different climate change scenarios indicate that droughts are likely to intensify over many parts of the world in the next 20–50 years (Dai, 2011), suggesting the need to assess drought impacts more accurately and develop appropriate mitigation strategies.

When a drought event occurs, moisture deficits are identified from many hydrologic variables such as precipitation, streamflow, soil moisture, snow pack, ground water levels, and reservoir storage (Wilhite and Glantz, 1985). Because drought impacts are experienced differently across the world, no universally accepted definition of drought exists. However, three types of droughts are commonly featured in scientific literature (Dracup et al., 1980):

1. Meteorological droughts are caused due to deficits in precipitation amounts when compared to the long-term average for a region. This shortage in precipitation can develop quickly and also end abruptly.
2. Agricultural drought conditions prevail when available soil moisture is insufficient to replace evapotranspiration losses in the root zone (WMO, 1975). The timing of soil moisture deficit plays a critical role, because deficiencies during the growing season can adversely impact crop yields. During droughts, plants are under stress and cannot fight off pests. Fertilizers and pesticides are also not effective in the absence of moisture resulting in failure of crops. The onset of agricultural droughts depends on antecedent soil moisture conditions and usually lags meteorological droughts.

3. Hydrologic droughts reflect shortages in water supply mainly in the form of reduced streamflows, reservoir and lake levels, and groundwater levels. Hydrologic droughts persist for longer durations when compared to meteorological and agricultural droughts because precipitation deficits translate to deficit in other hydrologic variables with significant time lags in some instances.

Precipitation deficits are not the only cause of droughts. Industrial and agricultural water demands have increased exponentially over the last few decades leading to water scarcity. With the increase in the emission of greenhouse gases, a steady rise in temperature has been observed over many parts of the globe. Increasing temperatures have affected the global hydrologic cycle leading to spatiotemporal variability of precipitation at different scales (Mishra and Singh, 2011; Solomon et al., 2007), thereby affecting drought characteristics.

1.2 Drought Characterization and Monitoring

Drought indicators or indices are commonly used to characterize and monitor droughts and their impacts. Generally all drought indicators use some measure of water deficit for analysis. Multiple hydrometeorological variables can also be used in a single drought indicator to capture the complex interactions that lead to droughts. Some of the desirable properties of a drought index are: (1) it should be sensitive to the timescale appropriate for the problem at hand; (2) the index should be able to capture the characteristics of both shorter- and longer-duration droughts; (3) it should be applicable to the problem being studied; (4) it should be possible to identify historical droughts; (5) the index should be capable of monitoring droughts on a near real-time basis (Friedman, 1957; Heim, 2002); and (6) it should have drought forecasting capability.

Several studies provide comprehensive review of drought indices (Heim, 2002; Mishra and Singh, 2010). *Palmer Drought Severity Index* (PDSI; Palmer, 1965) is a popular meteorological drought index that uses precipitation and temperature for estimating demand and supply of soil moisture within a two-layer water balance model. PDSI provides outlooks of moisture conditions that are comparable across regions and over different months. PDSI values typically vary from −4.0 to +4.0, negative values indicating drought conditions, while positive values indicate wet conditions. Another drought index that is popular because of its computational simplicity and forecasting ability at different time scales is the *standardized precipitation index*

(SPI; McKee et al., 1993). The SPI first fits a probability distribution to historic precipitation time-series data, and then normalizes the fitted distribution using the standard inverse Gaussian function to compute the drought index. SPI values are dimensionless with negative values indicating drought conditions, and the magnitudes of their departures from zero indicating the severity of the drought.

The *Crop Moisture Index* (CMI; Palmer, 1968) that monitors short-term moisture supply was developed to monitor agricultural droughts. With improvements in satellite remote sensing, monitoring crop and vegetation health over large spatial extents has become routine. For example, *Vegetation Condition Index* (VCI; Liu and Kogan, 1996) uses the advanced very high resolution radiometer radiance (AVHRR) data to study drought characteristics (early onset, intensity, frequency, and duration) and vegetation health. Along similar lines, the *Normalized Difference Water Index* (NDWI; Gao, 1996) uses near infrared (NIR) and short wave infrared (SWIR) channels to study the variation of moisture content and spongy mesophyll in vegetation canopies.

Monthly nonexceedance probability computed by compiling weighted values of variables such as reservoir storage, stream flow, snow pack, and precipitation resulted in the development of a hydrologic drought index called the *Surface Water Supply Index* (SWSI; Shafer and Dezman, 1982). Other popular hydrologic drought indices are Standardized streamflow index (SSI) and Standardized runoff index (SRI; Shukla and Wood, 2008). The SRI is computed and interpreted along similar lines as SPI.

Drought indices have been used for identifying droughts and their triggers (Steinemann, 2003), assessing drought status (Kao and Govindaraju, 2010), forecasting droughts (AghaKouchak, 2014), performing drought risk analysis (Hayes et al., 2004), and studying relationship of droughts with local-scale regional hydrological variables such as water quality (Sprague, 2005) and large-scale climate patterns like El Niño-Southern Oscillation (Cole and Cook, 1998; Liu and Juárez, 2001; Ryu et al., 2010). Drought indices are also used for classifying droughts and quantifying their temporal trends. These two applications of drought indices are reviewed in the following paragraphs.

1.3 Drought Classification

Drought classification schemes typically classify droughts based on their severity or intensity, and are often based on drought indices that measure degree of departure of hydrometeorological variables, such as precipitation and streamflow, from their long-term averages. Water resources planners rely on drought classification to select drought mitigation strategies. Hence weather agencies throughout the world routinely issue drought classification bulletins. For example, the *US Drought Monitor* releases a weekly update of drought status in USA. by classifying droughts into five classes—D0 to D4 with the latter representing exceptional drought. Likewise, India Meteorological Department (IMD) issues drought bulletins that classify droughts into three categories, namely, mild, moderate, and severe.

The common quantitative drought classification schemes work in two steps—first, by defining a drought index using hydrometeorological observations and next, by categorizing droughts based on predefined thresholds on the index value. Examples include IMD classification that uses departure of rainfall from its long-term average as a drought index, and US drought monitor classification that, along with other indices, uses SPI as a drought index. Among several drought classification schemes (Dai, 2011; Heim, 2002; Mishra and Singh, 2010), the scheme based on SPI (McKee et al., 1993) is very popular because of its computational simplicity and versatility in comparing different hydrometeorological variables at different time scales. In SPI, historical observations are used to compute the probability distribution of the monthly and seasonal (3-months, 6-months, and 12-months) precipitation totals. The fitted probability distributions are then normalized using the standard inverse Gaussian function to calculate SPI values. A negative value of SPI indicates precipitation less than the median rainfall, and the magnitude of departure from zero represents the severity of a drought.

Standard SPI-based drought classification, though popular, has many weaknesses (Mallya et al., 2013a). It provides discrete classification and ignores uncertainties arising from data errors, model assumptions, and parameter estimates. Thus, users are not aware of inherent uncertainties in drought classification often required for making informed decisions. Further, in the context of SPI, there is an ongoing debate on the selection of the parametric distribution for fitting the data. McKee et al. (1995) in their original paper on SPI recommend a gamma distribution. Lloyd-Huges and Saunders (2002) found gamma distribution to be an appropriate model for Europe. Guttman (1999) suggested Pearson-III distribution as the best universal model for SPI because it provides more flexibility than the gamma distribution. Rossi and Cancelliere (2003) found normal, lognormal, and gamma distributions to be suitable for different datasets in their study. Loukas and Vasiliades (2004) investigated different theoretical distributions using Kolmogorov–Smirnov (K–S) test and Chi-squared test and found Extreme Value-I distribution to be most suitable for studying droughts over Thessaly, Greece. Mishra et al. (2007) argue that different distributions may be appropriate for different drought durations (window size), and recommend the K–S test for choosing an appropriate distribution. Bonaccorso et al. (2013) used Lilliefors test to choose among normal, lognormal, and gamma distributions while Russo et al. (2013) used the three parameter generalized extreme value (GEV) distribution for SPI analysis. Thus there is no consensus on the choice of distribution for SPI analysis.

Mallya et al. (2013a) used hidden Markov model (HMM) for drought classification by conceptualizing hidden states in the model to represent drought states. Their model avoided the need for specifying thresholds for drought classification and provided probabilistic drought classification by accounting for model uncertainties; however, the number of hidden states (drought classes) was prespecified. To facilitate comparison of HMM drought index (HMM-DI) classification with standard methods, they specified 11 hidden states. Since the number of states is imposed on the model, it is possible that for datasets with short record length the model suffers from an overspecification problem, that is, the model structure is more complicated than supported by the dataset. Specifically, in the HMM context, overspecification would occur if the number of specified hidden states is more than that needed to model the data. Overspecification can result in parameter identification problems leading to unreliable results.

Mallya et al. (2015) proposed a method that adapts SPI drought classification methodology by employing gamma mixture model (Gamma-MM) in a Bayesian framework. The method alleviates the problem of selecting suitable distribution for SPI analysis, quantifies modeling uncertainties, and propagates them for probabilistic drought classification. Further, it avoids overspecification by using a Bayesian approach for optimally selecting the number of hidden states in the model.

1.4 Temporal Trends in Droughts

Temporal trends in droughts are identified by determining changes or sudden shifts in the distributional properties of the underlying hydrological variable. Classical drought indices such as SPI, or even probabilistic drought indices such as HMM-DI or probabilistic SPI, make several model assumptions about the hydrological variables used in their construct. Among them, the most important assumption is that the time-series of hydrological variable is stationary, that is, its distributional properties used to define droughts do not change over time. Thus, temporal trends in droughts cannot be estimated using classical drought indices. Nevertheless, hydrological time-series may exhibit nonstationarity due to changes in climate and land use or due to natural cycles that operate over a period of several years to decades and hence it is important to study temporal trends of droughts.

Several studies in the literature have proposed methods to perform drought analysis under nonstationary conditions. Mishra and Desai (2006) used autoregressive integrated moving average (ARIMA) models and variants of Artificial Neural Networks (ANNs), namely, recursive multi-step neural network approach (RMSNN) and direct multistep neural network approach (DMSNN) for drought forecasting in presence of nonstationarity. Coulibaly and Baldwin (2005) proposed the use of dynamic recurrent neural network (RNN) to model and forecast nonstationary hydrologic time-series. Belayneh et al. (2014) used wavelet analysis to first denoise the series, and then train ANNs or Support Vector Regressors on the decomposed signals to perform drought forecasting in arid regions of Ethiopia. Türkes and Tatlı (2009) studied droughts in nonstationary precipitation series by modifying the classical SPI using the concepts of empirical mode decomposition. Unlike the classical SPI, the modified SPI accounts for local or higher-order statistics in the precipitation time-series. Han et al. (2010) proposed the use of ARIMA models on potentially nonstationary remote sensing data to predict Vegetation Temperature Condition Index for drought forecasting. Verdon-Kidd and Kiem (2010) emphasized the need to evaluate risk to water resources systems during drought under a nonstationary climate over Australia. Mitra and Srivastava (2016) use the modified Mann-Kendall (MK) test on SPI and standardized precipitation evapotranspiration index (SPEI) series to study the spatiotemporal variability of meteorological droughts in southeast USA.

The literature review suggests that in the context of SPI, temporal changes in droughts are mostly studied by using one of the following two approaches. The first approach divides the study period into smaller intervals or epochs (~30 years) where the underlying rainfall series is assumed to be stationary, computes relative SPI for each epoch (Dubrovsky et al., 2009), and then compares drought characteristics between different epochs. The second approach allows the distribution of the hydrological variable to change with time, but the parameters of the distribution can follow only a prespecified temporal pattern. For example, the Standardized nonstationary Precipitation Index (SnsPI) proposed by Russo et al. (2013) assumes that the scale parameter of the gamma distribution for rainfall varies linearly with time.

The main objectives of this study are:

1. To investigate drought characteristics in Indiana by using a probabilistic drought classification approach that adapts SPI methodology by employing gamma mixture model (Gamma-MM) in a Bayesian framework (Mallya et al., 2015), and compare its results with classical SPI and HMM-based SPI.
2. To develop a methodology for studying temporal changes in droughts that does not require (1) making stationarity assumption about the precipitation time-series, and (2) prespecifying nature of temporal trend in the precipitation series.
3. To study temporal changes in the droughts in Indiana by the new methodology, and compare its results with the existing methods.

The remainder of the chapter is structured as follows. The next section describes the study area, Indiana, and provides an account of its historical droughts. The datasets and drought classification tools available for the study area are described in Section 3. Section 4 presents the mathematical formulation of the methods used for classifying droughts and quantifying temporal variation. These methods will be applied to precipitation data over Indiana and the results will be presented and discussed in Section 5. The chapter ends with a set of concluding remarks.

2. STUDY AREA

Indiana is located in the midwest United States, west of the Appalachian Mountains. While most of Indiana's topography consists of flat plains, regions with high elevations are located along the east central region of the state. The lowest elevation point is at the confluence of the Ohio River and the Wabash River at the southeastern corner of the state, at 97.5 m above mean sea level. The highest elevation point is at the Hoosier Hill in Wayne County at 383 m. The average elevation for Indiana is about 229 m.

The Wabash River is the most prominent river system and drains nearly 62,150 km^2 of Indiana. Other river systems include the Maumee River at the northeast corner of the state draining into Lake Erie, the St. Joseph River at the north draining into Lake Michigan, and the Kankakee River draining along the western border into the state of Illinois.

As Indiana is an agriculturally dominant economy, the type of soil plays a critical role in planning and management of agricultural activities. West central and east central Indiana have fertile prairie soils, and witness extensive farming activities that produce highest crop yields when compared to rest of the state. Northern and southern parts of the state are dominated by sandy soils and limestone deposits. As a result they have poor water retention capacity, and farms need to be irrigated during dry spells. East central Indiana and some parts of northeast Indiana have clayey soils, which have high water retention capacity and experience ponding.

Indiana has complex climate patterns with four distinct seasons—winters are cold, springs are characterized by thunderstorms and tornadoes, summers are very humid with high temperatures, and autumns are sunny with low humidity.

Indiana mostly receives precipitation when southerly winds that are warm and moisture laden from the Gulf of Mexico meet the continental polar air entering the state from central and western Canada. Moisture-laden winds from the Pacific also occasionally bring precipitation, although the Rocky Mountains stand as a barrier to this wind system. As Lake Michigan is located at the northern boundary of Indiana, this region receives heavy precipitation in the form of snowfall due to lake effect. Lake effect also ensures that the northern parts of the state are relatively cooler during the summer. The state of Indiana receives an average annual precipitation of 1020 mm. May is typically the wettest month of the year (with an average statewide precipitation of 130 mm), while February is the driest month of the year (with an average statewide precipitation of about 80 mm). Although Indiana receives adequate precipitation annually, its spatial and temporal variations result in droughts and floods. Indiana temperature and precipitation are also influenced by El Nino/La Nina and the Southern Oscillation (ENSO). During an El Nino event, the two jet streams which typically bring precipitation over Indiana, shift their paths. The Polar jet stream moves further north, and the Pacific jet stream moves further south, resulting in warmer and drier winters. Below normal precipitation is also experienced during spring and early summer months. During La Nina events, the Polar jet stream moves south, and the Pacific jet stream moves north resulting in above-normal precipitation over Indiana. While the summer precipitation remains below normal, winter precipitation increases. February, which is typically the driest month, witnesses heavy precipitation.

2.1 Historical Droughts in Indiana

Some of the major droughts that the state has witnessed during the period of instrumental records are chronicled below.

2.1.1 *Droughts of 1900*

Moderate to extreme (D2–D4) drought conditions prevailed over most of the state between 1897 and 1901. During this period, droughts were also wide spread over southeast, midwest, and western states in the country (parts of California, Arizona, and Nevada). In the midwest, Indiana, Ohio, and Kentucky were the worst affected.

2.1.2 *Dust Bowl Drought of 1930s*

Heat waves and lack of precipitation over the Mississippi and Ohio River basin caused extreme drought conditions not just in Indiana, but across United States—specially affecting the midwest and west regions. This mega drought is popularly referred to as the Dust Bowl drought as extreme dry conditions often led to dust storms across the country. It not only affected human health and ecology, but caused extreme hardships for farmers—which in turn severely impacted the agriculturally driven economy, leading to the Great Depression. While there were minor short-term respites from the effects of drought (1932, 1936, 1937), specially over Indiana, the negative impacts were felt far and wide across the United States throughout the decade.

2.1.3 *Texas Drought of 1950s*

During 1950s most of the southwestern United States, Texas, and southeast United States experienced extreme drought conditions, with Texas being the worst affected. The impacts of droughts in the south were quickly felt in the central United States and the midwest. Indiana experienced extreme drought conditions during 1952–54.

2.1.4 *Northeastern Droughts of 1960s*

While states in the northeastern United States (Virginia, Pennsylvania, New York, and New Jersey) were the worst affected by droughts, the impacts were also felt in parts of midwest, including Indiana and the neighboring states—Ohio, Illinois, Michigan, and Wisconsin—between 1962 and 1964.

2.1.5 Drought of Late 1980s

Short but particularly intense impacts of drought were felt across Indiana and other midwest and Northern Great Plain regions during 1987–88. Intense heat wave and wild fires were common during this drought event.

2.1.6 Drought During Summer 2012

Indiana had been experiencing above normal precipitation since the 2000s. The most recent drought conditions were felt during the summer of 2012 when impacts of a particularly extreme drought in the South and Western parts of United States spread quickly to the states in the midwest, including Indiana. Record high temperatures and water shortages for farming activities were felt during this period.

3. DATASETS

3.1 COOP Precipitation Data

Precipitation data were obtained from daily surface dataset (TD 3200) of cooperative (COOP) stations from National Climatic Data Center (NCDC). A total of 89 COOP stations were available with data record length greater than 30 years. If the data were missing for an entire month, they were replaced by the historic mean of that specific month (Kao and Govindaraju, 2010). The earliest precipitation record beginning in the year 1888 is available at COOP station (ID: 126001) in Mt. Vernon, IN. This station also has been the longest operating station with 128 years of data. Fig. 6.1 shows the study area along with the COOP stations for which the precipitation data were available.

3.2 Climate Division Data

Precipitation records are also available from NCDC at the nine climate divisions over Indiana (Vose et al., 2014). The climate divisions are shown in Fig. 6.1. Precipitation data are compiled based on Global Historical Climatology Network (GHCN) data using 5-km gridded approach. The record length for the data is from 1895 to present (2016). An annual time-series of water-year precipitation was computed at each of the nine climate divisions. The histogram of this annual time-series is shown in Fig. 6.2 for each of the nine climate divisions. The mean and standard deviation of the water-year precipitation time-series is also reported in this figure. Among the nine climate divisions, climate division 3 located

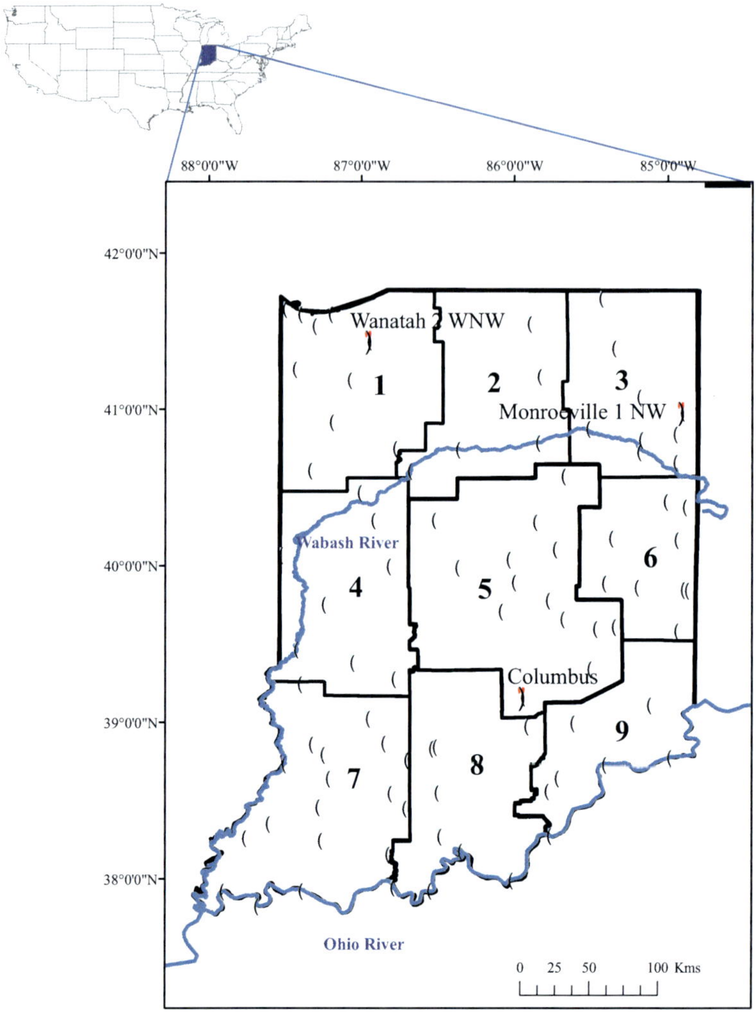

Figure 6.1 Map showing the study area along with the location of COOP precipitation stations that have a minimum record length of 30 years and have no missing data. Square markers denote the stations where results are discussed in subsequent sections. The nine climate divisions of Indiana are also shown.

in northeast has the lowest mean precipitation during the water-year (approximately 920 mm), and climate division 8 located in the south-central part of Indiana has the highest mean precipitation (approximately 1140 mm) during the water-year.

Several studies have investigated drought potential, vulnerability, and identification over Indiana. Rao and Voeller (1997) analyzed the use of

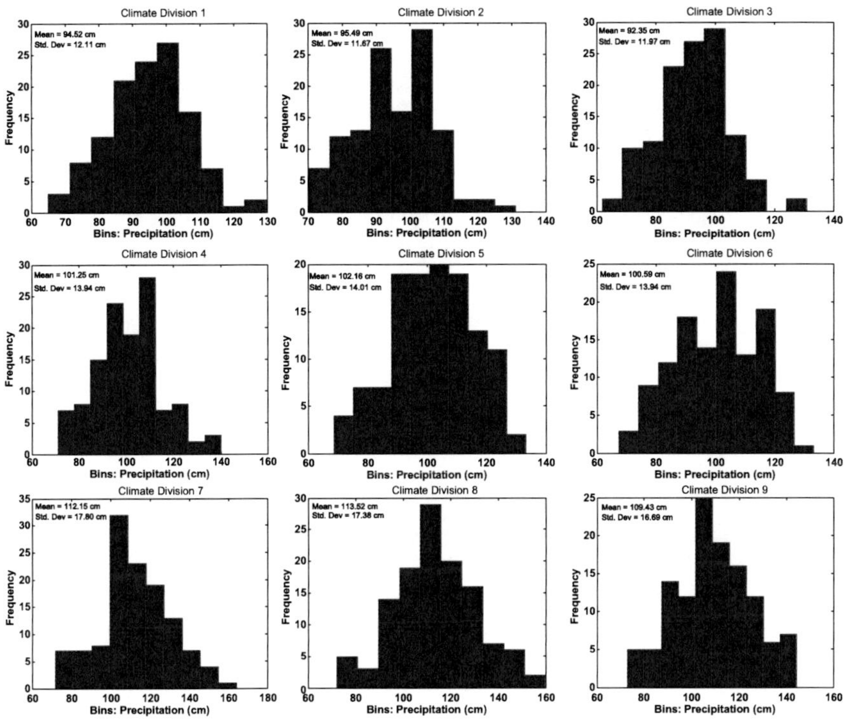

Figure 6.2 Histogram of cumulative water-year (October to September) precipitation at each climate division over Indiana.

several drought indicators for Indiana. Their study used both meteorological and hydrological variables as indicators for droughts. Some of the variables include precipitation anomaly, groundwater levels, streamflow, reservoir storage, and Palmer hydrologic drought index (PHDI). The study found streamflow variable 7Q10 and PHDI to be most suitable indicators for detecting droughts in Indiana. Lozano-Garcia et al. (1995) used Normalized Difference Vegetation Index (NDVI) computed from AVHRR data for monitoring agricultural droughts over Indiana. Charusombat and Niyogi (2011) found that some parts of southern Indiana are more vulnerable to droughts during the growing season. Mallya et al. (2013b) investigated potential causes and impacts of the 2012 drought over midwest United States.

An analysis of previous water-year drought events based on SPI values computed at each of the nine climate divisions over Indiana suggested that the most severe conditions persisted during the periods 1897–1900, 1933–40 (the Dust Bowl drought), 1952–53, 1962–65, and 1987–88. The most

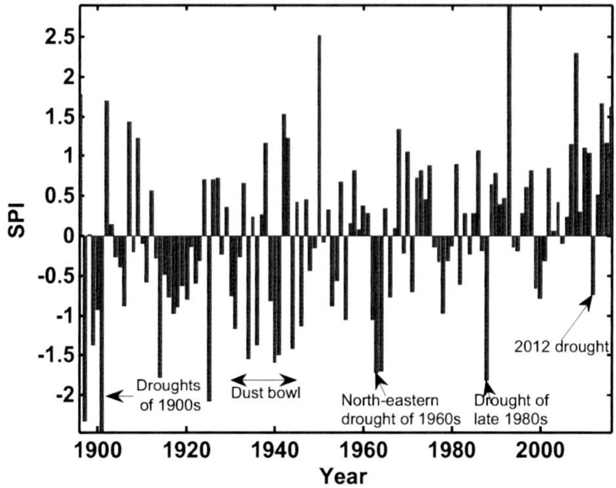

Figure 6.3 SPI time-series corresponding to water-year (October to September) at climate division 1 located in northwest Indiana.

recent drought conditions over Indiana occurred between 2012 and 13 (Mallya et al., 2013b). Fig. 6.3 shows historical SPI values for water-year ending in September of each year at climate division 1 located in northwest Indiana. For this climate division, the five most severe droughts were recorded during 1901, 1897, 1925, 1988, and 1914 (listed according to decreasing severity). The most recent water-year drought was during 2012.

3.3 Other Datasets

In addition to precipitation dataset which is mostly used in this study, other common variables used for drought analysis include streamflow (United States Geological Survey [USGS], Fig. 6.4), soil moisture data (Climate Prediction Centre [CPC] 0.5° gridded soil moisture dataset; Fan and van den Dool (2004)), temperature and humidity data from National Climatic Data Centre (NCDC), land-use information from National Land Cover Database 2011 (see Fig. 6.5), vegetation health data (NDVI), soil data (SSURGO/STATSGO), National Agricultural Statistics Service (NASS) Cropland Data Layer, and CropScape data.

Fig. 6.5 shows the NLCD 2011 land-use data over Indiana. The dataset was reclassified to a coarser resolution to represent four dominant classes—farm land, forest, urban, and residential zones, and water. Approximately 83% of Indiana comprises of farm land and forests. Therefore droughts over Indiana can have significant economic and social impact.

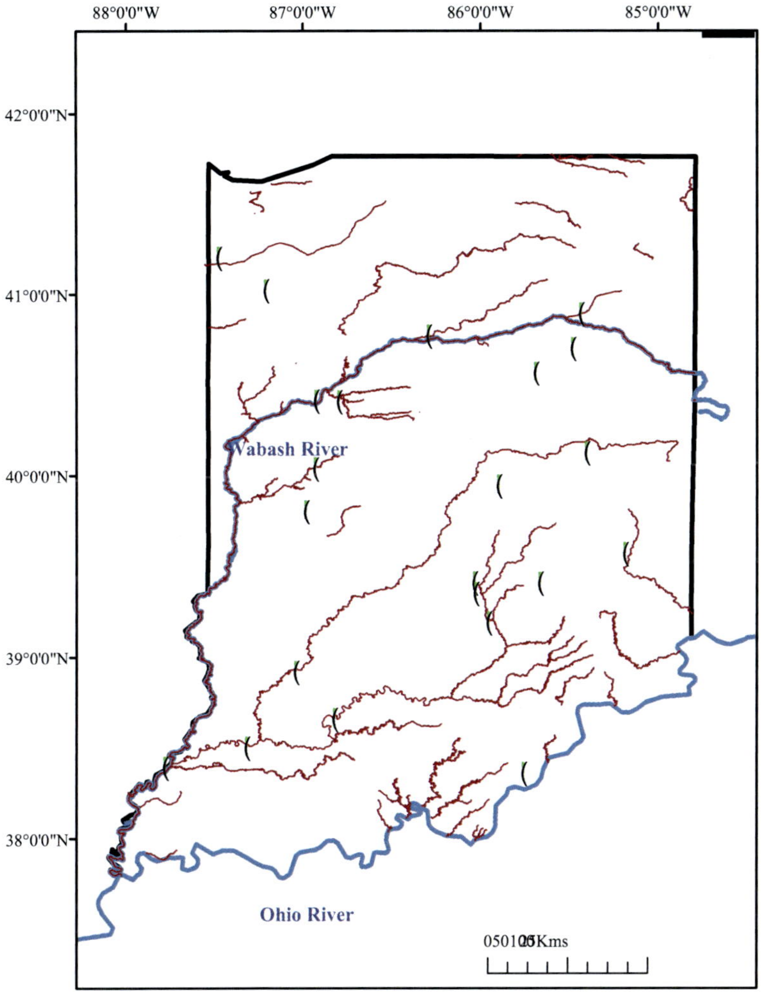

Figure 6.4 36 unregulated USGS stations over Indiana having at least 50 years of data are shown as circular markers.

3.4 Drought Research Initiative Network

A Drought Research Initiative Network (DRINET) was established in Indiana to provide cyber-infrastructure for drought research and community engagement. The network provides a platform for researchers to find and publish research data, develop tools that aid in drought characterization, forecasting, risk assessment, and decision making. It also fosters education initiatives and helps build partnerships between researchers; federal,

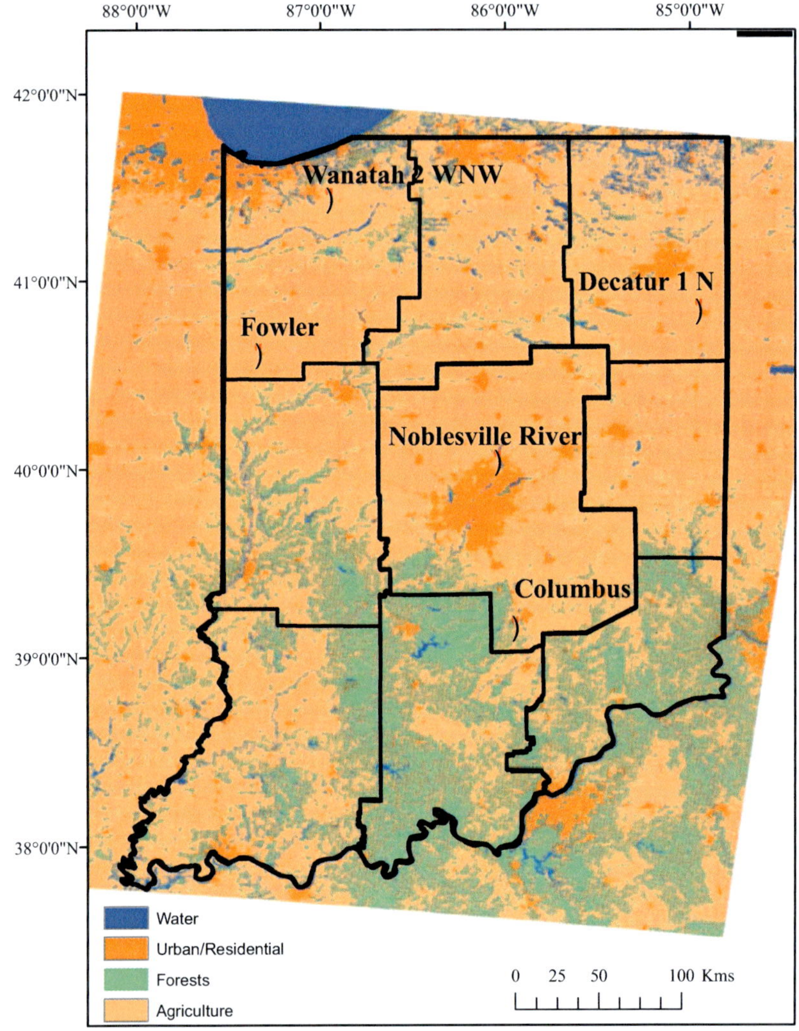

Figure 6.5 NLCD 2011 land-use data over Indiana. Color coding for the four dominant land-use classes are shown in the figure.

state, and local stakeholders; as well as the community. The DRINET website (http://drinet.hubzero.org) has several useful datasets including precipitation and temperature datasets; streamflow datasets; St. Joseph watershed streamflow and water quality data; CPC soil moisture, temperature, runoff, and evaporation data.

Several interactive drought assessment tools are also available on the DRINET website that are briefly summarized below.

- **The stream flow deficit viewer** provides hydrologic drought state of streams in Indiana using Joint Deficit Index (JDI) developed by Kao and Govindaraju (2010). Given a drought state, the tool also provides the probability of recovering to normal conditions. The JDI is capable of capturing both emerging and prolonged hydrologic droughts. The tool classifies drought states of a stream into five categories—dry, moderate, severe, extreme, and exceptional (D0–D4).
- **JDI-precipitation viewer** performs copula-based drought analysis described in Kao and Govindaraju (2010). The tool captures the spatiotemporal dependence relationships in the long-term monthly precipitation dataset over upper midwest United States. Their joint probability distribution was constructed by combining drought marginals and the dependence structure. The copula-based joint deficit index (JDI) provides an objective (probability-based) description of the overall drought status. The tool compares the results from JDI with PDSI. The JDI estimates obtained from the tool can help in drought identification, and allow a month-by-month assessment for future drought recovery.
- **Water deficit viewer** provides estimates of precipitation that are required to recover from an existing drought and the probability of such recovery for a particular window for series of months.
- **Water balance viewer** is a tool that performs dynamic data retrieval from the Consortium of Universities for the Advancement of Hydrologic Science, Inc. (CUAHSI) and NCDC web servers and performs water balance computations online at different subwatersheds over Indiana.
- **Drought impact viewer** uses NASS crop data layer (that consists of 13 categories of crops and land uses) for a specified year, a weekly drought index layer from the Midwestern Regional Climate Center, and a county boundaries layer for Indiana to provide visualization and summary statistics of drought impacts at a county scale on a week-to-week basis.
- **HMM-based drought index viewer** creates graphs showing probabilistic classification of droughts by HMM and discrete classification of droughts by SPI for user-defined time period at any chosen precipitation or streamflow station in Indiana.

4. METHODOLOGY

The mathematical formulations of the three drought classification methods, namely SPI, HMM-DI, and Gamma-MM, are first presented. Next, the methods used for studying temporal changes in droughts over Indiana are described.

4.1 Standardized Precipitation Index (SPI)

The method involves following steps:

1. Decide a drought duration (time-window) and estimate cumulative precipitation during that period. For example, to estimate droughts during a winter season, estimate cumulative precipitation during 3 months of the winter season (DJF) for each year. This will yield an annual time-series of cumulative precipitation. Likewise, for analyzing water-year droughts obtain an annual time-series of cumulative precipitation for 12 months starting on 1st October and ending on 30th September of following year.
2. Fit a gamma distribution to the cumulative precipitation series. The cumulative distribution function (CDF) of the gamma distribution is standardized using the standard inverse Gaussian function to compute the SPI drought index. As stated earlier, a negative value of SPI indicates drought conditions and the magnitude of its departure from zero indicates the severity of a drought.
3. Decide a threshold on CDF to determine drought class. To draw parallels with the US Drought Monitor, we have used the same thresholds as used by them for SPI drought classification (Table 6.1).

According to McKee et al. (1993), a drought event begins when SPI takes a value of −1.0 or less and it ends when SPI becomes positive. In this study, since drought classes are defined to coincide with designations in the US Drought Monitor, a drought event would begin when the SPI took a value of −0.5 or less and it ended when SPI was positive. Thus each

Table 6.1 US Drought Monitor Classification Scheme. SPI Ranges Are Prescribed for the Inverse of the Normal Distribution. Corresponding Thresholds on CDF are Given in the Last Column

Category	Description	SPI Range	Threshold on CDF
D0	Abnormally dry	−0.5 to −0.8	0.212 to 0.309
D1	Moderate drought	−0.8 to −1.3	0.097 to 0.212
D2	Severe drought	−1.3 to −1.6	0.055 to 0.097
D3	Extreme drought	−1.6 to −1.9	0.023 to 0.055
D4	Exceptional drought	−2.0 or less	0.023 or less

drought event had a duration defined by its beginning and end, and an intensity for each month the drought event prevailed. Based on the above definition, the following statistics were noted for SPI: the number of drought events; duration of the drought events; and number and average duration of droughts under each drought category (D0–D4).

4.2 Hidden Markov Model–Based Drought Index (HMM-DI)

The HMM (e.g., Rabiner, 1989) is a statistical model where observations from a system are assumed to be conditioned on the state of the system. The state is hidden (i.e., not observed) and satisfies the Markov property. The HMM was developed in late 1960s and early 1970s for speech recognition, and it has since been used successfully in many applications including hydrology and climate modeling (Robertson et al., 2004, 2003; Thyer and Kuczera, 2003). The mathematical formulation of the HMM used in this work is described in Tripathi and Govindaraju (2009) and the details of HMM-DI are given in Mallya et al. (2013a). A brief description of this method is presented in the following paragraphs.

Let the hydrologic variable of interest at time t be denoted by x_t, $t = 1 \ldots N$ $\{x_t \in R$ and $X=[x_1,\ldots, x_N]^T = x_{1:N}\}$. In HMM, the quantity x_t is assumed to depend on the state variable z_t,$\{Z=[z_1,\ldots, z_N]^T = z_{1:N}\}$ that denotes drought states, is hidden, and follows the first order Markov property. The state variable z_t is a discrete random variable with K values (drought states), $\{1, 2,\ldots, K\}$. HMM makes three assumptions about the underlying process being modeled:

1. The drought state z_t evolves according to (first-order) Markov property. Given the drought state at the previous month z_{t-1}, the drought states in the current and future months are independent of past drought states $(z_{t-2},\ldots, z_1)$, i.e., $P(z_t|z_{t-1}, z_{t-2},\ldots, z_1) = P(z_t|z_{t-1})$. The probabilities $P(z_t|z_{t-1})$ are referred to as transition probabilities.
2. Given the current drought state z_t, the monthly observation x_t for that month is assumed to be conditionally independent of the observations or drought states of other months, $P(x_t|x_{1:t-1}, z_{1:t}) = P(x_t|z_t)$. The probability distributions $P(x_t|z_t)$ are referred to as emission distributions.
3. The transition and emission probabilities depend only on the drought states and observations, and not on the time series index of the observation, $P(z_t = k|z_{t-1} = j) = P(z_l = k|z_{l-1} = j)$ and $P(x_t \leq x|z_t = k) = P(x_l \leq x|z_l = k)$.

Further, assuming the number of states K is known a priori, the joint distribution over the SPI categories and monthly hydrologic observations decomposes as a product

$$P(z_{1N}, x_{1N}) = P(z_1)\prod_{t=2}^{N} P(z_t|z_{t-1})\prod_{t=1}^{N} P(x_t|z_t) \tag{6.1}$$

An HMM can be completely described by:

1. The conditional distribution of the hydrologic variable given the drought state $P(x_t|z_t)$ referred to as emission distribution.
2. The conditional distribution of the present drought state given the previous drought state, that is, $P(z_t|z_{t-1})$. Because z_t is a K- valued discrete variable, the conditional distribution is given by a $K \times K$ transition matrix A with elements$A_{jk} = p(z_t = k|z_{t-1} = j)$.
3. Marginal distribution of the drought state at the first time step, $p(z_1)$,given by K-dimensional vector with $\pi_k = P(z_1 = k)$.

The posterior probability of being in a state $z_t = k$ at time t is given by

$$P(z_t = k|X, \lambda) = \frac{\alpha_t(k)\beta_t(k)}{\sum_{j=1}^{K}\alpha_t(j)\beta_t(j)} \tag{6.2}$$

where $\alpha_t(k) = P(x_1,\ x_2,\ldots,\ x_t,\ z_t = k|\lambda)$ and $\beta_t(k) = P(x_{t+1},\ x_{t+2},\ldots,\ x_N,\ z_t = k|\lambda)$, and λ represents the set of model parameters, namely the parameters of the emission distributions (θ), the transition matrix (A), and the initial distribution of the states (π).

For a drought index, a definition of drought states that remains unaltered irrespective of the location of a drought is desirable. To achieve this property, the following two steps are taken:

1. The data at any desired time scale (from 1 month out to several years) are transformed to departures from the mean. This step brings the data from different locations to a common baseline for comparison purposes. The HMM model is applied to the transformed data.
2. The probability density function for the emission distribution is chosen to be a Gaussian distribution of the form

$$p(x_t|z_t = k) = N\left(x_t|\mu_k, \sigma_k^2\right). \tag{6.3}$$

where μ_k and σ_k^2 are the mean and the variance of a Gaussian distribution, respectively. Because the states are hidden (i.e., not observed), the true nature of emission distributions cannot be determined a priori. The choice of Gaussian emission distribution is primarily for mathematical convenience. Many complex processes combine to create droughts, and one may expect their combined influence, expressed through deviations from the mean, to be Gaussian. Finally, if there is no temporal dependence in the drought states, the HMM automatically collapses to GMM for which theories are well developed (Reynolds and Rose, 1995). Since the results of the developed drought index are compared with SPI, the number of states (components in the Gaussian mixture) K was set to 11 (D0−D4, N, W0−W4) as described earlier. Unlike SPI where thresholds are fixed (see Table 6.1) for drought classification, HMM-DI utilizes a data-driven approach to estimate the parameters of the emission distributions. The μ_k's and σ_k's for all the components of the emission distribution were learnt from the data in a maximum likelihood framework using the Baum-Welch algorithm as described in Rabiner (1989).

As HMM-DI is a probabilistic drought index, for the purpose of this study a drought event was assumed to begin when total probability of drought was greater than 0.5, and assumed to end when the total probability of drought is less than 0.5. As in the case of SPI, a drought event has a duration defined by its beginning and end. Based on the above definition, the following two statistics were noted for HMM-DI: the number of drought events and duration of the drought events.

4.3 Gamma Mixture Model (Gamma-MM)

As discussed in the Introduction section, there is an ongoing debate on the choice of a suitable distribution for fitting data in SPI analysis. Mallya et al. (2015) addressed this problem by using the Gamma-MM. Given sufficient number of components in the mixture, the Gamma-MM is proven to provide arbitrarily close approximation to any general continuous distribution in the range $(0, \infty)$ (see, DeVore and Lorentz, 1993).

The use of Gamma-MM is not new in hydrology. To model data with multiple modes and different types of skewness, Evin et al. (2011) proposed the use of Gamma-MM for strictly positive hydrological data. In the assessment of hydrological droughts for Yellow River in China, Shiau et al. (2007) first fitted mixtures of exponential and gamma distributions to drought duration and drought severity, respectively, and then used the copula method to construct a bivariate drought distribution. In the following, we provide a brief description of the Gamma-MM. The readers are referred to Wiper et al. (2001) and Richardson and Green (1997) for details on mixture models. A summary of the mathematical details of the Gamma-MM as described in Mallya et al. (2015) is presented below.

Let the cumulative rainfall at time t be denoted by x_t, $t = 1,\ldots, N$ $\{x_t \in R \text{ and } X = [x_1,\ldots, x_N]^T\}$. If the total number of components of Gamma-MM, M, is known a priori, then the weighted sum of M mixtures of gamma is given by the equation

$$P(x_t|\lambda) = \sum_{i=1}^{M} w_i G\left(x_t \middle| \nu_i, \frac{\nu_i}{\mu_i}\right) \tag{6.4}$$

where w_i are the mixture weights or mixing ratios, and $G\left(x_t \middle| \nu_i, \frac{\nu_i}{\mu_i}\right)$ are the components of gamma densities of the form

$$G\left(x_t \middle| \nu_i, \frac{\nu_i}{\mu_i}\right) = \frac{\left(\frac{\nu_i}{\mu_i}\right)^{\nu_i}}{\Gamma(\nu_i)} x_t^{(\nu_i - 1)} \exp\left(-\frac{\nu_i}{\mu_i} x_t\right), \tag{6.5}$$

with mean μ_i and shape parameter ν_i. Further, the mixture weights satisfy the constraint $\sum_{i=1}^{M} w_i = 1$. The parameter set is represented as, $\lambda = \{\boldsymbol{w}, \boldsymbol{\mu}, \boldsymbol{\nu}\}$ where $\boldsymbol{w} = [w_1, w_2,\ldots, w_M]^T$, $\boldsymbol{\mu} = [\mu_1, \mu_2,\ldots, \mu_M]^T$ and $\boldsymbol{\nu} = [\nu_1, \nu_2,\ldots, \nu_M]^T$.

In the Bayesian framework, the model parameters are obtained by specifying prior distributions to model parameters. Parameter estimation is accomplished by introducing a latent variable $Z = [\boldsymbol{z}_1,\ldots, \boldsymbol{z}_N]^T$ for each time step. The variable $\boldsymbol{z}_t$ is an M-dimensional binary random variable, $\boldsymbol{z}_t = [z_{t1},\ldots, z_{tM}]^T$, in which a particular element is equal to 1 and all other elements are zero, that is, $\sum_{i=1}^{M} z_{ti} = 1$ and $z_{ti} \in \{0, 1\}$. The variable $\boldsymbol{z}_t$ denotes the component to which the data x_t belongs, and hence it is also called an *indicator variable*. The conditional distribution of x_t given $\boldsymbol{z}_t$ is

$$P(x_t|z_{ti} = 1) \sim G\left(x_t \middle| \nu_i, \frac{\nu_i}{\mu_i}\right) \tag{6.6}$$

The posterior probability of the model parameters and latent variables is obtained by applying Bayes' Rule as

$$P(\lambda|X) \propto P(X|\lambda)P(\lambda) \tag{6.7}$$

where the parameter set λ includes the latent variable as well. The *likelihood function* given the latent variable is

$$P(X|\lambda) = P(X|Z, \boldsymbol{\mu}, \boldsymbol{\nu}) = \prod_{t=1}^{N}\prod_{i=1}^{M}\left(\mathrm{G}\left(x_t\middle|\nu_i, \frac{\nu_i}{\mu_i}\right)\right)^{z_{ti}} \tag{6.8}$$

Following Wiper et al. (2001) *the prior distribution* over the model parameter is given as $P(\lambda) = P(Z|\boldsymbol{w})P(\boldsymbol{w})P(\boldsymbol{\mu})P(\boldsymbol{\nu})$ with

$$P(Z|\boldsymbol{w}) = \prod_{t=1}^{N}\prod_{i=1}^{M} w_i^{z_{ti}},$$

$$P(\boldsymbol{w}) = \mathrm{Dir}(\boldsymbol{w}|\boldsymbol{\Phi}) = C(\boldsymbol{\Phi})\prod_{i=1}^{M} w_i^{\phi_i - 1},\ \boldsymbol{\Phi} = [\phi_1, \ldots, \phi_M]^T,$$

$$P(\boldsymbol{\nu}) = \mathrm{Exp}(\boldsymbol{\nu}|\boldsymbol{\theta}) = \prod_{i=1}^{M}\frac{1}{\theta_i}\exp(-\theta_i\nu_i),\ \boldsymbol{\theta} = [\theta_1, \ldots, \theta_M]^T, \text{ and}$$

$$P(\boldsymbol{\mu}) = \mathrm{GI}(\boldsymbol{\mu}|\boldsymbol{\alpha}, \boldsymbol{\beta}) = \prod_{i=1}^{M}\frac{\beta_i^{\alpha_i}}{\Gamma(\alpha_i)}\mu_i^{-\alpha_i - 1}\exp\left(-\frac{\beta_i}{\mu_i}\right), \boldsymbol{\alpha}$$
$$= [\alpha_1, \ldots, \alpha_M]^T \text{ and } \boldsymbol{\beta} = [\beta_1, \ldots, \beta_M]^T$$

where Dir, Exp, and GI represent Dirichlet, exponential, and inverted gamma distributions respectively, and $C(\boldsymbol{\Phi})$ is a normalizing constant. The prior distribution is made noninformative by assigning following values to the hyperparameters.

$$\phi_i = 1; \theta_i = 0.01; \alpha_i = \beta_i = 1 \text{ for } i = 1, \ldots, M.$$

The posterior distribution $P(\lambda|X)$ does not have a closed form and has to be estimated by either deterministic approximation (variational Bayes' methods) or stochastic approximation (MCMC; Markov chain Monte Carlo methods). In this study the posterior distribution is estimated using stochastic approximation by sampling posterior distribution with Gibbs sampler, an MCMC algorithm (Geman and Geman, 1984).

The Gibbs sampling algorithm samples posterior distribution of the parameters by sequentially sampling from the conditional distribution of a parameter given all other parameters. The sampling starts with an initial value and proceeds as follows.

1. Set iteration number $j = 0$, and parameters to their initial value $\boldsymbol{\lambda}^{(0)} = [\boldsymbol{w}^{(0)}, \boldsymbol{\mu}^{(0)}, \boldsymbol{\nu}^{(0)}]$. The initial value is obtained by randomly sampling from the prior distribution of the parameters.
2. Sample from $P\left(\boldsymbol{z}_t^{(j+1)} \middle| X, \boldsymbol{w}^{(j)}, \boldsymbol{\mu}^{(j)}, \boldsymbol{\nu}^{(j)}\right) \sim \text{Multinomial}\ (\boldsymbol{z}_t|\boldsymbol{r}_t)$ where $\boldsymbol{r}_t = [r_{t1}, \ldots, r_{tM}]^T$, $r_{ti} = \frac{s_{ti}}{\sum_{i=1}^{M} s_{ti}}$ and $s_{ti} = w_i G\left(x_t \middle| \nu_i, \frac{\nu_i}{\mu_i}\right)$ and Multinomial represents multinomial distribution.
3. Sample from $P\left(\boldsymbol{w}^{(j+1)} \middle| X, Z^{(j+1)}, \boldsymbol{\mu}^{(j)}, \boldsymbol{\nu}^{(j)}\right) \sim \text{Dir}\left(\boldsymbol{w}\middle|\widehat{\boldsymbol{\Phi}}\right)$ where $\widehat{\boldsymbol{\Phi}} = [\phi_i + n_i, \ldots, \phi_M + n_M]^T$ and $n_i = \sum_{t=1}^{N} z_{ti}$.
4. Sample from $P\left(\boldsymbol{\mu}^{(j+1)} \middle| X, Z^{(j+1)}, \boldsymbol{w}^{(j+1)}, \boldsymbol{\nu}^{(j)}\right) \sim \text{GI}\left(\boldsymbol{\mu}\middle|\widehat{\boldsymbol{\alpha}}, \widehat{\boldsymbol{\beta}}\right)$ where $\widehat{\boldsymbol{\alpha}} = [\alpha_i + n_i\nu_i, \ldots, \alpha_M + n_M\nu_M]^T$ and $\widehat{\boldsymbol{\beta}} = \left[\beta_i + \nu_i \sum_{t=1}^{M} x_t z_{ti}, \ldots, \beta_M + \nu_M \sum_{t=1}^{M} x_t z_{tM}\right]^T$.
5. Sample from $P(\boldsymbol{\nu}^{(j+1)}|X, Z^{(j+1)}, \boldsymbol{w}^{(j+1)}, \boldsymbol{\mu}^{(j+1)})$. This conditional distribution does not have a closed form. Hence samples are generated using Metropolis-Hasting algorithm from a proposal distribution $P(\widetilde{\nu}_i|\nu_i) \sim G(h, h|\nu_i)$ and is accepted with a probability

$$\min\left\{1, \frac{f(\widetilde{\nu}_i)P(\nu_i|\widetilde{\nu}_i)}{f(\nu_i)P(\widetilde{\nu}_i|\nu_i)}\right\}$$

where $f(\nu_i) \propto \frac{\nu_i^{n_i\nu_i}}{\Gamma(\nu_i)^{n_i}} \exp\left(-\nu_i\left(\theta_i + \frac{\sum_t x_t z_{ti}}{\mu_i} + n_i \log \mu_i - \log\left(\prod_{t=1; z_{ti}=1}^{N} x_t\right)\right)\right)$.

If the new sample $\widetilde{\nu}_i$ is rejected, the current value of ν_i is retained. The above procedure is repeated to sample ν_i for all components $i = 1, \ldots, M$. In this study the parameter of the proposal distribution h is set to 2.
6. Set $j = j + 1$ and go to Step 2 until convergence. In this study, 15,000 samples are generated after ignoring initial 500 samples (*burn-in* period). Trace plots of the samples are monitored for convergence.

To keep the notations uncluttered, the iteration number is omitted from the parameters of the conditional distributions.

In the above formulation of Gamma-MM, we have assumed that the number of mixture components, M, is known. However, in a general context, M is not known and should be estimated from data. One approach for estimating M is to consider it as a model parameter, assign prior distribution to it, and estimate posterior distribution by MCMC method. Since changing M will result in a different model structure, usual MCMC algorithms such as Gibbs sampler cannot be applied. Instead reversible jump MCMC (RJMCMC; Green (1995) and Richardson and Green (1997)) may be used. In this study we implemented RJMCMC for Gamma-MM as described by Richardson and Green (1997) and Wiper et al. (2001). The results suggested that RJMCMC algorithm requires significantly higher number of iterations for convergence compared to a model where M is specified. We found that if we start with a model having sufficiently large number of components, M the Bayesian algorithm automatically prunes the components that are not relevant for modeling by making the mixing ratio (w) very small, thereby determining optimum number of components. We recommend the latter approach for hydrological applications where the number of components is usually limited to 2 or 3.

In the Bayesian framework, mixture models have *identifiability* problem, that is, an M component mixture model will have a total of $M!$ equivalent solutions. The problem can be avoided by introducing asymmetricity in the likelihood function. For example, in the context of Gamma-MM, Wiper et al. (2001) recommended the following restriction on the means of the mixture components, $\mu_1 < \mu_2 < \ldots < \mu_M$. However, for finding a good density model, as required in the present application, the problem of identifiability is not relevant because any of the equivalent solutions is as good as another (Bishop, 2006).

As in the case of HMM-DI, a drought event begins when the total probability of drought is greater than 0.5, and ends when the probability is less than 0.5. Based on this definition, the number of drought events and their average durations were recorded for each station in the study area.

4.4 Temporal Trends in Droughts

Evaluation of temporal trends associated with retrospective drought events provides a basis to understand regional patterns of severity and duration of droughts. It also provides insight into the nature of possible future droughts and potential vulnerabilities over the study region. In this study, we first investigated trends in precipitation series at each COOP station and also at each climate division. A modified MK trend test that accounts for autocorrelation in time-series (Kulkarni and von Storch, 1995; Hamed and Rao, 1998) was used to detect trends in seasonal and water-year precipitation.

The trends were tested at 5% significance level. The effect of spatial correlations in the data (Burn and Elnur, 2002; Yue and Wang, 2002) on the trend results were accounted by using false discovery rate (FDR) (Benjamini and Hochberg, 1995; Ventura et al., 2004).

If the precipitation series exhibits a trend, drought classification methods such as SPI, HMM-DI, and Gamma-MM are not applicable for drought analysis because they assume precipitation time-series to be stationary. In this study, we propose a method for drought analysis that overcomes this problem by explicitly removing trends from the precipitation series. The new method is compared with relative SPI (Dubrovsky et al., 2009) and SnsPI (Russo et al., 2013) for performing drought analysis of a nonstationary precipitation series. The following paragraphs describe the new method and provide a summary of relative SPI and SnsPI methods.

4.4.1 Proposed Method for Drought Analysis of a Non-stationary Precipitation Series

A precipitation series may exhibit nonstationarity because of any of several reasons (Fuller, 1996)—(1) the mean is a function of time, (2) the variance or other higher-order moments are functions of time, and (3) the stochastic mechanism generating time series is nonstationary. In this study, we consider nonstationarity arising from first kind. Specifically, we consider a *trend stationarity process* in which the underlying trend is solely a function of time and can be removed to obtain a stationary process. A trend stationarity process y_t is expressed as

$$y_t = f(t) + z_t \tag{6.9}$$

where t represents times, z_t denotes a zero mean stationary process, and $f(t)$ is a function of time representing trend of the process at time t. The trend can be determined either extrinsically by specifying a linear or nonlinear functional form for $f(t)$ (e.g., regression models) or intrinsically by using the data without prespecifying a functional form (e.g., empirical mode decomposition (Wu et al., 2007)).

Drought classification for nonstationary time-series can be approached in one of the following two ways: (1) assuming that the "normal" conditions for a station are evolving and hence the drought thresholds for different categories are changing with time (as in SnsPI), and (2) assuming that the "normal" conditions for a station are fixed (with respect to a reference period and hence the thresholds for drought classification are fixed), but the frequency of droughts are changing with time (as in relative SPI). The proposed method classifies droughts for nonstationary time-series by using both the approaches. The steps of the proposed methods are as follows:

1. Identify trend in the time-series $f(t)$ by using either an extrinsic approach (regression model) or an intrinsic approach (empirical mode decomposition).
2. Determine z_t by removing trend from the time-series and fit a suitable distribution to obtain the cumulative distribution function $F_Z(z_t) = P(Z \leq z_t)$, where Z is a random variable belonging to a family of stationary stochastic processes.
3. Determine the CDF of the rainfall time-series at time t as

$$F_{Y_t}(y_t) = P(Y_t \leq y_t) = P(Z \leq y_t - f(t)) = F_Z(y_t - f(t)). \qquad (6.10)$$

Estimate drought thresholds at each time step by using $F_{Y_t}(y_t)$ and the SPI drought definitions given in Table 6.1.

Select a reference year (t_0), and determine $F_{Y_{t_0}}$ and corresponding drought thresholds. Estimate drought class at time t based on these fixed thresholds for the reference year.

4.4.2 Standardized NonStationary Precipitation Index (SnsPI)

The SnsPI method fits a nonstationary model to the precipitation data by linearly varying the scale parameter (s_t) of the gamma distribution with time. Following the notations used in Section 4.3, the gamma distribution is represented as $G(x_t|\nu, s_t)$ with scale parameter $s_t = \mu_t/\nu$ and $E(x_t) = \mu_t = b_1 + b_2 t$, where b_1 and b_2 are constants. In this study, the parameters ν, b_1, and b_2 are estimated by the maximum likelihood method.

4.4.3 Relative SPI

The relative SPI is defined with respect to a reference period in which precipitation time series is assumed to be stationary. For estimating relative SPI, the gamma distribution is first fitted to the reference period and to the period for which the temporal changes have to be analyzed. The two distributions are then compared to determine changes in the concerned period with respect to the reference period.

5. RESULTS AND DISCUSSION

This section is divided into three subsections. Section 5.1 presents the results for drought classification by using SPI, HMM-DI, and Gamma-MM drought indices. Section 5.2 describes the results for precipitation trend analysis. Section 5.3 presents the results of the proposed methodology for the drought classification of nonstationary precipitation series (detrended—SPI) and compares them with the results of classical SPI, relative SPI, and SnsPI for synthetic and real-world precipitation series.

5.1 Drought Classification

The drought indices described in Section 4 are applied to study seasonal (3-month time window) and water-year (12-month time window) droughts in Indiana. Indiana has four distinct seasons, each spanning about 3 months: winter (December to February), spring (March to May), summer (June to August), and fall (September to November). The water-year in Indiana extends from October to September of the following year. First, an annual time-series of cumulative precipitation during any chosen season or water-year are computed. Next, droughts are classified using SPI and Gamma-MM methods. The latter method is also referred to as probabilistic SPI. Both the methods assume that the cumulative precipitation time-series are stationary, and consist of independent and identically distributed samples. In the following paragraphs, winter and water-year droughts from the two methods are presented for a selected station. The two methods are also compared with HMM-based probabilistic drought classification for the same station during winter months.

5.1.1 Winter Droughts

The results are presented for COOP station Wanatah 2 WNW (ID: 129222) located in northwest Indiana. The station is located close to Lake Michigan. Fig. 6.6 shows the empirical cumulative distribution function (CDF) obtained by using Weibull plotting position formula (Chow et al., 1988) along with CDFs of fitted gamma distribution (fitted using maximum likelihood approach) and Gamma-MM for winter precipitation (December—February). The CDF of Gamma-MM is closer to empirical CDF than the CDF of gamma distribution, particularly for the smaller rainfall values [$F(X) < 0.25$], which are critical for drought classification. The Gamma-MM owes its better fit to the large number of tuning parameters ($3M$-1, where M is number of components in Gamma-MM) compared to two parameter gamma distribution.

Increasing the number of mixture components (M) in Gamma-MM method ensures that the model provides a better fit to the data. However, it may also result in overfitting. The Gamma-MM model addresses this problem by using a Bayesian framework that avoids overfitting by marginalizing over the model parameters instead of making point estimates. Fig. 6.7 shows the mixing ratio of a 5-component Gamma-MM fitted to cumulative winter precipitation at this station. The model identifies that three of the five components have negligible contribution and are

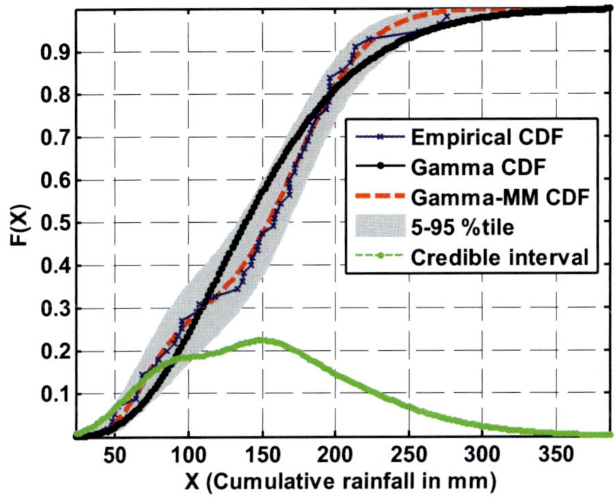

Figure 6.6 Empirical CDF along with CDFs obtained by fitting gamma distribution (Gamma CDF) and gamma mixture model (Gamma-MM CDF) to the 3-month cumulative precipitation in winter (December to February) at COOP station (ID: 129222) Wanatah 2 WNW, IN. The grey band shows 5th and 95th percentile of the Gamma-MM CDF and the *green dotted line* shows width of its credible interval.

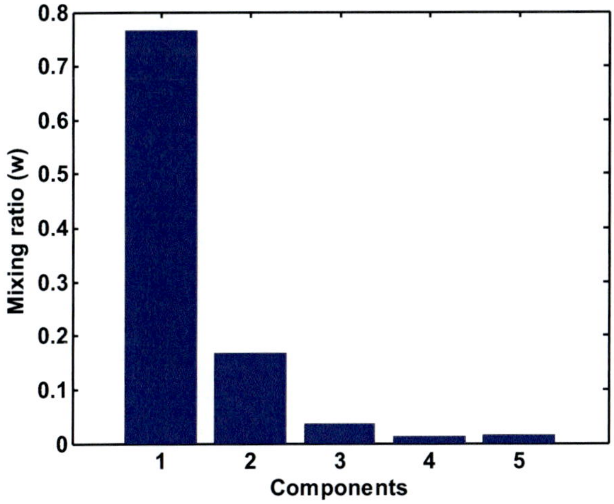

Figure 6.7 Mixing ratios of the components of the Bayesian Gamma-MM. Two components are identified as significant for characterizing winter (December to February) droughts at COOP Station (ID: 129222) Wanatah 2 WNW, IN.

effectively pruned from the model. Thus, the Bayesian framework identifies optimal number of mixture components needed to fit the data.

The Bayesian framework also allows quantification of model uncertainties and their propagation to model estimates. In the context of Gamma-MM, the posterior distribution of model parameters is estimated from which the CDF is obtained. Unlike maximum likelihood approach that yields a point estimate of CDF, the Bayesian approach treats CDF as a random variable and yields distribution of CDFs for a given value of precipitation. The grey shaded band in Fig. 6.6 represents 90% credible interval (5th and 95th percentile). The width of the credible interval is not constant but varies with the magnitude of precipitation. It has a maximum value of 0.22 near the median precipitation (150 mm), a plateau near the intersection of two components (~110 mm; Fig. 6.8), and a monotonic decreasing trend on either side of the median.

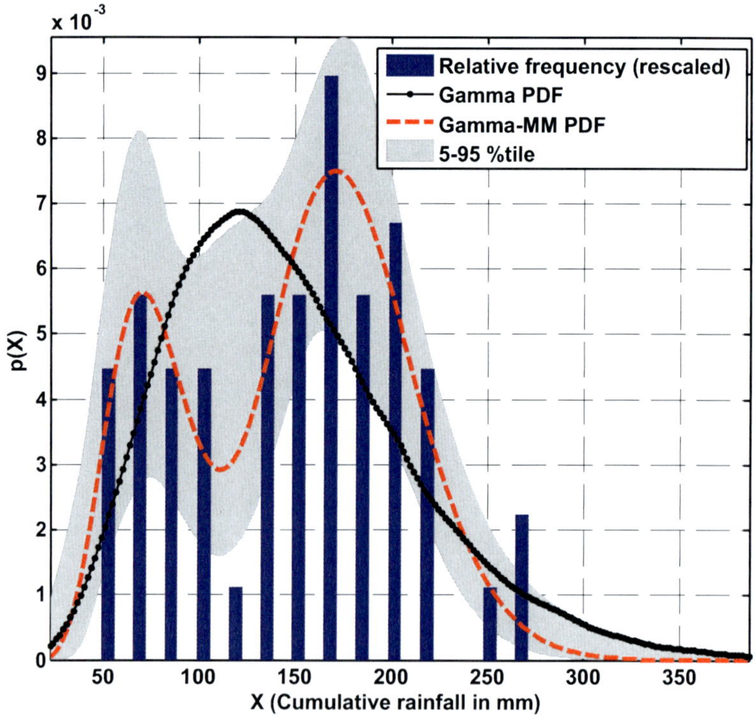

Figure 6.8 Relative frequency of the 3-month cumulative precipitation during winter (December to February) at COOP station (ID: 129222) Wanatah 2 WNW, IN, and probability density functions of the fitted gamma distribution (Gamma PDF) and gamma mixture model (Gamma-MM PDF). The grey band shows 90% credible interval (5th and 95th percentile) of the Gamma-MM PDF.

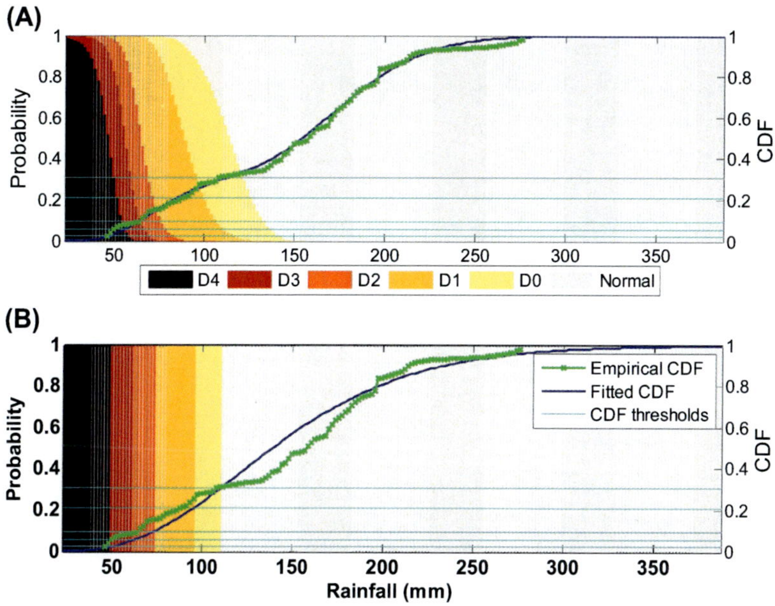

Figure 6.9 Drought classification using 3-month cumulative precipitation during winter (December to February) at COOP station (ID: 129222) Wanatah 2 WNW, IN by (A) the probabilistic SPI (Gamma-MM) (*top panel*) and (B) standard SPI (Gamma distribution) (*bottom panel*). The colored patches represent drought classes, the *light horizontal lines* denote thresholds on CDF specified by US Drought Monitor, and the *solid curves* represent empirical and fitted CDFs.

The width of the credible interval is large even for smaller values of CDFs that decide drought classes in the SPI methodology. In this study, we attempted to engage credible interval of CDF for drought classification. Fig. 6.9B shows the drought classification using standard SPI method. The empirical CDF along with the fitted CDF and drought classification thresholds are shown in the figure. The SPI drought classification uses fixed thresholds, hence the boundaries separating two drought classes are vertical lines on the panel. Fig. 6.9A shows probabilistic drought classification by using Gamma-MM. The classification uses the same thresholds on CDF as SPI but engages uncertainty in the estimate of CDF resulting in probabilistic drought classification. Unlike standard SPI, the demarcating boundaries in the probabilistic SPI are curves denoting varying classification probabilities.

The probabilities associated with drought classification represent uncertainties in determining drought classes. For example, the D4 category drought represents drought conditions where nonexceedance probability of the cumulative rainfall is less than 0.023 [F(X) < 0.023; Table 6.1]. The probabilistic drought classification acknowledges that given limited data and model assumptions such a threshold cannot be determined uniquely but can be estimated probabilistically. The method honors model uncertainty and provides results in a format that could be useful for drought managers.

Fig. 6.10 shows historical drought classes during winter at COOP station 129222 using standard SPI (Fig. 6.10B) and probabilistic SPI (Fig. 6.10A). The droughts classified by the methods are similar; however, the advantages of probabilistic classification are evident in some years. For

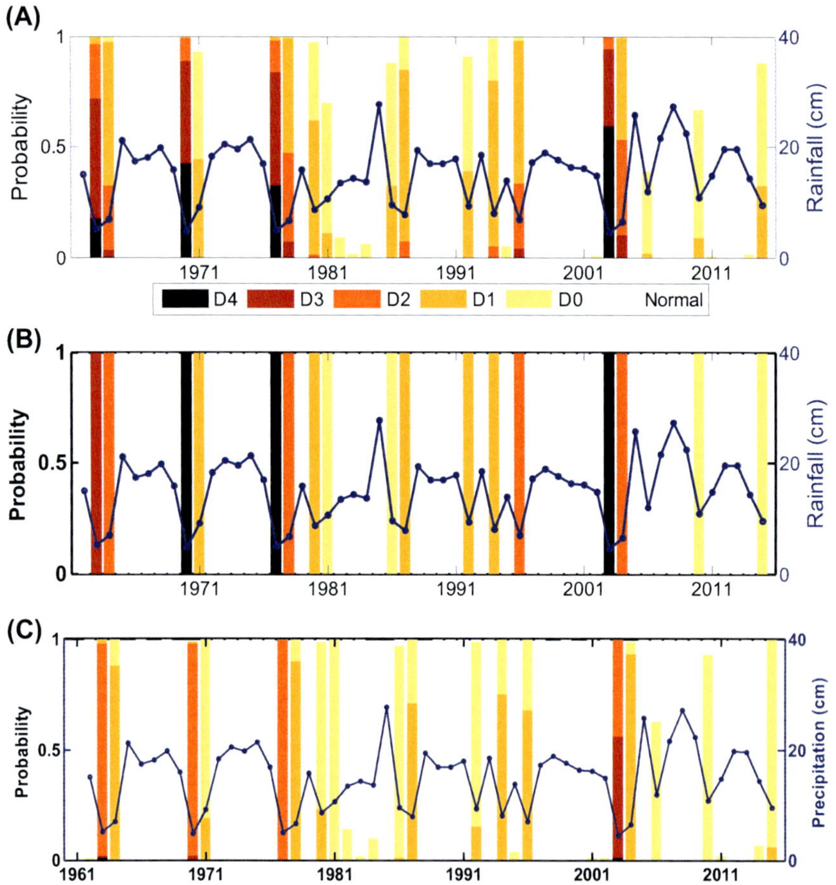

Figure 6.10 Classification of historical 3-month time window droughts during winter (December to February) at COOP station (ID: 129222) Wanatah 2 WNW, IN, using (A) probabilistic SPI, (B) standard SPI, and (C) HMM-DI approaches. The *solid blue line* represents cumulative precipitation during winter months.

example, in 1963 and 1970 the cumulative rainfall values were 52 and 48 mm, respectively. Considering that the difference in cumulative rainfall between these years is less than 7% of their standard deviation (59 mm), one would not expect them to belong to two different drought classes as categorized by SPI (1963 in D3 and 1970 in D4). The probabilistic SPI classifies 1963 and 1970 to D3 class with probability 56% and 47%, and to D4 class with probabilities 17% and 43%, respectively (the remaining probability masses being assigned to other drought classes). The historical drought classes at COOP station 129222 using HMM-DI are shown in Fig. 6.10C. Compared to the probabilistic SPI results (Fig. 6.10A), drought classes obtained using HMM are less conservative. This is evident for the years 1963, 1964, 1970, and 1971, where droughts are classified into less severe category by HMM-DI compared to drought classification using probabilistic SPI. However, it appears that HMM-DI is more conservative when classifying normal conditions as seen for years 1982–84, and 2010. An HMM with 11 hidden states may suffer from an overspecification problem.

5.1.2 Water-Year Droughts

Water-year droughts were analyzed at all COOP stations in the study area. The advantage of using a Gamma-MM model was also validated, specially in cases where precipitation data does follow the gamma distribution.

For brevity, the results are presented only for COOP Station (ID: 125815) in Monroeville 1 NW, IN. This station is located in northwest Indiana in climate division 3. The cumulative precipitation records extend from 1940 to 2000, a total of 61 years of data. The mean water-year precipitation at this location is 845 mm with a standard deviation of 176 mm. Fig. 6.11 shows the relative frequency of cumulative water-year precipitation along with the PDFs for gamma distribution and the proposed Gamma-MM, and they appear to have almost identical shape and spread. The relative frequency of precipitation indicates that there is only one mode, which was captured by both gamma distribution and Gamma-MM. The CDFs of both gamma and Gamma-MM also closely match with the empirical CDF (see Fig. 6.12) of the data obtained by using Weibull plotting position formula for most values of precipitation, with few exceptions. Near 750 and 1500 mm CDFs of gamma distribution and the proposed Gamma-MM do not match the empirical CDF (see Fig. 6.12).

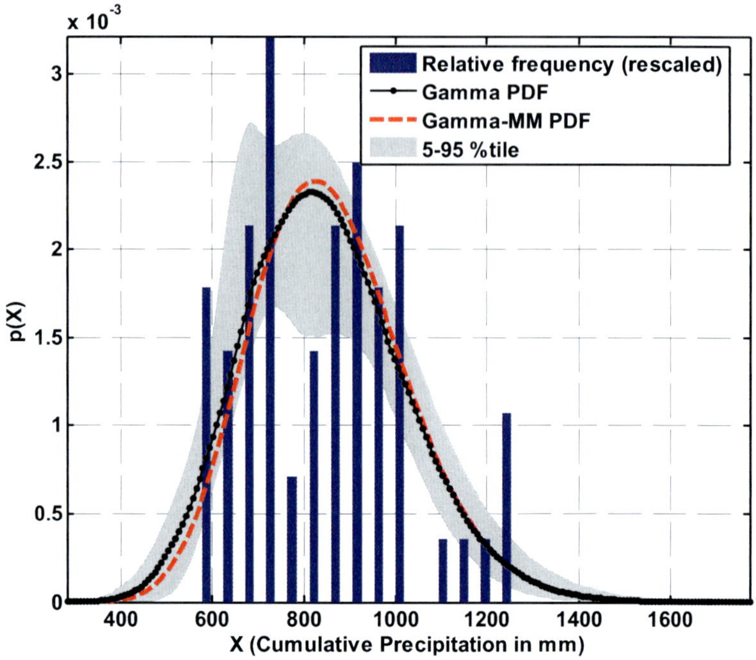

Figure 6.11 Relative frequency of the cumulative precipitation during water-year (October to September of following year) at COOP Station (ID: 125815) in Monroeville 1 NW, IN, and probability density functions of the fitted gamma distribution (Gamma PDF) and gamma mixture model (Gamma-MM PDF). The grey band shows 90% credible interval (5th and 95th percentile) of the Gamma-MM PDF.

The width of the credible interval is maximum near precipitation value of 750 mm, and monotonically decreases on either side.

The demarcating boundaries for the drought classes in the case of Gamma-MM model (shown in Fig. 6.13) are slightly curved for D0–D2 classes indicating less uncertainty. However, the demarcating boundaries are more curved for D3 and D4 drought classes indicating higher uncertainty in their classification.

Fig. 6.14 shows the historical drought classification for water-year droughts at Monroeville 1 NW station. The probabilistic SPI (Fig. 6.14B) provides drought classification similar to the standard SPI (Fig. 6.14A). The standard SPI is more conservative for less severe droughts, and probabilistic SPI is more conservative for severe droughts (D4 and D3 classes). As in the case of seasonal droughts (see Fig. 6.10), the advantages of probabilistic classification are evident for some years. For example, in 1941 and 1999 the cumulative rainfall values were 584, and 603 mm,

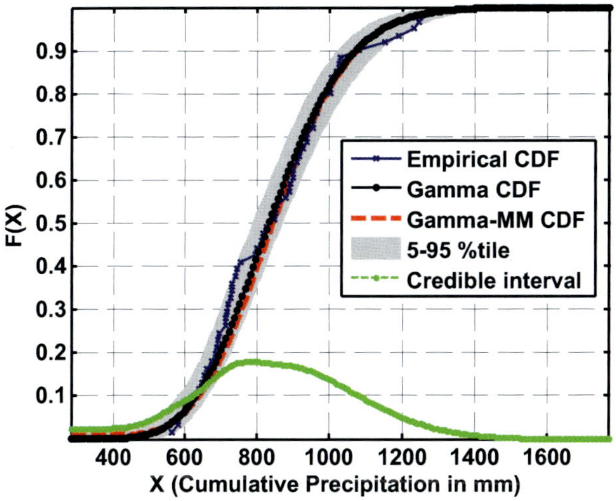

Figure 6.12 Empirical CDF along with CDFs obtained by fitting gamma distribution (Gamma CDF) and gamma mixture model (Gamma-MM CDF) to cumulative precipitation during water-year (October to September of following year) at COOP Station (ID: 125815) Monroeville 1 NW, IN. The grey band shows 5th and 95th percentile of the Gamma-MM CDF and the *green dotted line* shows width of its credible interval.

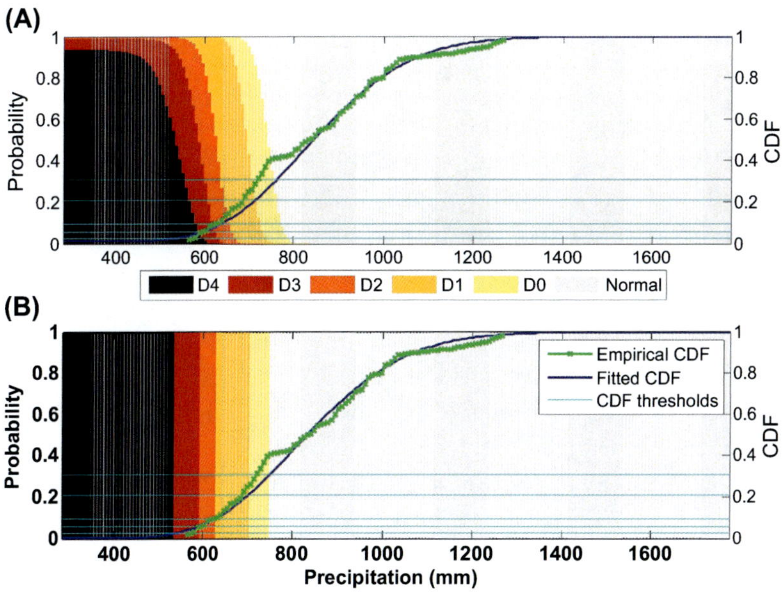

Figure 6.13 Drought classification using cumulative precipitation during water-year (October to September of following year) at COOP Station (ID: 125815) Monroeville 1 NW, IN, by (A) the probabilistic SPI (Gamma-MM) (*top panel*) and (B) standard SPI (Gamma distribution) (*bottom panel*). The colored patches represent drought classes, the *light horizontal lines* denote thresholds on CDF specified by US Drought Monitor, and the *solid curves* represent empirical and fitted CDFs.

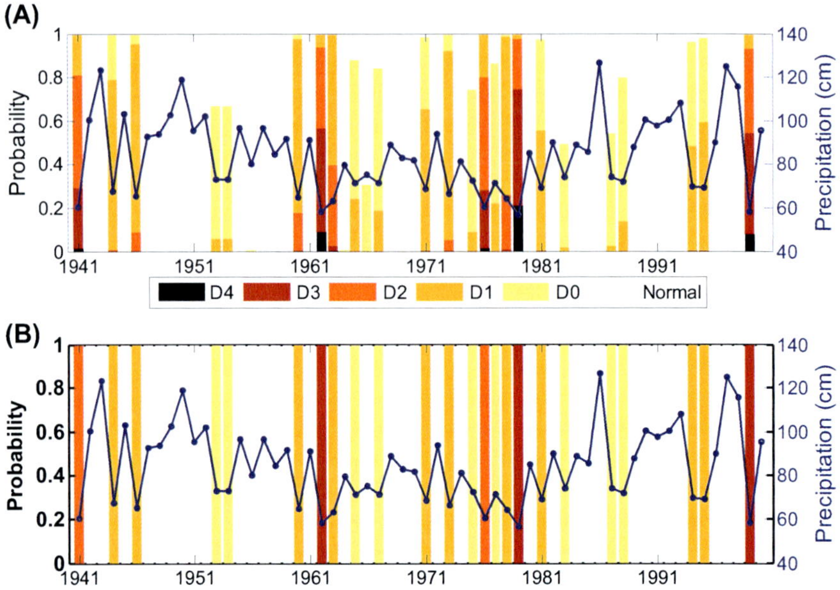

Figure 6.14 Classification of historical droughts during water-year (October to September of following year) at COOP station (ID: 125815) Monroeville 1 NW, IN, using (A) probabilistic SPI and (B) standard SPI approaches. The *solid blue line* represents cumulative precipitation during water-year.

respectively. Considering that the difference in cumulative rainfall between these years is 19 mm, that is, less than 11% of their standard deviation (176 mm), one would hope that these years would have similar drought classes. However, SPI classifies them as D2 (1941) and D3 (1999) classes, respectively. The probabilistic SPI, on the other hand, classifies 1941 and 1999 to D2 class with probability 52% and 39%, and to D3 class with probabilities 28% and 46%, respectively (the remaining probability masses being assigned to other drought classes).

Analyses of historical droughts over the climate divisions in Indiana (see Fig. 6.15) reveals that the drought characterizations by both standard SPI and Gamma-MM are almost identical for the major events, except for some minor differences. For example, in September 1940, climate division 6 is classified as D3 drought (−1.9) by standard SPI whereas Gamma-MM estimates average drought index as −2.1(based on 10,000 posterior estimates of the CDF) thereby classifying 1940 in D4 drought. By using fixed thresholds to characterize droughts (as in case of SPI) or when reporting drought class based on average Gamma-MM, we come to different conclusions. Such

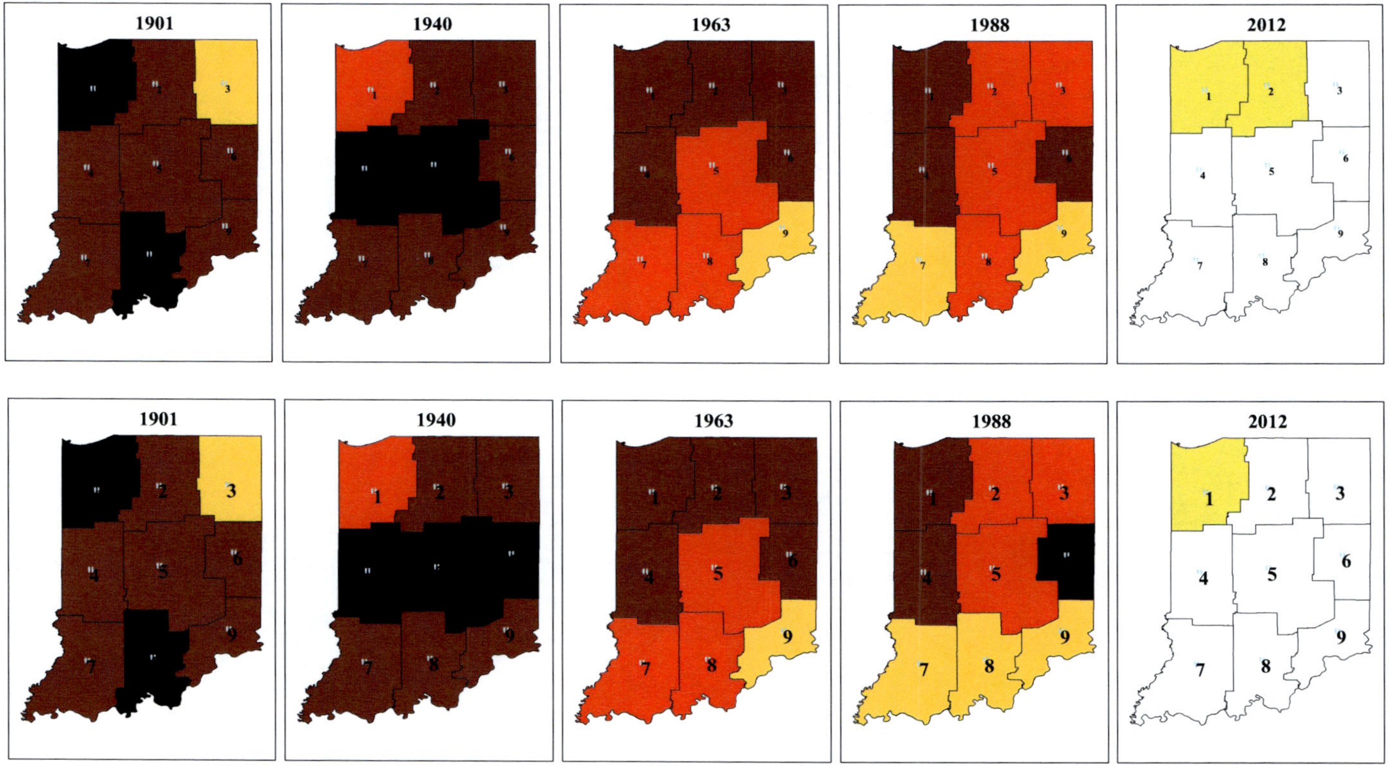

Figure 6.15 Spatial variation of historical water-year drought classes at each climate division during major drought events over Indiana according to standard SPI (top panel) and Gamma-MM (bottom panel). Color coding for drought classes similar to Fig. 6.9.

differences can have significant impact on decision making. It can result in incorrect assessment of drought impacts and lead to under/over allocation of resources and affect triggering of mitigation measures. A probabilistic drought classification provides estimates of uncertainty in drought classification, and thus helps decision makers to take informed actions.

5.2 Precipitation Trends

The seasonal and water-year precipitation series are analyzed for trends. Fig. 6.16 shows the results of MK trend test of seasonal precipitation series (i.e., 3-month time window) for each ending month at all COOP stations over Indiana. Results indicate that trends in seasonal precipitation series are not statistically significant ($\alpha = 0.05$) at most stations (denoted by circles). Majority of COOP stations that have statistically significant trends have positive slope, indicating that precipitation magnitudes are increasing with time. These stations are denoted by upward pointing triangles in Fig. 6.16. Stations with statistically significant negative trend in precipitation magnitudes are denoted by downward pointing triangular markers. Very few stations have negative trends, and these are predominantly present during winter and spring months (February to May).

Similar analysis was performed for 12-month cumulative precipitation time-series at all COOP stations over Indiana, for each ending month. While majority of the stations (56 COOP stations for water-year precipitation series) have no statistically significant trends, 32 stations exhibited statistically significant increase in cumulative precipitation, while only one station showed significant negative trend. The results for other ending months were similar, except for minor differences, and are not shown here for brevity.

5.3 Drought Classification for Nonstationary Precipitation Series

Fig. 6.17 shows the results of MK trend test of summer precipitation series at COOP station in Columbus, IN. The trend is found to be significant ($\alpha = 0.05$), with a positive Sen's slope of 0.52 mm/year, indicating that summer precipitation totals are increasing over time. Fig. 6.17 also shows the histogram of summer precipitation along with autocorrelation at several lags.

As the precipitation series exhibits nonstationarity, classical drought indices such as SPI may not be appropriate. Alternate methods, namely, relative SPI and SnsPI are applied to the precipitation series. Fig. 6.18 shows

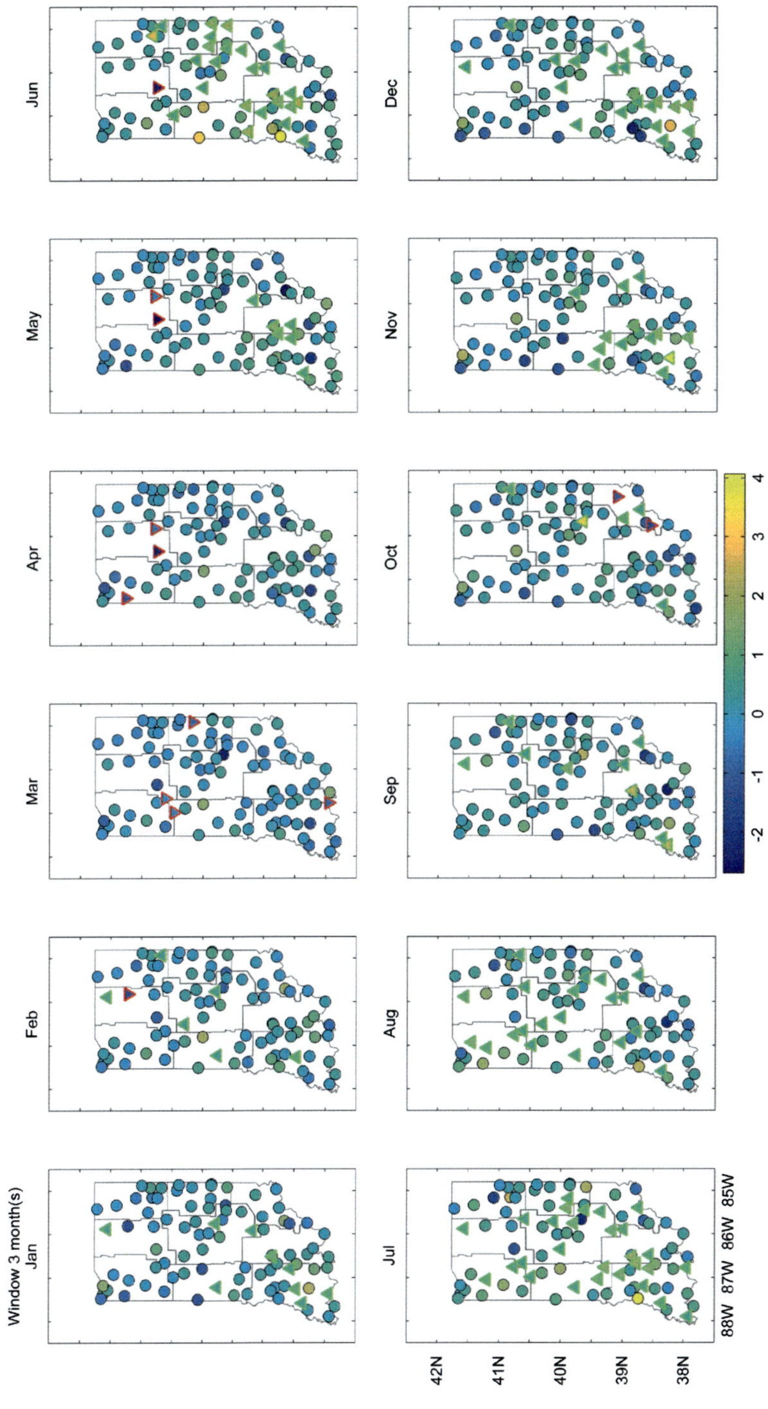

Figure 6.16 MK trend test (with $\alpha = 0.05$) of seasonal precipitation series (3-month time window) for each ending month (January to December) at COOP stations over Indiana. Markers are color coded based on the magnitude of Sen's slope. Stations with statistically significant positive slope are denoted by *upward pointing triangular markers*, while those with significant negative slope are denoted as *downward pointing triangular markers*.

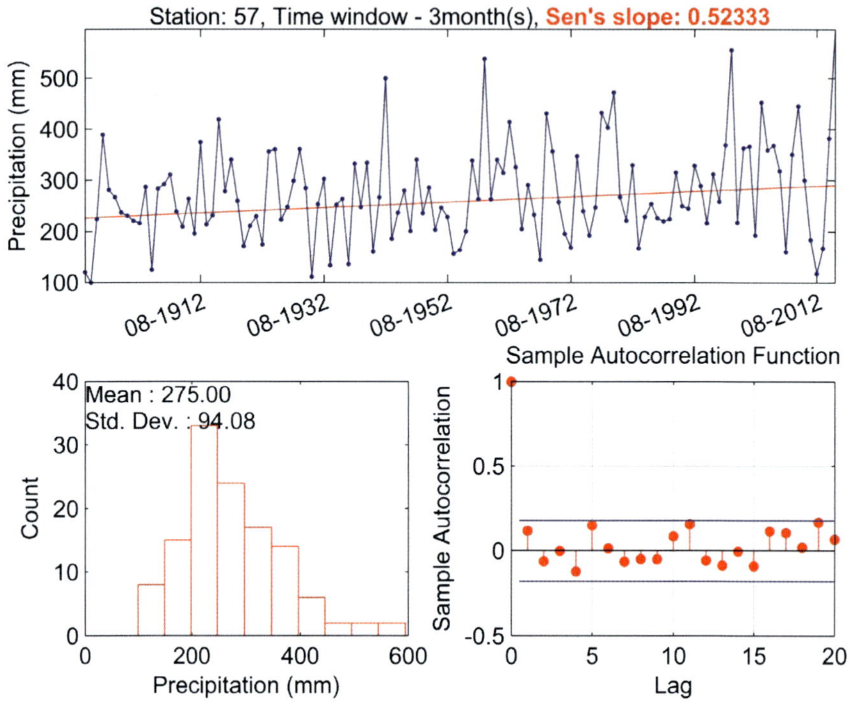

Figure 6.17 MK trend test result (with $\alpha = 0.05$) of summer precipitation series (June to August) at COOP station (ID: 121747) at Columbus, IN.

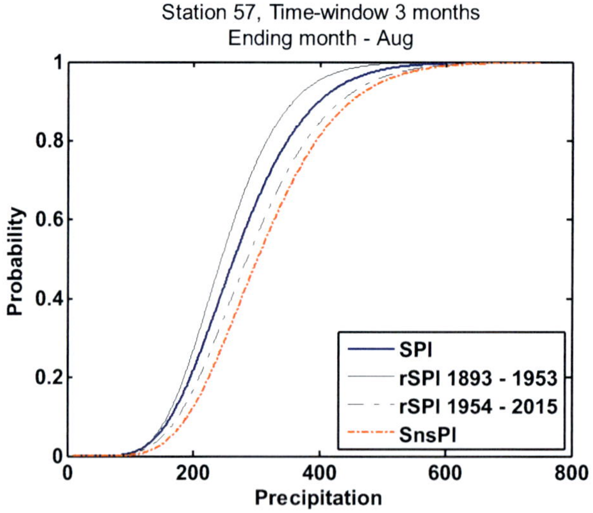

Figure 6.18 Comparison of CDFs of classical SPI, relative SPI (rSPI), and SnsPI summer precipitation series (June to August) at COOP station (ID: 121747) at Columbus, IN.

the comparison of CDFs of three drought analysis methods. The differences in the CDF for different epochs (1893–1953 and 1954–2015) indicate that recent periods are becoming wetter. For example, a cumulative precipitation of 200 mm corresponds to relative SPI of 0.25 and 0.15 for the first (1893–1953) and second half (1954–2015) of the study period, respectively. The corresponding values from standard SPI and SnsPI are 0.2 and 0.1, respectively. These different values of drought index suggest that when precipitation series is nonstationary, standard SPI and its variants that assume stationarity are not reliable.

The proposed method for classifying nonstationary precipitation series is first applied on two synthetic series and then to the summer precipitation series at COOP station in Columbus, IN.

The first step of the proposed method involves detrending the time series (Section 4.4). One approach for detrending is to first decompose the time series using Empirical Mode Decomposition which separates both deterministic and stochastic trends as Intrinsic Mode Functions (IMFs; Wu et al., 2007). The IMF consists of amplitude and frequency modulations and are extracted from the original time-series in a stepwise manner, beginning with high frequency local oscillations that are super-imposed on low frequency part of the data. These high-frequency oscillations are subtracted from the data to obtain the residual. Then the next IMF representing highest frequency oscillations present in the residual is extracted. The process is repeated until we obtain a residual with no oscillation. Typically, if we have a dataset of length N, we can extract about $log_2 N$ IMFs from the data. Then, those IMFs that are statistically different from white noise are subtracted from the original series to obtain a detrended series.

The second approach, and the one that is used in this study, is to fit polynomials of several orders to the data, and obtain the best polynomial by using the Akaike Information Criteria (AIC). Trend line is then obtained from the coefficients of the best fitting polynomial.

5.3.1 Synthetic Series

First, 100 samples are drawn from the gamma distribution $G(x_t|\nu,s_t)$ with scale parameter $s_t = 8.33$ and shape parameter $\nu = 36.0$. The time-series of the 100 samples are shown in Fig. 6.19. The mean and standard deviation of the series are around 300 and 50 mm, respectively.

For the first synthetic example, a linear trend is added to the mean of the synthetic series. The resulting series is shown in Fig. 6.20. Drought characterization is then performed using the following methods:

- SPI—A gamma distribution is fitted to the synthetic series and SPI values are obtained as described in Section 4.1. The drought classes for classical SPI are shown in the top panel of Fig. 6.21.
- SnsPI (Russo et al., 2013)—The SnsPI drought indices are obtained by using the method presented in Section 4.4. In this method, the scale parameter of the gamma distribution is assumed to vary linearly with time. Such a model specifies a linear trend for the mean of the series and a quadratic trend for its variance. Drought classes obtained for SnsPI are shown in the second panel of Fig. 6.21.
- Detrended SPI—first, trend in the mean of the series is obtained by fitting a polynomial. Next, the detrended series is obtained by subtracting the estimated trend from the data. Several candidate distributions (beta, Birnbaum-Saunders, exponential, extreme value, gamma, generalized extreme value, generalized pareto, inverse Gaussian, logistic,

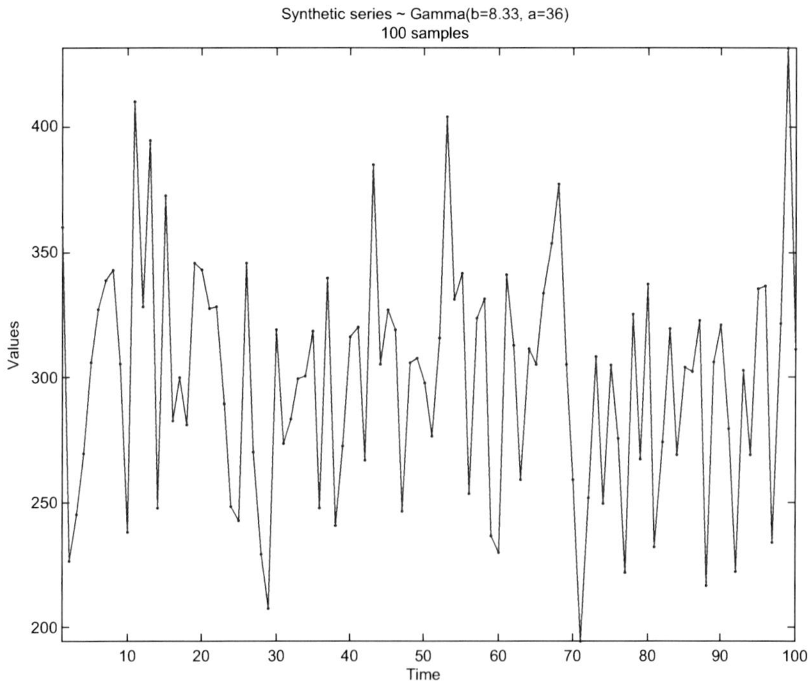

Figure 6.19 Synthetic time series generated from a gamma distribution with mean 300 mm and standard deviation equal to 50 mm.

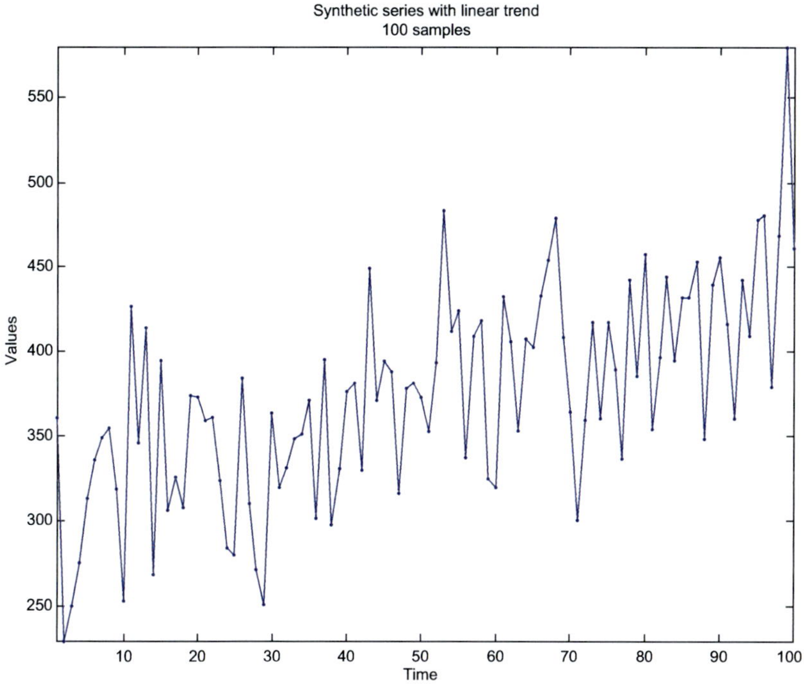

Figure 6.20 Synthetic series with linear trend in mean.

loglogistic, lognormal, Nakagami, normal, Rayleigh, Rician, t location-scale, and Weibull distribution) are fitted to the detrended series, and the best distribution is selected based on AIC score (see Fig. 6.22). The method identifies normal distribution to be the best distribution. The selected distribution is then used for obtaining drought indices shown in the third panel of Fig. 6.21.

As expected, the standard SPI overestimates drought severities for the earlier time steps, and underestimates it for the later time steps. This is evident from the top panel of Fig. 6.21 and also by observing that the threshold that defines any drought class is fixed over the entire time period (dashed line in Fig. 6.23A and first panel of Fig. 6.23B). For example, the threshold for D4 drought ($F(X) < 0.023$) is about 260 mm, and remains unchanged over the entire period. The SnsPI treats the scale parameter as a time-varying component of the model, and is able to capture the linear trend in the synthetic data. This is evident from the dotted lines of Fig. 6.23A. For example, the threshold for D4 category drought is close to 200 mm at time-step 1, and close to 350 mm at time-step 100.

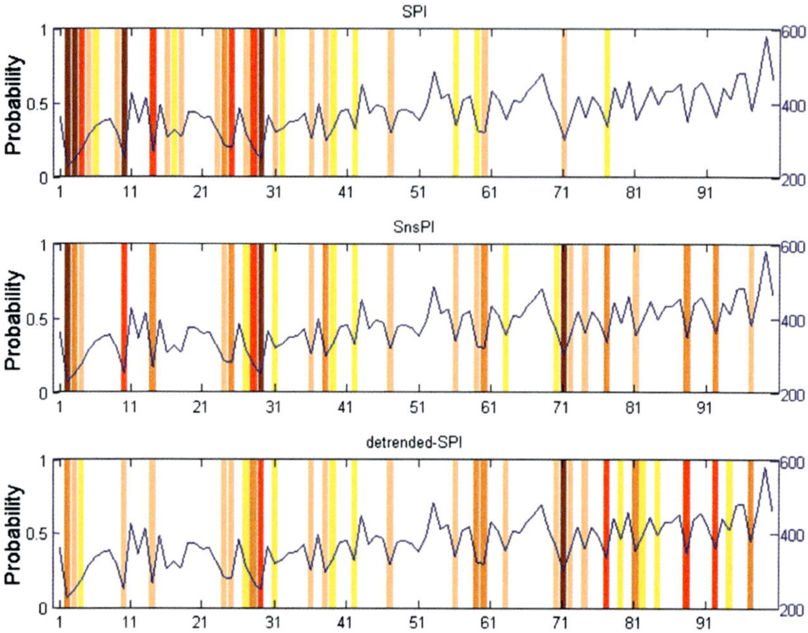

Figure 6.21 Droughts classified into D0–D4 classes according to US Drought Monitor classification scheme for SPI (*first panel*), SnsPI (*second panel*), and proposed detrended-SPI method (*third panel*). The *blue line* represents the original synthetic series that has linear trend.

The threshold varies linearly between these points (Fig. 6.23A). As a consequence, the drought classes obtained are markedly different compared to standard SPI approach. In SnsPI the drought occurrences are spread evenly across the study period (second panel of Fig. 6.21), rather than being concentrated in the earlier years like standard SPI (first panel of Fig. 6.21).

SnsPI prescribes linear trend for mean and a quadratic trend for variance. This property is evident from the second panel of Fig. 6.23B. The proposed detrended-SPI method, however, specifies linear trend for mean and no trend for variance (see third panel of Fig. 6.23B).

The second approach for drought analysis of nonstationary series assumes that the "normal" conditions for a station are fixed (with respect to a reference period), but the frequency of droughts are changing with time. Fig. 6.24 shows the drought classes by the new approach when the reference point is time-step 1. As expected, the droughts become less severe with time because of the positive trend in the series.

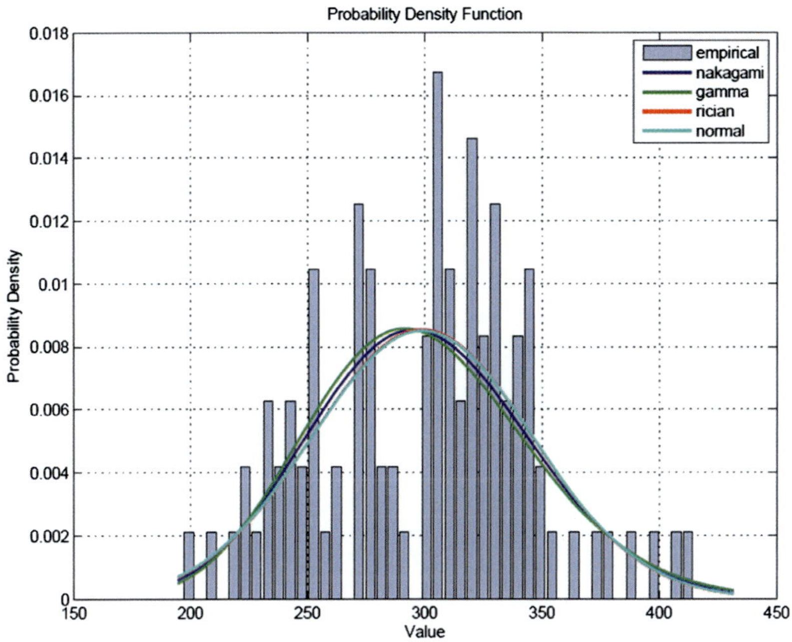

Figure 6.22 Top 4 out of 17 candidate distributions based on AIC scores that fit the detrended series (original synthetic series had linear trend).

For the second synthetic example, a quadratic trend is imposed on a series generated from the same gamma distribution as used in the first example (Fig. 6.25). The three methods, standard SPI, SnsPI, and detrended-SPI, are applied again. For the detrended-SPI, normal distribution was again selected as the best distribution based on AIC score. The results obtained are shown in Fig. 6.26.

The first panel of Fig. 6.26 shows the drought classes obtained from standard SPI. Given the significantly positive nonlinear trend of the time series, most drought occurrences are concentrated in the first half of the time period. The results obtained from the detrended-SPI (third panel of Fig. 6.26) are significantly different from SnSPI (second panel of Fig. 6.26), specifically in the second half of the series. These differences can be attributed to the fact that SnsPI can only account for linear trends in mean.

The advantages of proposed detrended-SPI are evident in Fig. 6.27A and B. Fig. 6.27A shows the variation of precipitation thresholds over time for different drought categories. The SPI thresholds remain constant over time (dashed lines in Fig. 6.27A), SnsPI thresholds vary linearly (dotted lines

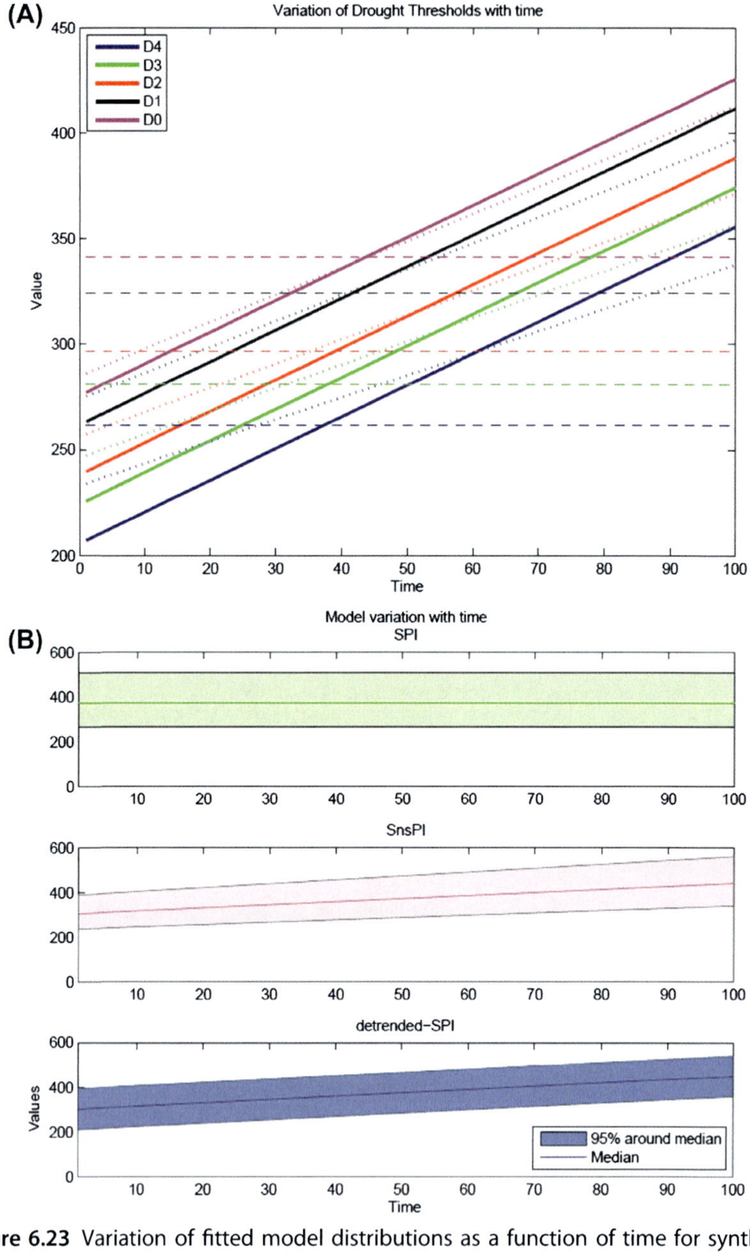

Figure 6.23 Variation of fitted model distributions as a function of time for synthetic data with linear trends. (A) shows the variation of data thresholds that define drought classes (D4 to D0) according to SPI (*dashed lines*), SnsPI (*dotted lines*), and proposed detrended-SPI (*solid lines*), (B) shows the median of the distribution and 95% distribution interval around the median for SPI (*first panel of B*), SnsPI (*second panel of B*), and detrended-SPI approach (third panel of B).

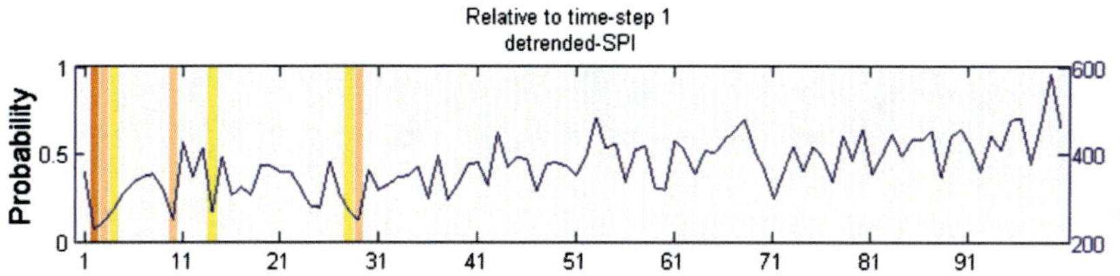

Figure 6.24 Drought classes calculated relative to distributional properties at the first time-step using detrended-SPI.

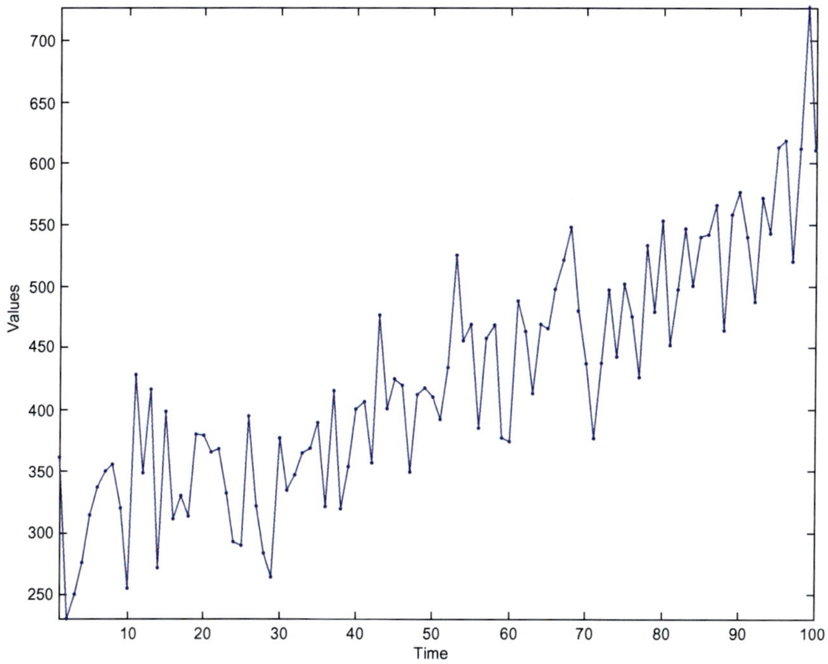

Figure 6.25 Synthetic time series with quadratic trend in mean.

in Fig. 6.27A), while the drought thresholds for the proposed method show a quadratic trend (solid lines in Fig. 6.27A).

Fig. 6.28 shows the drought classes over the study period according to detrended-SPI when the reference period is selected as time-step 1. Again as expected, the drought severities decrease with time, indicating wetter conditions (or positive trend in data).

5.3.2 Observed Precipitation Series

The proposed detrended-SPI method was tested on observed precipitation data at COOP station 121747 in Columbus, IN. The precipitation

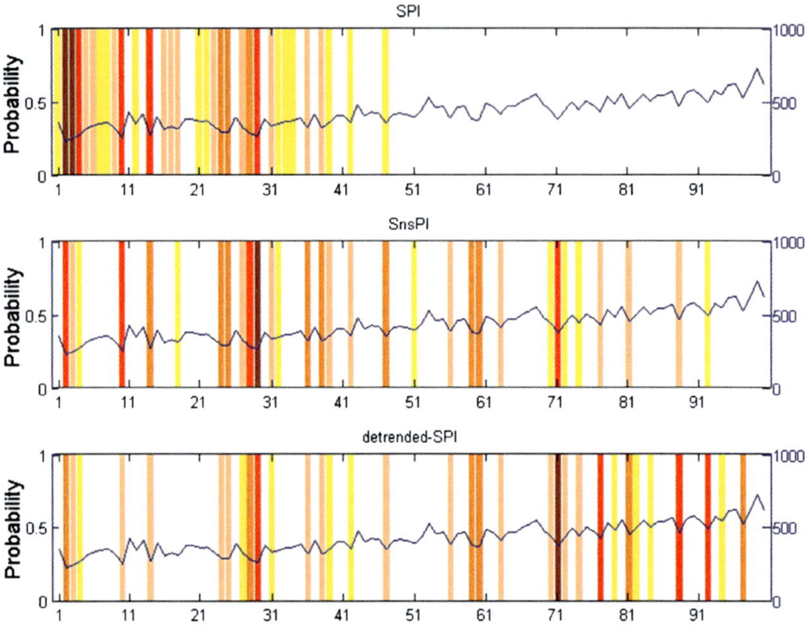

Figure 6.26 Droughts classified into D0–D4 classes, according to the US Drought Monitor classification scheme for SPI (first panel), SnsPI (second panel), and proposed detrended-SPI method (*third panel*). The *blue line* represents the original synthetic series that has nonlinear trend.

time-series (Fig. 6.17) corresponds to 3-month time window during summer (June to August). The red line indicates positive Sen's slope, indicating increasing precipitation over time. The series exhibits a statistically significant positive trend according to MK trend test ($\alpha = 0.05$).

The results obtained from the three methods, SPI, SnsPI, and detrended-SPI, are shown in Fig. 6.29. The detrended-SPI identifies linear trend in the precipitation series and selects normal distribution for the detrended series. As in the synthetic case, the SPI and SnsPI results are markedly different during the recent years. Further, the SnsPI owing to quadratic increase in the variance, assigns low severity to the recent droughts.

The consequences of model assumptions in the three methods become clear in Fig. 6.30. The dashed lines in Fig. 6.30A denote the thresholds of D4–D0 classes according to standard SPI. These dashed lines are horizontal

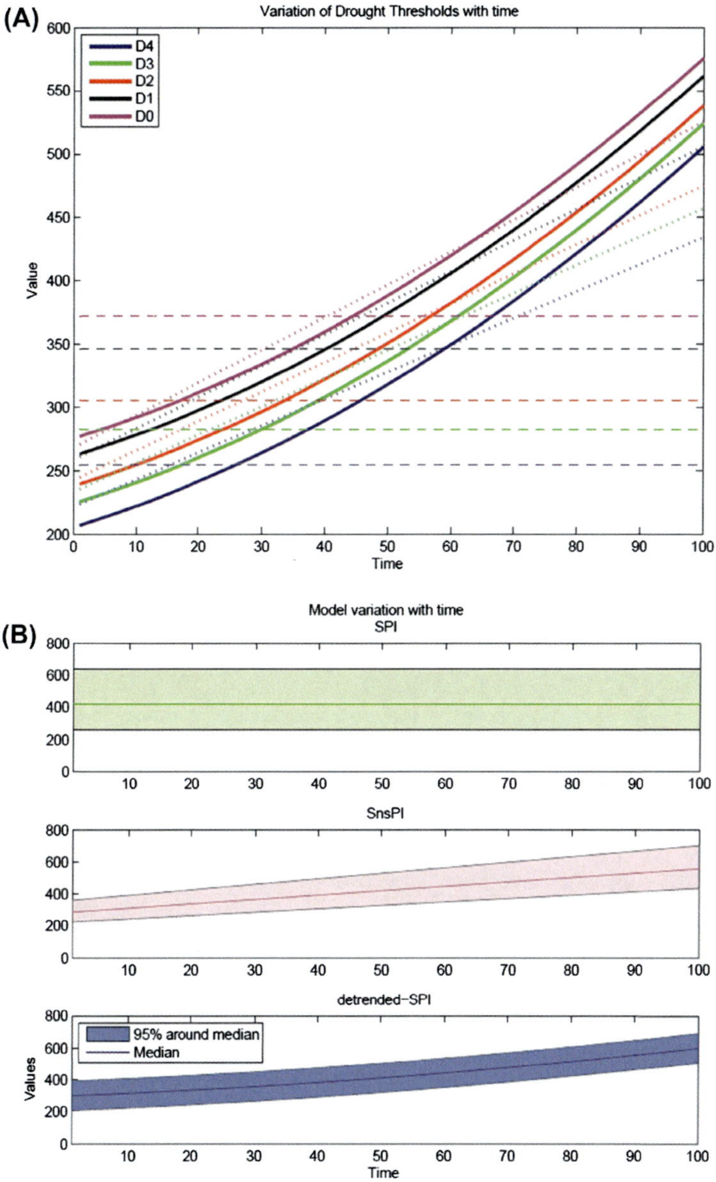

Figure 6.27 Variation of fitted model distributions as a function of time for synthetic data with nonlinear trends. (A) shows the variation of data thresholds that define drought classes (D4 to D0) according to SPI (*dashed lines*), SnsPI (*dotted lines*), and proposed detrended-SPI (*solid lines*); (B) shows the median of the distribution and 95% distribution interval around the median for SPI (*first panel of B*), SnsPI (*second panel of B*), and detrended-SPI approach (third panel of B).

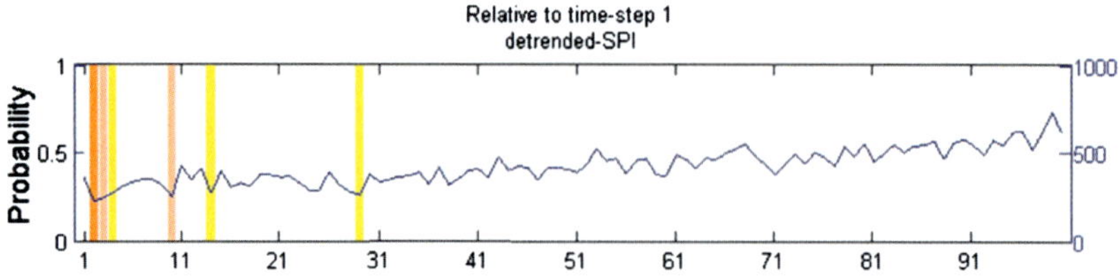

Figure 6.28 Drought classes for synthetic series with nonlinear trend in its mean, calculated relative to distributional properties at the first time-step using detrended-SPI.

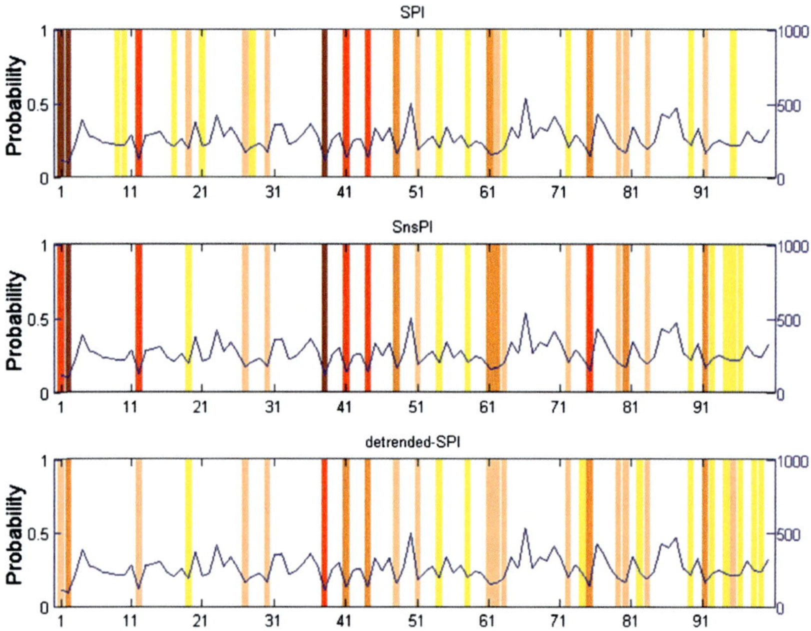

Figure 6.29 Droughts classified into D0–D4 classes according to the US Drought Monitor classification scheme for SPI (*first panel*), SnsPI (*second panel*), and proposed detrended-SPI method (*third panel*). The *blue line* represents the cumulative summer precipitation series at COOP station 121747 in Columbus, IN.

indicating that the thresholds on precipitation for different drought classes do not change over time (stationarity assumption). This property of SPI model also becomes evident in the median and variance in the first panel of Fig. 6.30B. For SnsPI (second panel of Fig. 6.30B) the median varies linearly with time, whereas the variance varies quadratically. For detrended-SPI (third panel of Fig. 6.30B), the median varies linearly, while the variance is constant.

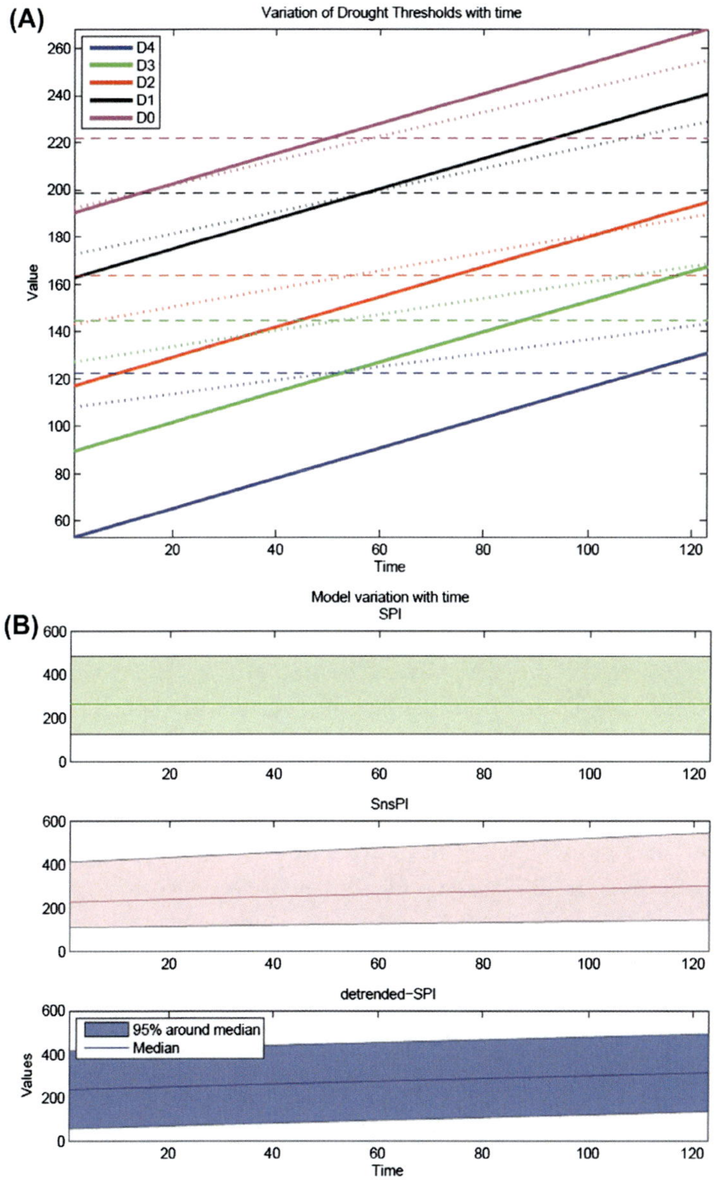

Figure 6.30 Variation of fitted model distributions as a function of time for nonstationary cumulative summer precipitation (3-month time window, June to August) at COOP station 121747 in Columbus, IN. (A) shows the variation of data thresholds that define drought classes (D4 to D0) according to SPI (*dashed lines*), SnsPI (*dotted lines*), and proposed detrended-SPI (*solid lines*), (B) shows the median of the distribution and 95% distribution interval around the median for SPI (*first panel of B*), SnsPI (*second panel of B*), and detrended-SPI approach (*third panel of B*).

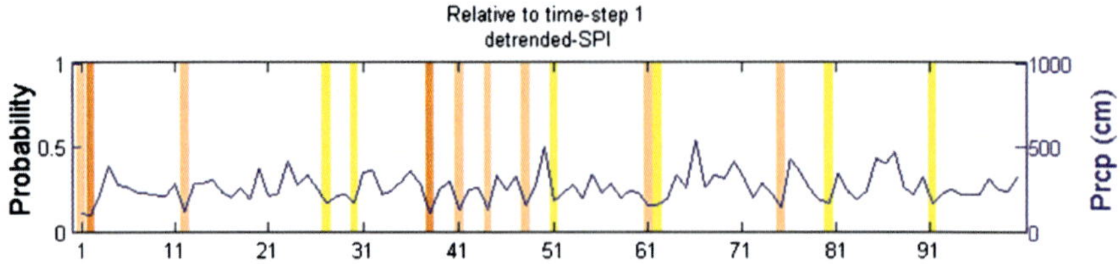

Figure 6.31 Drought classes calculated relative to distributional properties of cumulative summer precipitation series (3-month time window, June—August) at the first time step according to detrended-SPI method. Precipitation data is from COOP station 121747 in Columbus, IN.

Relative drought severity classes with respect to the first time-step are shown in Fig. 6.31. Since the series has positive trend (top panel Fig. 6.17), the summer droughts relative to the first year becomes milder over time.

6. SUMMARY AND CONCLUDING REMARKS

1. Meteorological droughts over Indiana are studied by using three drought classification methods, namely, standardized precipitation index (SPI), hidden Markov model based drought index (HMM-DI), and gamma mixture model—based drought index (Gamma-MM). The latter two provide probabilistic classification of droughts.
2. The Gamma-MM alleviates the problem of adopting a prescribed distribution for SPI analysis by modeling the data with a mixture of gamma distributions. Given sufficient components in the mixture, the Gamma-MM can give arbitrarily close approximation to any general continuous distribution in the range $(0, \infty)$. The problem of overfitting the data is avoided by using a Bayesian framework that determines optimum number of components for the model.
3. The results of droughts over Indiana during the water year (October—September) and for each of the four seasons (winter, spring, summer, and fall) suggest that the drought classification by the Gamma-MM is similar to SPI when the data satisfies SPI assumptions. However, the results of the two methods are significantly different when data violate SPI assumptions. Further, the drought classification obtained by Gamma-MM is less conservative compared to the HMM classification.

4. Mann-Kendall trend test of precipitation series indicated that many stations in Indiana have statistically significant trends. Positive trends are observed at many stations during the summer months and water-year, while negative trends are observed only in a few stations during winter and spring months.
5. A methodology is proposed for drought analysis of trend stationary precipitation series. The method first explicitly removes trend from the data, and then analyzes detrended series to estimate changes in drought frequency and drought classification thresholds.
6. The application of the proposed method on synthetic and real-world data suggests that the new method offers flexibility in modeling nonstationary time-series compared to standardized non-stationary precipitation index (SnsPI) and relative SPI.
7. The proposed method can be applied only if the trends are deterministic and can be explicitly estimated as a function of time. Future research should explore possibilities to relax this requirement.

REFERENCES

AghaKouchak, A., 2014. A baseline probabilistic drought forecasting framework using standardized soil moisture index: application to the 2012 United States drought. Hydrol. Earth Syst. Sci. 18, 2485–2492.

Belayneh, A., Adamowski, J., Khalil, B., Ozga-Zielinski, B., 2014. Long-term SPI drought forecasting in the Awash River Basin in Ethiopia using wavelet neural network and wavelet support vector regression models. J. Hydrol. 508, 418–429. https://doi.org/10.1016/j.jhydrol.2013.10.052.

Benjamini, Y., Hochberg, Y., 1995. Controlling the false discovery rate: a practical and powerful approach to multiple testing. J. R. Stat. Soc. Ser. B Methodol. 57, 289–300.

Bishop, C.M., 2006. Pattern Recognition and Machine Learning. Springer, New York.

Bonaccorso, B., Peres, D.J., Cancelliere, A., Rossi, G., 2013. Large scale probabilistic drought characterization over Europe. Water Resour. Manag. 27, 1675–1692. https://doi.org/10.1007/s11269-012-0177-z.

Burn, D.H., Elnur, M.A.H., 2002. Detection of hydrologic trends and variability. J. Hydrol. 255, 107–122. https://doi.org/10.1016/S0022-1694(01)00514-5.

Charusombat, U., Niyogi, D., 2011. A hydroclimatological assessment of regional drought vulnerability: a case study of Indiana droughts. Earth Interact. 15, 1–65. https://doi.org/10.1175/2011EI343.1.

Chow, V.T., Maidment, D.R., Mays, L.W., 1988. Applied hydrology. McGraw-Hill Ser. Water Resour. Environ. Eng. 396–397.

Cole, J.E., Cook, E.R., 1998. The changing relationship between ENSO variability and moisture balance in the continental United States. Geophys. Res. Lett. 25, 4529–4532. https://doi.org/10.1029/1998GL900145.

Coulibaly, P., Baldwin, C.K., 2005. Nonstationary hydrological time series forecasting using nonlinear dynamic methods. J. Hydrol. 307, 164–174.

Dai, A., 2011. Drought under global warming: a review. Wiley Interdiscip. Rev. Clim. Change 2, 45–65. https://doi.org/10.1002/wcc.81.

DeVore, R.A., Lorentz, G.G., 1993. Constructive Approximation. Springer.

Dracup, J.A., Lee, K.S., Paulson Jr., E.G., 1980. On the definition of droughts. Water Resour. Res. 16, 297–302. https://doi.org/10.1029/WR016i002p00297.

Dubrovsky, M., Svoboda, M.D., Trnka, M., Hayes, M.J., Wilhite, D.A., Zalud, Z., Hlavinka, P., 2009. Application of relative drought indices in assessing climate-change impacts on drought conditions in Czechia. Theor. Appl. Climatol. 96, 155–171. https://doi.org/10.1007/s00704-008-0020-x.

Evin, G., Merleau, J., Perreault, L., 2011. Two-component mixtures of normal, gamma, and Gumbel distributions for hydrological applications (W08525). Water Resour. Res. 47 https://doi.org/10.1029/2010WR010266.

Fan, Y., van den Dool, H., 2004. Climate Prediction Center global monthly soil moisture data set at 0.5° resolution for 1948 to present. J. Geophys. Res. Atmos. 109, D10102. https://doi.org/10.1029/2003JD004345.

Federal Emergency Management Agency, 1995. National Mitigation Strategy. FEMA, Washington, DC.

Friedman, D.G., 1957. The prediction of long-continuing drought in south and southwest Texas. Ocas. Pap. Meteorol. 1, 182.

Fuller, W.-A., 1996. Introduction to Statistical Time Series. In: Wiley series in probability and statistics, 2nd ed. Wiley, New York.

Gao, B.-C., 1996. NDWI—a normalized difference water index for remote sensing of vegetation liquid water from space. Remote Sens. Environ. 58, 257–266.

Geman, S., Geman, D., 1984. Stochastic relaxation, Gibbs distributions, and the Bayesian restoration of images. IEEE Trans. Pattern Anal. Mach. Intell. PAMI 6, 721–741. https://doi.org/10.1109/TPAMI.1984.4767596.

Green, P.J., 1995. Reversible jump Markov chain Monte Carlo computation and Bayesian model determination. Biometrika 82, 711–732. https://doi.org/10.1093/biomet/82.4.711.

Guttman, N.B., 1999. Accepting the standardized precipitation index: a calculation algorithm. JAWRA 35, 311–322. https://doi.org/10.1111/j.1752-1688.1999.tb03592.x.

Hamed, K.H., Rao, A.R., 1998. A modified Mann-Kendall trend test for autocorrelated data. J. Hydrol. 204, 182–196. https://doi.org/10.1016/S0022-1694(97)00125-X.

Han, P., Wang, P.X., Zhang, S.Y., Zhu, D.H., 2010. Drought forecasting based on the remote sensing data using ARIMA models. Math. Comput. Model. 51, 1398–1403. https://doi.org/10.1016/j.mcm.2009.10.031.

Hayes, M.J., Wilhelmi, O.V., Knutson, C.L., 2004. Reducing drought risk: bridging theory and practice. Nat. Hazards Rev. 5, 106–113.

Heim, R.R., 2002. A review of twentieth-century drought indices used in the United States. Bull. Am. Meteorol. Soc. 83, 1149.

Kao, S.-C., Govindaraju, R.S., 2010. A copula-based joint deficit index for droughts. J. Hydrol. 380, 121–134. https://doi.org/10.1016/j.jhydrol.2009.10.029.

Kulkarni, A., von Storch, H., 1995. Monte Carlo experiments on the effect of serial correlation on the Mann-Kendall test of trend. Meteorol. Z. 4, 82–85.

Liu, W., Kogan, F., 1996. Monitoring regional drought using the vegetation condition index. Int. J. Remote Sens. 17, 2761–2782.

Liu, W.T., Juárez, R.I.N., 2001. ENSO drought onset prediction in northeast Brazil using NDVI. Int. J. Remote Sens. 22, 3483–3501. https://doi.org/10.1080/01431160010006430.

Lloyd-Huges, B., Saunders, M.A., 2002. A drought climatology for Europe. Int. J. Climatol. 22, 1571–1592.

Loukas, A., Vasiliades, L., 2004. Probabilistic analysis of drought spatiotemporal characteristics in Thessaly region, Greece. Nat. Hazards Earth Syst. Sci. 4, 719–731.

Lozano-Garcia, D.F., Fernandez, R.N., Gallo, K.P., Johannsen, C.J., 1995. Monitoring the 1988 severe drought in Indiana, U.S.A. using AVHRR data. Int. J. Remote Sens. 16, 1327–1340. https://doi.org/10.1080/01431169508954479.

Mallya, G., Tripathi, S., Govindaraju, R.S., 2015. Probabilistic drought classification using gamma mixture models. J. Hydrol. 526, 116–126.

Mallya, G., Tripathi, S., Kirshner, S., Govindaraju, R., 2013a. Probabilistic assessment of drought characteristics using Hidden Markov Model. J. Hydrol. Eng. 18, 834–845. https://doi.org/10.1061/(ASCE)HE.1943-5584.0000699.

Mallya, G., Zhao, L., Song, X., Niyogi, D., Govindaraju, R., 2013b. 2012 Midwest drought in the United States. J. Hydrol. Eng. 18, 737–745. https://doi.org/10.1061/(ASCE)HE.1943-5584.0000786.

McKee, T.B., Doesken, N.J., Kleist, J., 1995. Drought monitoring with multiple time scales. In: Proceedings of the 9th Conference on Applied Climatology. American Meteorological Society Dallas, Boston, MA, pp. 233–236.

McKee, T.B., Doesken, N.J., Kleist, J., 1993. The relationship of drought frequency and duration to time scales. In: Conf. Appl. Climatol. Am. Meteorol. Soc. Anaheim CA.

Mishra, A.K., Desai, V.R., 2006. Drought forecasting using feed-forward recursive neural network. Ecol. Model. 198, 127–138. https://doi.org/10.1016/j.ecolmodel.2006.04.017.

Mishra, A.K., Desai, V.R., Singh, V.P., 2007. Drought forecasting using a hybrid stochastic and neural network model. J. Hydrol. Eng. 12, 626–638. https://doi.org/10.1061/(ASCE)1084-0699(2007)12:6(626).

Mishra, A.K., Singh, V.P., 2011. Drought modeling—a review. J. Hydrol. 403, 157–175.

Mishra, A.K., Singh, V.P., 2010. A review of drought concepts. J. Hydrol. 391, 202–216. https://doi.org/10.1016/j.jhydrol.2010.07.012.

Mitra, S., Srivastava, P., 2016. Spatiotemporal variability of meteorological droughts in southeastern USA. Nat. Hazards 1–32. https://doi.org/10.1007/s11069-016-2728-8.

Palmer, W.C., 1968. Keeping track of crop moisture conditions, nationwide: the new crop moisture index. Weatherwise 21, 156–161. https://doi.org/10.1080/00431672.1968.9932814.

Palmer, W.C., 1965. Meteorological Drought, Research Paper No. 45. U.S. Weather Bureau, Washington, DC.

Rabiner, L.R., 1989. A tutorial on hidden Markov models and selected applications in speech recognition. Proc. IEEE 77, 257–286. https://doi.org/10.1109/5.18626.

Rao, A.R., Voeller, T.L., 1997. Development and testing of drought indicators. Water Resour. Manag. 11, 119–136. https://doi.org/10.1023/A:1007933725122.

Reynolds, D.A., Rose, R.C., 1995. Robust text-independent speaker identification using Gaussian mixture speaker models. IEEE Trans. Speech Audio Process. 3, 72–83. https://doi.org/10.1109/89.365379.

Richardson, S., Green, P.J., 1997. On Bayesian analysis of mixtures with an unknown number of components (with discussion). J. R. Stat. Soc. Ser. B Stat. Methodol. 59, 731–792. https://doi.org/10.1111/1467-9868.00095.

Robertson, A.W., Kirshner, S., Smyth, P., 2004. Downscaling of daily rainfall occurrence over northeast Brazil using a hidden Markov model. J. Clim. 17, 4407–4424. https://doi.org/10.1175/JCLI-3216.1.

Robertson, A.W., Kirshner, S., Smyth, P., 2003. Hidden Markov Models for Modeling Daily Rainfall Occurrence over Brazil. University of California.

Rossi, G., Cancelliere, A., 2003. At-site and regional drought identification by Redim model. In: Rossi, G., Cancelliere, A., Pereira, L.S., Oweis, T., Shatanawi, M., Zairi, A. (Eds.), Tools for Drought Mitigation in Mediterranean Regions. Water Science and Technology Library. Springer, Netherlands, pp. 37–54.

Russo, S., Dosio, A., Sterl, A., Barbosa, P., Vogt, J., 2013. Projection of occurrence of extreme dry-wet years and seasons in Europe with stationary and nonstationary standardized precipitation indices. J. Geophys. Res. Atmos. 118, 7628–7639. https://doi.org/10.1002/jgrd.50571.

Ryu, J.H., Svoboda, M.D., Lenters, J.D., Tadesse, T., Knutson, C.L., 2010. Potential extents for ENSO-driven hydrologic drought forecasts in the United States. Clim. Change 101, 575–597. https://doi.org/10.1007/s10584-009-9705-0.

Shafer, B.A., Dezman, L.E., 1982. Development of a Surface Water Supply Index (SWSI) to Assess the Severity of Drought Conditions in Snowpack Runoff Areas.

Shiau, J.-T., Feng, S., Nadarajah, S., 2007. Assessment of hydrological droughts for the Yellow River, China, using copulas. Hydrol. Process. 21, 2157–2163. https://doi.org/10.1002/hyp.6400.

Shukla, S., Wood, A.W., 2008. Use of a standardized runoff index for characterizing hydrologic drought. Geophys. Res. Lett. 35, 7. https://doi.org/10.1029/2007GL032487.

Solomon, S., Qin, D., Manning, M., Marquis, M., Averyt, K., Tignor, M., Miller Jr., H.L., Chen, Z., 2007. Climate Change 2007: The Physical Science Basis.

Sprague, L.A., 2005. Drought effects on water quality in the South Platte River Basin, Colorado1. JAWRA 41, 11–24. https://doi.org/10.1111/j.1752-1688.2005.tb03713.x.

Steinemann, A., 2003. Drought Indicators and Triggers: A Stochastic Approach to Evaluation. Wiley Online Library.

Thyer, M., Kuczera, G., 2003. A hidden Markov model for modelling long-term persistence in multi-site rainfall time series 1. Model calibration using a Bayesian approach. J. Hydrol. 275, 12–26. https://doi.org/10.1016/S0022-1694(02)00412-2.

Tripathi, S., Govindaraju, R.S., 2009. On the identification of intra-seasonal changes in the Indian summer monsoon. In: Proceedings of the Third International Workshop on Knowledge Discovery from Sensor Data, SensorKDD '09. ACM, New York, NY, USA, pp. 62–70.

Türkes, M., Tatlı, H., 2009. Use of the standardized precipitation index (SPI) and a modified SPI for shaping the drought probabilities over Turkey. Int. J. Climatol. 29, 2270–2282. https://doi.org/10.1002/joc.1862.

Ventura, V., Paciorek, C.J., Risbey, J.S., 2004. Controlling the proportion of falsely rejected hypotheses when conducting multiple tests with climatological data. J. Clim. 17, 4343–4356. https://doi.org/10.1175/3199.1.

Verdon-Kidd, D.C., Kiem, A.S., 2010. Quantifying drought risk in a nonstationary climate. J. Hydrometeorol. 11, 1019–1031. https://doi.org/10.1175/2010JHM1215.1.

Vose, R.S., Applequist, S., Squires, M., Durre, I., Menne, M.J., Williams, C.N., Fenimore, C., Gleason, K., Arndt, D., 2014. Improved historical temperature and precipitation time series for U.S. climate divisions. J. Appl. Meteorol. Climatol. 53, 1232–1251. https://doi.org/10.1175/JAMC-D-13-0248.1.

Wilhite, D.A., Glantz, M.H., 1985. Understanding: the drought phenomenon: the role of definitions. Water Int. 10, 111–120. https://doi.org/10.1080/02508068508686328.

Wiper, M., Insua, D.R., Ruggeri, F., 2001. Mixtures of gamma distributions with applications. J. Comput. Graph Stat. 10, 440–454. https://doi.org/10.1198/106186001317115054.

WMO, 1975. "Drought and Agriculture," WMO Technical Note No. 138, Report of the CAgM Working Group on the Assessment of Drought, Geneva, Switzerland, p. 127.

Wu, Z., Huang, N.E., Long, S.R., Peng, C.-K., 2007. On the trend, detrending, and variability of nonlinear and nonstationary time series. Proc. Natl. Acad. Sci. 104, 14889–14894.

Yue, S., Wang, C.Y., 2002. Applicability of prewhitening to eliminate the influence of serial correlation on the Mann-Kendall test. Water Resour. Res. 38, 1068. https://doi.org/10.1029/2001WR000861.

CHAPTER 7

Variations and Trends in Global and Regional Sea Levels

Ramesh S.V. Teegavarapu, Alejandra R. Schmidt
Department of Civil, Environmental and Geomatics Engineering, Florida Atlantic University, Boca Raton, FL, United States

Contents

Trends and Changes in Hydroclimatic Variables
ISBN 978-0-12-810985-4
https://doi.org/10.1016/B978-0-12-810985-4.00007-4

1. INTRODUCTION

Sea level rise (SLR) is considered as one of the most controversial and recognized areas of research study in the world. Global sea level is known to have been affected by different factors namely climate change and climate variability. From 1993 to 2008 the mean sea level has increased on average by 0.13 inches per year (Ablain et al., 2009). However, by 2100, it has been estimated that the mean global SLR will vary on average from 1 foot to 6 feet (Nicholls and Cazenave, 2010). Global warming, one of the main causes of climate change, has known to have played a major role in the rise of the world's mean surface temperature (Morice et al., 2012), and the heat stored in the sea (Hobbs and Willis, 2013). Melting of ice that exists over land contributes to the increase of volume of water in the oceans, thus contributing to sea level rise. However, melting of sea ice does not contribute to increase in sea levels. When the ocean temperature rises, thermal expansion occurs in the sea waters. This thermal expansion ultimately causes the rise of sea levels. Likewise, glaciers that melt and eventually contribute to rise in sea levels (Cazenave and Cozannet, 2014). Many scientific studies have focused on trends linked to climate change to forecast future SLR. However, climate variability is known to have impact on local sea levels when smaller time frames are considered. Climate variability then must be considered when preparing a prevention plan against the effects of SLR (Hamilton et al., 2015). Thermal expansion, melting of glaciers and polar ice caps, and ice loss from Greenland and West Antarctica are the three main factors that influence the rise in sea levels.

SLR is measured using tide gauges that estimate sea level fluctuations with reference to a secured benchmark (Warrick and Oerlemans, 1993). Each secured benchmark is also monitored for any ground motions (or movements) that could be induced by any tectonic or volcanic activities, land subsidence, and human exercises such as the removal of groundwater and oil or gas (Cazenave and Cozannet, 2014). Because ground motions directly influence the measurements from tide gauges, adjustments were required to analyze tide gauge measurements regarding absolute sea level change (Cazenave and Llovel, 2010). As the adjustment quantities are not guaranteed to be 100% correct or error-free, the historically recorded data from many stations may not be completely accurate. The inaccuracy of the measurements can lead to wrong projections/predictions of sea level trends, as well as inaccurate sea level forecasts. The analysis of sea levels presented in this chapter focuses only on observed (i.e., historical data) measurements and not on future sea levels attributed to projections linked to climate change.

2. SEA LEVEL RISE: CONTRIBUTING FACTORS

The major contributing factors to present-day global mean SLR are thermal expansion of oceans, land ice melt, and the addition of fresh water to the oceans from land water reservoirs. These factors vary depending on natural changes and global warming caused by anthropogenically green-house gases (Cazenave and Cozannet, 2014). Sea levels are currently increasing and are predicted to increase for many years to come, even if greenhouse gas releases are restricted (Church and White, 2011). Since 1960, thermal expansion has caused nearly 25% of the recorded SLR (Domingues et al., 2008), and almost 50% from 1993 to 2003 (Solomon et al., 2007). However, thermal expansion seemed to slow down after this time span because only 30% has caused SLR from 1993 to 2007 (Cazenave and Llovel, 2010). Throughout the 1900s mountain glaciers and small ice caps have receded, and these processes have increased rapidly by the beginning of the 1990s (Meier et al., 2007). From 1993 to 2007, it was estimated that 30% of the recorded SLR was due to the melting of land ice (Cazenave and Llovel, 2010). On the other hand, the storage of fresh water is considered as a minor factor that affects the global sea levels, since it contributes to less than 10% influence on the existing sea level variations (Church et al., 2010). Additionally, dam building on rivers has decreased sea level by 0.5 mm/year throughout the mid-20th century (Chao et al., 2008).

3. VARIATIONS OF SEA LEVELS: LINKS TO CLIMATE VARIABILITY

Different coupled oceanic and atmospheric oscillations (El Niño Southern Oscillation, Atlantic Multidecadal Oscillation, Pacific Decadal Oscillation, North Atlantic Oscillation, etc.) or teleconnections are known to have varying influences on several hydroclimatic variables and sea levels in different regions across the globe. The annual and seasonal influences of these oscillations on sea levels are not clearly understood. The analysis reported in this chapter can help our understanding of these influences at regional and global levels. El Niño Southern Oscillation (ENSO) combined cycles influence the atmosphere and ocean differently, affecting sea levels in a complex way. There are two different phases of ENSO, warm phase and cool phase, with a cycle of two to seven years (Tootle et al., 2005) and therefore ENSO is referred to as interannual oscillation. Each phase of ENSO impacts different regions of the world differently; El Niño (warm phase) and La Niña (cool phase) events make modifications in airstreams, precipitation, and sea surface temperatures (SST) (Pacific Climate Futures, 2017). Main providers to Global Mean Sea Level (GMSL) change and ability to fluctuate are ocean thermal growth and sea mass. Sea mass changes when land ice masses, land water storage, or water vapor content in the atmosphere shifts (Llovel et al., 2011). Researchers have found that ENSO interannual ability to fluctuate the worldwide water cycle greatly affects water vapor in the atmosphere, land water storage, and therefore sea mass and GMSL (Cazenave et al., 2012).

Similarly, the Pacific Decadal Oscillation (PDO), Atlantic Multidecadal Oscillation (AMO), and North Atlantic Oscillation (NAO) are also considered to be carriers of climate variability. PDO effects are similar to ENSO's, though PDO acts in a larger time cycle and has greater influence in North Pacific and North American region (Trenberth and Hurrell, 1994). PDO cool and warm phases naturally last for around 25 years (Mantua et al., 1997; Mantua and Hare, 2002) and its activity is usually denoted to as decadal. AMO drives different climate abnormalities around the world which include the irregularity of rainfall in the northeast Brazilian region (Folland et al., 2001), the alterations in the summer climate for North America and Europe (Sutton and Hodson, 2005), and the periodicity of hurricanes in the Atlantic (Goldenberg et al., 2001). AMO cool and warm phases persist for around 65–80 years (Gray et al., 2004; Kerr, 2000) and its activity is usually denoted to as multidecadal. Finally, NAO in its positive phase contributes to warmer environments in northern and central Europe and in the eastern United States; and cooler environments over the Mediterranean, eastern Canada, and Greenland (Scaife et al., 2008). The negative phase of NAO results in average heights and pressure beyond the high altitudes of the North Atlantic and below average heights and pressure beyond the central North Atlantic, the eastern United States, and western Europe (NOAA, 2017). NAO's periodicity from the 1800s has shown to be quasibiennial and quasidecadal (Hurrell and Loon, 1995).

4. VARIATIONS OF SEA LEVELS: LINKS TO CLIMATE CHANGE

Global warming is known to increase the temperature in the oceans. When the temperature in the ocean increases, the seawaters expand, and consequently the levels of the seas rise; a process identified as thermal expansion (Cazenave and Llovel, 2010). The precise quantity of thermal expansion relies on several factors: the amount of heat absorbed; water temperature, pressure, and salinity. Consequently, the future SLR due to global warming will be based on the amount of greenhouse gas emissions and whether these emissions can be diminished or stabilized (Janin and Mandia, 2012). Studies that were found to be similar to this work, in terms of how the data was analyzed, and the climate variability patterns used, were examined initially and are recognized. Information related to these studies is presented in Table 7.1, and the most crucial findings are also briefly discussed in this section.

The influence of ENSO on the variation of the annual sea level cycles in the South China Sea was evaluated in a recent study by Cheng et al. (2016). The impact of ENSO on sea levels was analyzed by using tide gauge data, satellite altimetry data, and Ocean General Circulation Model (OGCM) outputs. The findings of the study (Cheng et al., 2016) showed the impact of ENSO on historical yearly sea level variations. In El Niño phase, a less

Table 7.1 List of Research Studies That Evaluated Links Between Sea Levels and Different Coupled Oceanic and Atmospheric Oscillations (Teleconnections)

Teleconnection(s)	Region	Study
ENSO	Australia	White et al. (2014)
ENSO	China	Cheng et al. (2016)
ENSO and PDO	United States	Hamilton et al. (2015)
NAO and AMO	The United Kingdom and western Europe	Ezer et al. (2016)
NAO	Eastern North Atlantic and the Mediterranean Sea	Calafat et al. (2012)
NAO	North Sea	Chen et al. (2014)
PDO	Pacific Ocean	Si and Xu (2014)
PDO	Bohai, yellow, and east China	Han and Huang (2008)
ENSO and PDO	Pacific Ocean	Zhang and Church (2012)
PDO	Global	Hamilton et al. (2013)

than average yearly sea level cycle was noted from all datasets, whereas during La Niña phase the opposite effect was found. Study conducted in the United States by Hamilton et al. (2015) investigated the combined impact of ENSO and PDO on sea levels by using long reconstructed datasets and tide gauges. The results showed that ENSO could provide as far as half a meter of sea level variation on short time intervals in some areas, and as far as 20 cm along the West Coast of the United States (Hamilton et al., 2015).

Other findings (Hamilton et al., 2013) included contributions of PDO on global sea levels to be as high as 0.49 ± 0.25 mm/year in the past 20 years and discovered that eliminating the PDO contribution decreases the global sea levels for the last 60 years. These findings were based on reconstructed altimeter records. Another study conducted on the western European coast discovered that the greatest sea level variability impact for NAO occurs in the North Sea and the Norwegian coasts. The AMO greatest sea level variability occurs in the subpolar Gyre and the Labrador Sea (Ezer et al., 2016).

The main objectives of any study involving analysis of time series of sea levels at single or multiple sites should attempt to:

- Understand and confirm the existence (or nonexistence) of statistically significant changes in sea level observations by adopting parametric and nonparametric statistical hypothesis tests for evaluation of variations in space and time.
- Develop and evaluate new methods or variants of existing trend evaluation methods considering issues related to homogeneity, data length, and others factors.
- Decipher and confirm any attributable reasons (natural climate variability influences) to the changes or trends.
- Evaluate the possibility of flooding of coastal areas due to sea level rise.
    - Determine the frequency and duration of flood events
    - Understand climatic patterns and their influences on sea levels based on year-to-year differences
    - Assess seasonal occurrences of flood events due to sea level variations and hide tides
    - Evaluate nuisance flood (i.e., high tide flooding) (Sweet et al., 2014) occurrences

A study by Sweet and Marra (NOAA, 2015) highlighted the following observations about nuisance flooding in the coastal regions of the United States. The observations are provided verbatim from a national climate report supplement (https://www.ncdc.noaa.gov/sotc/national/2015/8/supplemental/page-1):

- The occurrence of nuisance flooding continues to increase over years and decades, as relative sea levels continue to climb. Relative sea level rise is driven by long-term rise in mean sea level, which is strongly tied to climate change, and by subsidence of coastal land, which is significant at some locations.
- The number of nuisance flooding events is also quite sensitive to the El Niño/Southern Oscillation patterns.
- The year 2014 saw daily occurrences of nuisance flooding in the mid-Atlantic in line with the region's long-term and accelerating trends toward more nuisance flooding events. The West Coast saw coastal flooding events more in line with long-term trends. The Northeast and Gulf Coast regions saw below-trend numbers of nuisance flooding events.
- The projected totals for nuisance flooding events in 2015 indicate the potential for a record number of nuisance flooding events at several locations in the mid-Atlantic coast.

5. MEASUREMENT OF MEAN SEA LEVEL VARIATIONS

The Centre for Operational Oceanographic Products and Services (COOPS) has been measuring sea level for over 150 years, with tide gauge stations of the National Water Level Observation Network (NWLON) operating on all US coasts. Changes in mean sea level (MSL), either a sea level rise or sea level fall, have been computed at 142 long-term water level stations using a minimum span of 30 years of observations at each location. These measurements have been averaged by month to remove the effect of higher frequency phenomena. The MSL trends are measured by tide gauges and are local relative MSL trends as opposed to the global sea level trend. Tide gauge measurements are made with respect to a local fixed reference level on land; therefore, if there is some long-term vertical land motion occurring at that location, the relative MSL trend measured there is a combination of the global sea level rate and the local vertical land motion. The global sea level trend has been recorded by satellite altimeters since 1992, and the latest calculation of the trend can be obtained from NOAA's Laboratory for Satellite Altimetry, along with maps of the regional variation in the trend. The University of Colorado's Sea Level Research Group compares global SLRs calculated by different research organizations and provides detailed explanations about the issues involved.

In the past, tide gauges were placed in marine harbors to assist profitable shipping and to alleviate the unpredictability of flooding and retreating tides (Beenstock et al., 2015). These gauges were not designed for measuring sea levels, but at the beginning of the 20th century, they began to be used for this purpose (Emery and Aubrey, 1991). The allocations of tide gauges throughout the world are mainly found near sea sites leaving islands and open sea with very little coverage (Pirazolli, 1993). During the early 1950s, the number of tide gauges increased. However, many countries worldwide do not have a single gauge station. This is especially true for the southern Mediterranean area, while several regions of Africa and South America are underrepresented. The spatial absence of tide gauge position is intensified by a lack of historical data due to the restricted number of tide gauge with extended records (Beenstock et al., 2015).

In the 1960s, sea level data began to be gathered, stored, and documented by computers. Inspectors preserved and regulated gauges with the tide readings. Tide stations were visited every year to ensure the stations were well preserved and without any debris (NOAA, 2017). From the early 1990s, sea levels began to be measured by using high precision altimeters. Altimeters work by emitting microwave radiation against the ocean crust that partially reflects the signal back to a satellite. The magnitude of the cycle travel time provides the height of the satellite over the immediate ocean crust and an estimation of sea level (Nicholls and Cazenave, 2010). Currently, NOAA, which oversees over 300 gauges throughout the world, utilizes innovative acoustic and electronic instruments as recording systems. These recording systems, which are very similar to the satellite altimeters, send audio signals, instead of microwave radiation, down to a sounding tube and calculate the time it takes for the returned sign to travel back from the ocean's surface. The stations collect data every 6 min and use a Geostationary Operational Environmental Satellite (GOES) to broadcast their measurements hourly to NOAA headquarters. Additionally, all the raw and processed data are available on the NOAA website (NOAA, 2017).

5.1 Issues with Sea Level Measurements

Sea level measurements suffer from numerous limitations, and they include:

- The lack of gauges throughout the world limits the understanding of the global sea level behavior or regional variations. For instance, southern Europe and Africa are considered as underrepresented regions because only thirty-two (32) stations have been installed in these two areas.

- Sea level records cover different time periods in each region. In the coastal regions of the United States, most of the stations have data starting from 1949. Regions with lowest data are South America, Central America, and North America, whereas North Europe has the highest number of data.
- Tide gauges, which are fixed to the ground, supply measurements relative to the earth's crust, which could change over time. Because the benchmarks could move, sea level data can be interpreted incorrectly if the vertical land motion is not considered and corrections are not made to the observations.

6. DATA DESCRIPTION: LENGTH OF DATA AND AVAILABILITY

Monthly average sea level data reported in meters obtained from NOAA are used for the analysis reported in this chapter. The global sea level data from a total of 374 stations around the world is used for the analysis. The locations of these stations are shown in Fig. 7.1. The oldest dataset found from the database is from a gauge located in Brest, France; this dataset ranges from 1807 to 2010. The most recent dataset is from a tidal gauge located in Prudhoe Bay, Alaska; this dataset ranges from 1988 to 2016. On an average, most of the stations have around 60 years of data.

7. GLOBAL AND REGIONAL TRENDS

Sea levels at different sites (i.e., tidal gauge locations or stations or gauge stations) around the world are evaluated for trends using nonparametric

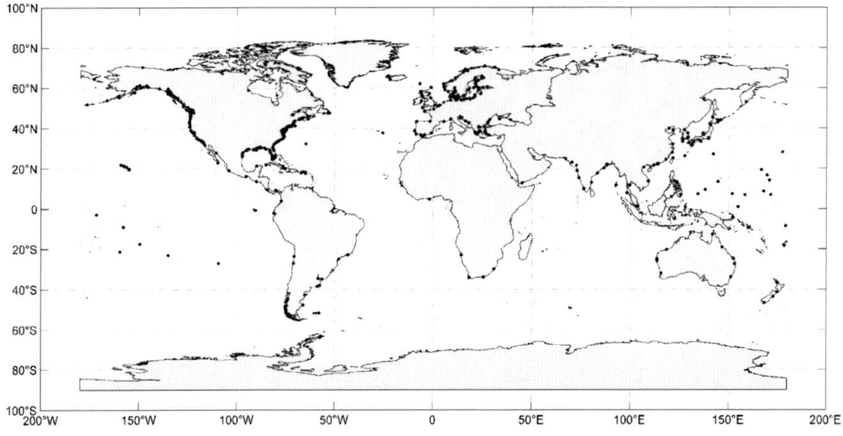

Figure 7.1 Location of tide gauges across the globe used for sea level measurements.

methods discussed in Chapter 1. Nonparametric trend analysis methods were applied to the average monthly and yearly maximum sea level data from 374 tidal gauges to detect any trends in sea level time series at each station.

8. VARIATION OF SEA LEVELS: CLIMATE CHANGE

Climate change has been affecting the sea levels for the past several hundred thousand years. The length of currently available data at each gauge station varies. Since the focus of the analysis is to understand the global sea level trends, larger periods of data are needed to acquire a basic understanding of sea level fluctuations. Annual extremes at each gauge are analyzed using Spearman's Rho (SR) and Mann-Kendall (MK) trend tests and results from these tests are reported in this chapter.

8.1 Spearman's Rho Trend Test Analysis Results

The SR test is used to confirm the existence or nonexistence of trends. SR test is applied to annual maximum and annual monthly sea level data. A hypothesis test is used for all evaluations, at a 5% significance level. The annual maximum is obtained for all stations for the entire period of record and SR tests are conducted. Fig. 7.2 shows the locations where sea levels are increasing with time depicted by the red colored dots, sea levels decreasing with time, depicted by the blue colored dots, and neither increasing nor decreasing with time depicted by the clear dots. Furthermore, the global results show that a total of 263 stations (~70.3%) have increasing or decreasing trends that are statistically significant. A total of 216 stations

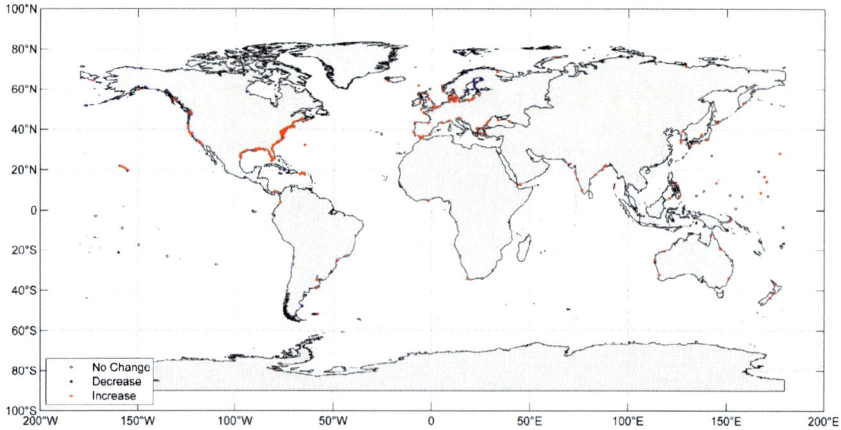

Figure 7.2 Variation of global sea level trends based on Spearman's Rho test.

(~82.1%) show an increasing trend, whereas, only a total of 47 stations (~17.9%) show a decreasing trend and the remaining 111 stations (~29.7%) show no change in sea level for the time series. Sea level trends are analyzed for the United States using the SR test. The results show that a total of 128 stations (~90.1%) have increasing or decreasing trends that are statistically significant. A total of 109 stations (~85.2%) show an increasing trend, whereas a total of 19 (~14.8%) stations show a decreasing trend with time. Only 14 stations (~9.9%) show no change in sea level for the time series. SR test is used to evaluate the trends at these tidal gauges, and the results are reported in Fig. 7.2. It is evident that majority of sites around the world are experiencing increasing trends. The SR test is conducted at 5% significance level.

8.2 Mann-Kendall Trend Test Analysis Results

The MK test at a 5% significance level is used to evaluate the trends in annual extremes of sea levels at each gauge. Fig. 7.3 shows an overall view of the locations where sea levels are increasing with time depicted by the red dots, decreasing with time depicted by the blue dots, and neither increasing nor decreasing with time depicted by the clear dots. Additionally, the global results show that a total of 278 stations (~74.3%) have increasing or decreasing trends and these trends are statistically significant. A total of 226 stations (~81.3%) show an increasing sea level trend with time, whereas only a total of 52 stations (~18.7%) show a decreasing sea level trend with time. The remaining 96 stations (~25.7%) show no trends in sea levels.

In case of coastal sites of the United States, results show that a total of 128 stations (~90.1%) have an increasing or decreasing trends and are

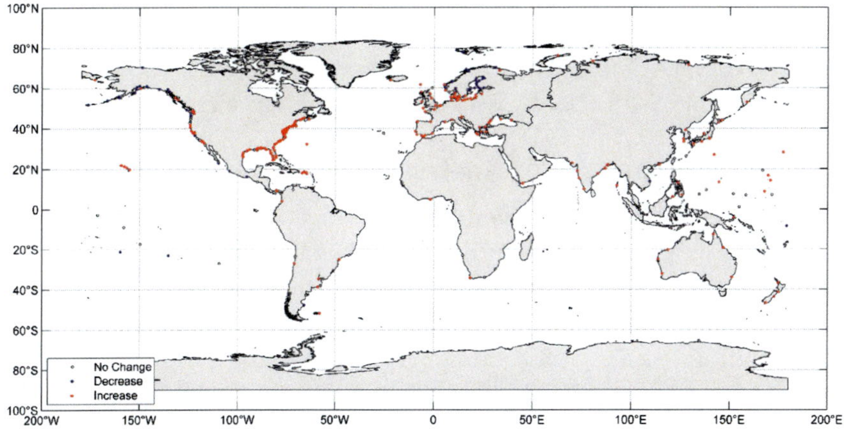

Figure 7.3 Variation of global sea level trends based on Mann-Kendall test.

statistically significant. A total of 109 stations (~85.2%) show an increasing trend with time, whereas, a total of 19 (~14.0%) stations show a decreasing trends with time. Only 14 stations (~9.9%) show no change in sea level for the time series.

8.3 Spearman's Rho and Mann-Kendall Trend Analysis Results

The results from SR and MK tests discussed in Sections 8.1 and 8.2 led to similar findings for trends. For instance, the number of stations in the United States for both nonparametric tests (i.e., SR and MK tests) show the same number of sites with statistically significant trends (128 stations) and nonstatistically significant trends (18 stations). However, since these two statistical methods use different approaches to calculate correlations based on associations between two independent variables, differences were found in the results for the global stations. Results from MK tests indicate 15 more stations with statistically significant trends (278 stations) compared to 263 stations globally from SR tests.

9. VARIATION OF SEA LEVELS: INFLUENCES DUE TO CLIMATE VARIABILITY

Four different analyses are used to estimate sea level fluctuations throughout the world. Additionally, all the gauges from the United States are analyzed separately to understand which oscillations have a higher impact on the east, west, and gulf coasts of the United States. The entire data was separated into temporal windows coinciding with two different phases of different oscillations (ENSO, AMO, PDO, and NAO). The data separation process is described in Section 9.1 Once data is separated based on warm and cool phases of all oscillations, a nonparametric hypothesis test, Mann–Whitney U test, described in Chapter 1, is utilized to test statistically significant changes in sea levels in two different phases.

9.1 Data Preparation for Analysis

The annual maximum sea level data values for the entire period of record for all stations are derived from the available historical data. Additionally, the median sea level data values were also obtained for each year for all stations. The temporal windows that coincide with warm and cool phases for the four-coupled oceanic-atmospheric oscillations used in this study are obtained from previous studies (Birk, 2009; Birk et al., 2009). The temporal windows associated with warm and cool phases of different oscillations and they are listed as interannual (ENSO) in Table 7.2, decadal (PDO) in Table 7.3, multidecadal (AMO) in Table 7.4, and quasi-decadal (NAO) in Table 7.5. Once the temporal windows are established for all oscillations, the sea level data for the entire period of record for every station are separated into their respective cool and warm phases.

Coupled influences of two oscillations can be evaluated by identifying different combinations of two phases of these oscillations. A procedure

Table 7.2 List of Years That Coincide With Warm and Cool Phases of ENSO

Phase	Years
El Niño (Warm)	1902, 1904, 1905, 1911, 1913, 1918, 1925, 1929, 1930, 1940 1951, 1957, 1963, 1965, 1969, 1972, 1976, 1982,1986, 1987 1991, 1997, 2002, 2006, 2009
La Niña (Cool)	1903,1906,1908, 1909, 1910, 1916, 1922, 1924, 1938, 1942, 1944, 1949, 1954,1955, 1956, 1964, 1967, 1970, 1971, 1973 1974, 1975, 1988, 1998, 1999, 2007, 2010

Table 7.3 List of Years That Coincide With Warm and Cool Phases of PDO

Phase	Period
Cool	1920–1925
Warm	1926–1945
Cool	1946–1976
Warm	1977–1999
Cool	2000–2010

Table 7.4 List of Years That Coincide With Warm and Cool Phases of AMO

Phase	Period
Cool	1900–1925
Warm	1926–1969
Cool	1970–1994
Warm	1995–2010

Table 7.5 List of Years That Coincide With Warm and Cool Phases of NAO

Phase	Period
Warm	1950–1951
Cool	1952–1972
Warm	1973–1976
Cool	1977–1980
Warm	1981–2001

similar to the one discussed previously in this section is required for separating the data based on time period for each oscillation phase. The procedure is briefly described here. The data coupling consisted of the separation of the data of ENSO with PDO, ENSO with AMO, and ENSO with NAO. The coupling of the data was based on the cool phase and warm phase of AMO, PDO, and NAO, such that each oscillation coincides with ENSO years; for example, warm La Niña years, warm El Niño years, cool La Niña years, cool El Niño years. Table 7.6, Table 7.7, Table 7.8 show temporal windows for two phases of different oscillations.

Table 7.6 List of Years That Coincide With Warm and Cool Phases of Coupled AMO and ENSO

Phase	AMO	El Niño	La Niña
Cool	1920–1925 2000–2010	1938,1942,1944, 1982 1986, 1987, 1991, 1997	1929, 1930, 1940 1988, 1998, 1999
Warm	1926–1945 1977–1999	1925, 2002, 2006, 2009	1922, 1924, 2007, 2010

Table 7.7 List of Years That Coincide With Warm and Cool Phases of Coupled PDO and ENSO

Phase	PDO	El Niño	La Niña
Cool	1926–1945	1925, 1951, 1957, 1963, 1965	1922, 1924, 1949, 1954,1955
	1977–1999	1969, 1972, 1976, 2007, 2010	1956, 1964, 1967, 1970, 1971 1973, 1974, 1975, 2007, 2010
Warm	1920–1925	1929, 1930, 1940, 1982, 1986	1938, 1942, 1944, 1988, 1998
	1946–1976 2000–2010	1987, 1991, 1997	1999

Table 7.8 List of Years That Coincide With Warm and Cool Phases of Coupled NAO and ENSO

Phase	NAO	El Niño	La Niña
Cool	1952–1972 1977–1980	1957, 1963, 1965, 1969 1972	1954, 1955, 1956, 1964 1967, 1970, 1971
Warm	1950–1951 1973–1976 1981–2001	1951, 1976, 1982, 1986 1987, 991, 1997	1973, 1974, 1975, 1988 1999

9.2 El Niño Southern Oscillation (ENSO)

For this analysis, the monthly values are used for all stations for the entire period of record. The median values for each phase are computed to find whether these values are equal between the paired samples, as well as testing whether these results are statistically significant by using a nonparametric hypothesis test (Mann Whitney U test). Fig. 7.4 shows the locations where sea levels have higher median values during La Niña phase depicted by the blue colored dots, El Niño phase depicted by the red colored dots, and sea levels having the same medians depicted by the clear dots. Furthermore, a total of 180 stations throughout the world (~48.1%) are influenced by ENSO. A total of 107 stations (~59.4%) are influenced by El Niño phase, whereas, a total of 73 stations (~40.6%) are influenced by La Niña phase. The remaining 194 stations throughout the world (~51.9%) are not influenced by ENSO.

In the United States, a total of 56 stations (~39.4%) are influenced by ENSO. A total of 43 stations (~76.8%) are influenced by El Niño phase, whereas only 13 stations (~23.2%) are influenced by La Niña phase. The remaining 86 stations in the United States (~60.5%) are not influenced by ENSO. The results show that a total of 23 stations in the east coast are influenced by the ENSO oceanic–atmospheric oscillation. A total of 14 stations (~60.9%) are influenced by El Niño phase, whereas only 9stations (~39.1%) are influenced by La Niña phase. Moreover, 25 stations on the west coast are said to be influenced by ENSO. A total of 22 of these stations

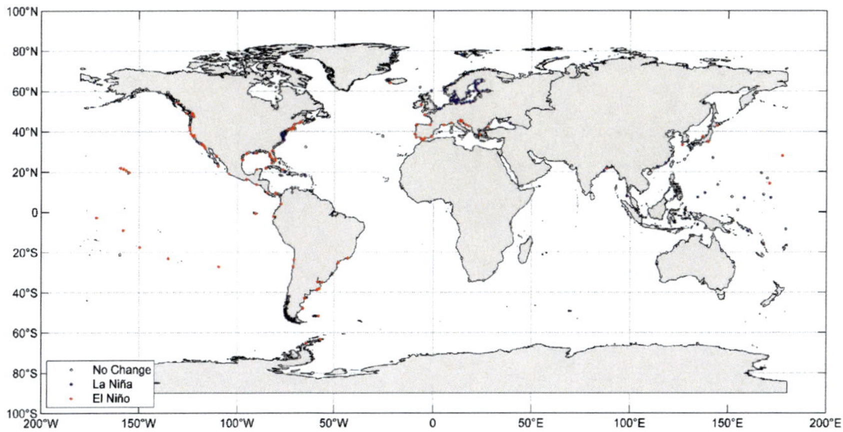

Figure 7.4 Changes in global sea levels attributed to two phases of ENSO based on Mann–Whitney U test.

(~88.0%) are influenced by El Niño phase. Whereas, only 3 stations (~12.0%) are influenced by La Niña phase. A total of 8 stations in the gulf coast are influenced by ENSO. A total of 7 stations (~87.5%) are influenced by El Niño phase, and only one station (~12.5%) is influenced by La Niña phase. Even though ENSO only influences sea levels in the United Sates by 39%, it is important to emphasize that during El Niño phase there is a possibility of higher sea levels than La Niña.

9.3 Atlantic Multidecadal Oscillation (AMO)

For this analysis, the same methodology discussed in Section 9.2 is applied to all the monthly sea level values. Fig. 7.5 shows the locations where sea levels have higher median sea level values during the cool phase depicted by the blue colored dots, the warm phase depicted by the red colored dots, and where the sea levels have the same median values depicted by the clear dots. Furthermore, a total of 280 stations throughout the world (~74.9%) are influenced by AMO. A total of 222 stations (~79.3%) are influenced by the warm phase, whereas, only 58 stations (~20.7%) are influenced by the cool phase. The remaining 94 stations throughout the world (~25.1%) are not influenced by AMO. In the United States, a total of 111 stations (~78.2%) are influenced by the AMO. A total of 101 stations (~91.0%) are influenced by the warm phase, whereas only 10 stations (~9.0%) are influenced by the cool phase. The remaining 31 stations in the United States (~21.8%) are not influenced by AMO.

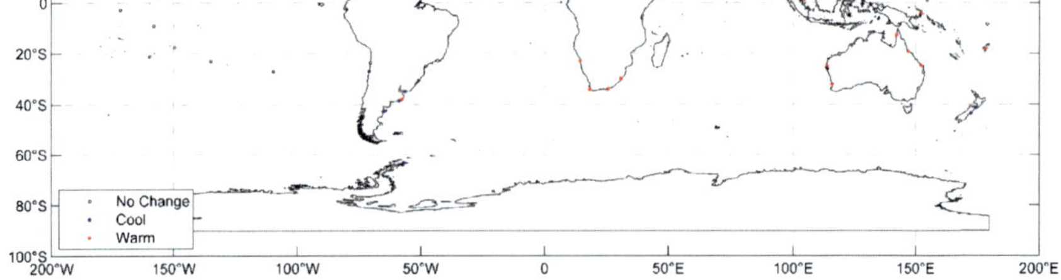

Figure 7.5 Changes in global sea levels attributed to two phases of AMO based on Mann–Whitney U test.

The results show that a total of 56 stations on the east coast are influenced by AMO. All stations on the east coast and gulf coast are found to be influenced by the AMO warm phase. Moreover, 33 stations on the west coast are said to be influenced by AMO. A total of 23 of these stations (~69.7%) are influenced by the warm phase. Whereas, a total of 10 stations (~30.3%) are influenced by the cool phase. From the results, it is observed that around 75% of global sea levels are influenced by AMO; specially during AMO's warm phase. The greatest finding in this section is that over 78% of stations in the United States are influenced by AMO, particularly on the east coast and gulf coast where all the stations are found to be 100% influenced by AMO's warm phase.

9.4 Pacific Decadal Oscillation (PDO)

Changes in monthly sea levels in warm and cool phases of PDO are evaluated and the results are summarized in this section. Fig. 7.6 shows the locations where sea levels have higher median sea level values during the cool phase. Variation of global sea levels of PDO based on Mann–Whitney U test is depicted by blue colored dots, the warm phase is depicted by the red colored dots, and where the sea levels have the same median values, they are depicted by the clear dots. Furthermore, a total of 261 stations throughout the world (~69.8%) are influenced by PDO. A total of 104 stations (~39.9%) are influenced by the warm phase, whereas a total of 157 stations (~60.1%) are influenced by the cool phase. The remaining 113 stations throughout the world (~30.2%) are not influenced

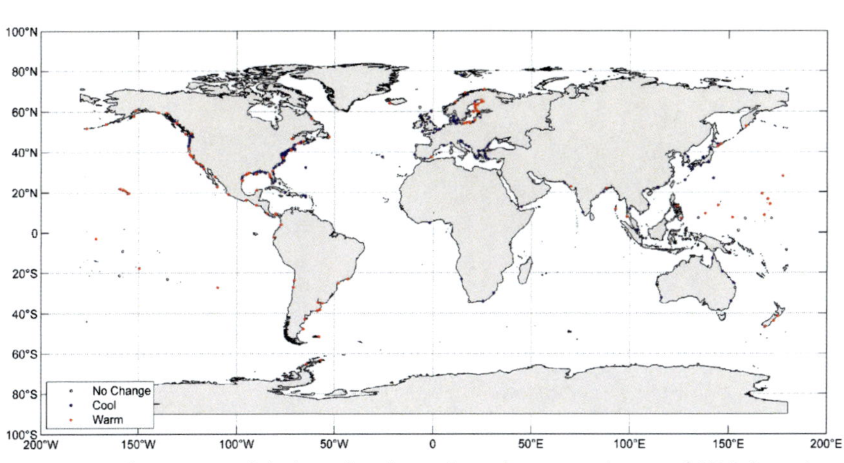

Figure 7.6 Changes in global sea levels attributed to two phases of PDO based on Mann–Whitney U test.

by PDO. In the United States, a total of 103 stations (~72.5%) are influenced by PDO, and a total of 35 stations (~34.0%) are influenced by the warm phase, whereas a total of 68 stations (~66.0%) are influenced by the cool phase. The remaining 39 stations in the United States (~27.5%) are not influenced by PDO.

The results show that a total of 47 stations on the east coast are influenced by PDO. A total of 4 stations (~8.5%) are influenced by the warm phase, whereas a total of 43 stations (~91.5%) are influenced by the cool phase. Moreover, 38 stations on the west coast are said to be influenced by PDO. A total of 24 stations (~63.2%) are influenced by the warm phase, whereas a total of 14 stations (~36.8%) are influenced by the cool phase. In the gulf coast, a total of 18 stations are found to be influenced by PDO. A total of 7 stations (~38.9%) are influenced by the warm phase, and the remaining 11 (~61.1%) are influenced by the cool phase. From the results, it can be observed that around 70% global sea levels are influenced by PDO; specially during PDO's cool phase. Similarly, 73% of stations in the United States are influenced by PDO; the east coast and gulf coast are highly influenced by PDO's cool phase.

9.5 North Atlantic Oscillation (NAO)

Evaluation of NAO impacts on sea level variations is discussed in this section. Fig. 7.7 shows the locations where sea levels have higher median sea level values during the cool phase, depicted by the blue colored dots, the warm phase depicted by the red colored dots, and where sea levels have the same

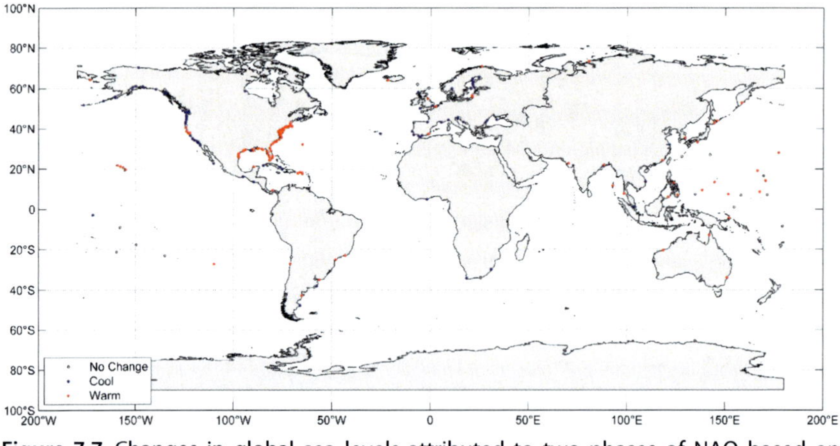

Figure 7.7 Changes in global sea levels attributed to two phases of NAO based on Mann–Whitney U test.

median values depicted by the clear dots. A total of 210 stations throughout the world (~56.2%) are influenced by NAO. A total of 126 stations (~60.0%) are influenced by the warm phase, whereas, a total of 84 stations (~40.0%) are influenced by the cool phase. The remaining 164 stations throughout the world (~43.8%) are not influenced by NAO.

In the United States, a total of 125 stations (~88.0%) are influenced by NAO oceanic-atmospheric oscillation. A total of 90 stations (~72.0%) are influenced by the warm phase, whereas, a total of 35 stations (~28.0%) are influenced by the cool phase. The remaining 17 stations in the United States (~12.0%) are not influenced by NAO.

Furthermore, the results show that a total of 56 stations on the east coast are influenced by NAO. A total of 55 stations (~98.2%) are influenced by the warm phase, whereas, only 1 station (~1.8%) is influenced by the cool phase. A total of 49 stations on the west coast are said to be influenced by NAO. From these 49 stations, 16 of them (~32.7%) are influenced by the warm phase, whereas the remaining 33 stations (~67.3%) are influenced by the cool phase. In the gulf coast, a total of 20 stations are influenced by NAO. A total of 19 (~95.0%) are influenced by the warm phase, and only one station (~5.0%) is influenced by the cool phase. Although around 56% of the global sea levels are influenced by NAO, most coastal regions are influenced by NAO's warm phase. However, it is important to emphasize that during NAO's cool phase the west coast in the United States has a possibility of higher sea levels than during NAO's warm phase.

10. EVALUATION OF COMBINED INFLUENCES OF COUPLED OCEANIC ATMOSPHERIC OSCILLATIONS

Three coupled oscillations are used to evaluate which ENSO phase has a higher influence on global sea levels provided that other multidecadal, decadal or quasi-decadal oscillation is also in effect. The coupled oscillations include PDO and ENSO, AMO and ENSO, and NAO and ENSO. All coupled effects evaluations consist of four periods:

- AMO cool phase La Niña, AMO cool phase El Niño, AMO warm phase La Niña, AMO warm phase El Niño
- PDO cool phase La Niña, PDO cool phase El Niño, PDO warm phase La Niña, PDO warm phase El Niño
- NAO cool phase La Niña, NAO cool phase El Niño, NAO warm phase La Niña, NAO warm phase El Niño

The entire data was separated based on the temporal windows for the four periods described in Section 9.1. Once the data were separated into its respective temporal windows, Mann—Whitney U test was used to evaluate statistically significant changes in the sea levels from two temporal windows. If the results are statistically significant, then the null hypothesis is rejected (the medians for the cool La Niña and cool El Niño phases are the same, or the medians for the warm La Niña and warm El Niño are the same), at a 5% significance level. Otherwise, the alternative hypothesis is not rejected indicating that the medians values are not the same.

10.1 AMO Cool Phase and ENSO

The monthly sea level values are used for this analysis for all stations. The median values for the AMO cool La Niña phase and AMO cool El Niño phase are calculated and used to test if the values are the same or not, and to find if the results are statistically significant by using a hypothesis test. Fig. 7.8 shows the locations where sea levels are dominated by AMO cool La Niña phase depicted by the blue dots, sea levels are dominated by AMO cool El Niño phase depicted by the red dots, and sea levels not being influenced by either phase are depicted by the clear dots. The results presented a total of 60 stations throughout the world (~16.0%) to be influenced by AMO cool ENSO coupled oscillation. A total of 45 stations (~75.0%) are influenced by AMO cool El Niño phase, whereas a total of 15 stations are influenced by AMO cool La Niña phase (~25.0%).

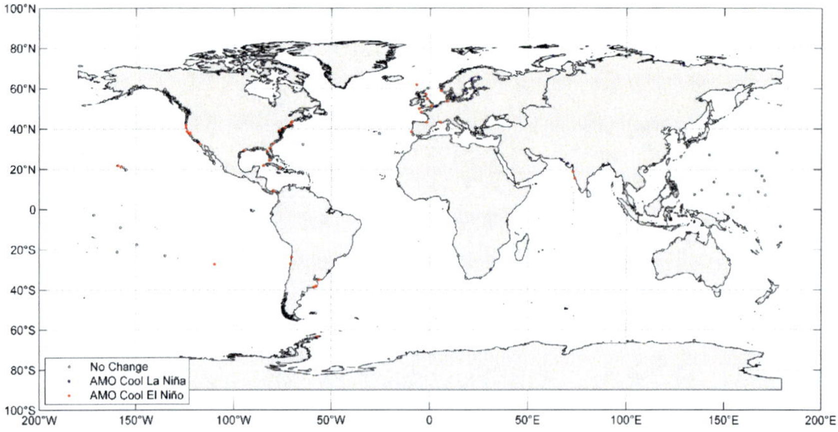

Figure 7.8 Changes in global sea levels attributed to AMO cool phase and two phases of ENSO based on Mann—Whitney U test.

The remaining 314 stations throughout the world (~84.0%) are not influenced by AMO cool ENSO coupled oscillation.

In the United States, a total of 21 stations (~14.8%) are influenced by AMO cool ENSO coupled oscillation. All statistically significant stations found to be influenced by AMO cool ENSO coupled oscillation, that is 21 stations, are influenced by AMO cool El Niño phase. The remaining 121 stations (~85.2%) are not impacted by this coupled oscillation. Although only 16% of the world is influenced by AMO cool ENSO coupled oscillation, it is important to emphasize that in the United States, AMO cool phase combined with El Niño phase increases the risk of sea level rise at all given 21 stations.

10.2 AMO Warm Phase and ENSO

Fig. 7.9 shows the locations where sea levels are dominated by AMO warm La Niña phase depicted by the blue dots, sea levels are dominated by AMO warm El Niño phase depicted by the red dots, and sea levels not being influenced by either phase depicted by the clear dots. The results presented a total of 195 stations throughout the world (~52.1%) to be influenced by AMO warm ENSO coupled oscillation. A total of 29 stations (~14.9%) are influenced by AMO warm El Niño phase, whereas a total of 166 stations are influenced by AMO warm La Niña phase (~85.1%). The remaining 179 stations throughout the world (~47.9%) are not influenced by AMO warm ENSO.

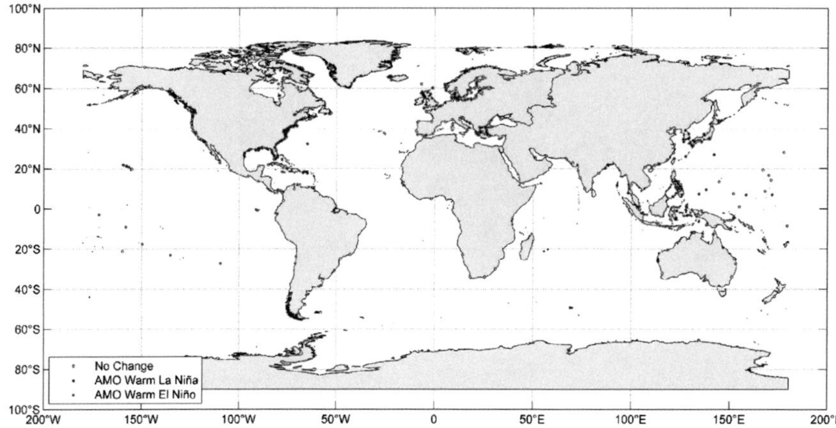

Figure 7.9 Changes in global sea levels attributed to AMO warm phase and two phases of ENSO based on Mann–Whitney U test.

In the United States, a total of 83 stations (~58.5%) are influenced by AMO warm ENSO coupled oscillation with La Niña phase. The remaining 59 stations (~41.5%) are not influenced by this coupled oscillation. The results show that more than 50% of stations throughout the world are influenced by this coupled oscillation. The AMO warm phase contributes to a higher sea levels when it is coupled with La Niña phase; especially in the United States.

10.3 PDO Cool Phase and ENSO

Fig. 7.10 shows the locations where sea levels are dominated by PDO cool La Niña phase depicted by the blue dots; PDO cool El Niño phase are depicted by the red dots, and sea levels not being influenced by either phase are depicted by the clear dots. The results presented a total of 150 stations throughout the world (~40.1%) to be influenced by PDO cool ENSO coupled oscillation. A total of 111 stations (~74.0%) are influenced by PDO cool El Niño phase whereas only 39 stations are influenced by PDO cool La Niña phase (~26.0%).

The remaining 224 stations throughout the world (~59.9%) are not influenced by PDO cool ENSO coupled oscillation. In the United States, a total of 50 stations (~35.2%) are influenced by PDO cool ENSO coupled oscillation with El Niño phase. The remaining 92 stations (~64.8%) are not influenced by this coupled oscillation. The results show that only 40% of stations throughout the world are influenced by this coupled oscillation.

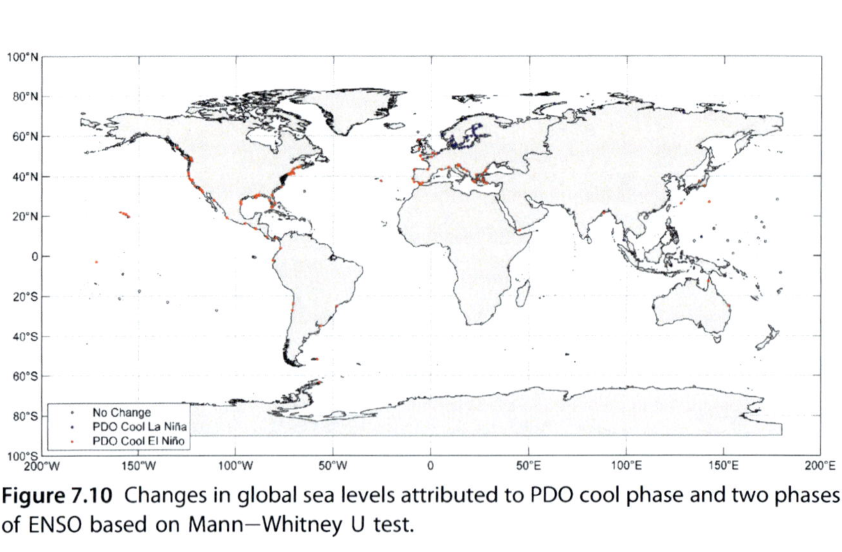

Figure 7.10 Changes in global sea levels attributed to PDO cool phase and two phases of ENSO based on Mann–Whitney U test.

However, when the PDO cool phase is combined with El Niño phase, the United States has a higher risk of sea level rise at 50 stations.

10.4 PDO Warm Phase and ENSO

Fig. 7.11 shows the locations where sea levels are dominated by PDO warm La Niña phase depicted by the blue dots, sea levels dominated by PDO warm El Niño phase depicted by the red dots, and sea levels not being influenced by either phase depicted by the clear dots.

The results indicate a total of 205 stations throughout the world (~54.8%) are influenced by PDO warm ENSO coupled oscillation. A total of 37 stations (~18.0%) are influenced by PDO warm El Niño phase, whereas, a total of 168 stations are influenced by PDO cool La Niña phase (~82.0%). The remaining 169 stations throughout the world (~45.2%) are not influenced by PDO warm ENSO coupled oscillation. In the United States, a total of 80 stations (~56.3%) are influenced by PDO warm ENSO coupled oscillation. A total of 6 stations (~7.5%) are influenced by PDO warm El Niño phase and the remaining 74 stations (~92.5%) are influenced by PDO warm La Niña phase. The rest of the stations in the United States, 62 stations (~43.7%), are not influenced by this coupled oscillation. The results show that over 50% of stations throughout the world are influenced by this coupled oscillation. PDO warm phase produces higher sea levels when it is coupled with La Niña phase, especially in the United states where more than 90% of stations that are susceptible to PDO warm phase present a higher risk when this phase is coupled with La Niña phase.

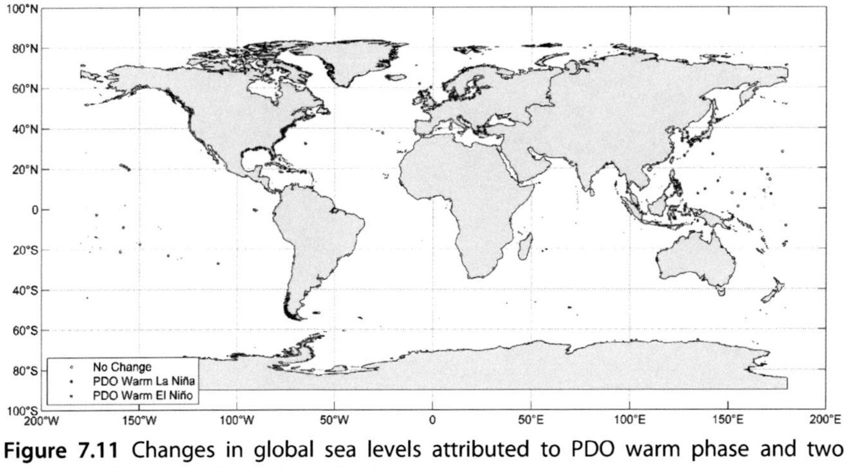

Figure 7.11 Changes in global sea levels attributed to PDO warm phase and two phases of ENSO based on Mann–Whitney U test.

10.5 NAO Cool Phase and ENSO

Fig. 7.12 presents an overall view of the locations where sea levels are dominated by NAO cool La Niña phase depicted by the blue dots, sea levels at sites that are dominated by NAO cool El Niño phase depicted by the red dots, and sea levels not being influenced by either phase depicted by the clear dots. The results presented a total of 163 stations throughout the world (~43.6%) to be influenced by NAO cool ENSO coupled oscillation. A total of 101 stations (~62.0%) are influenced by NAO cool El Niño phase, whereas, a total of 62 stations are influenced by NAO cool La Niña phase (~38.0%). The remaining 211 stations throughout the world (~56.4%) are not influenced by NAO cool ENSO coupled oscillation.

In the United States, a total of 53 stations (~37.3%) are influenced by NAO cool ENSO coupled oscillation. A total of 50 stations (~94.3%) are influenced by NAO cool El Niño phase, and only 3 stations (~5.7%) are influenced by NAO cool La Niña phase. The rest of stations in the United States, 89 stations (~62.7%), are not influenced by this coupled oscillation. The results show that only around 40% of stations throughout the world are influenced by this coupled oscillation, though most of the stations in the United States (94.3%) influenced by NAO cool phase possess a higher risk of sea level rise when it is coupled with El Niño phase.

10.6 NAO Warm Phase and ENSO

Fig. 7.13 shows the locations where sea levels are dominated by NAO warm La Niña phase depicted by the blue dots, sea levels dominated by

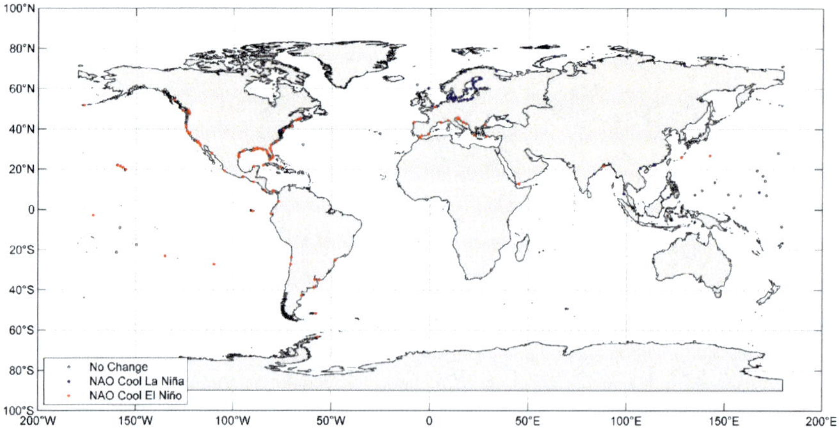

Figure 7.12 Changes in global sea levels attributed to NAO cool phase and two phases of ENSO based on Mann–Whitney U test.

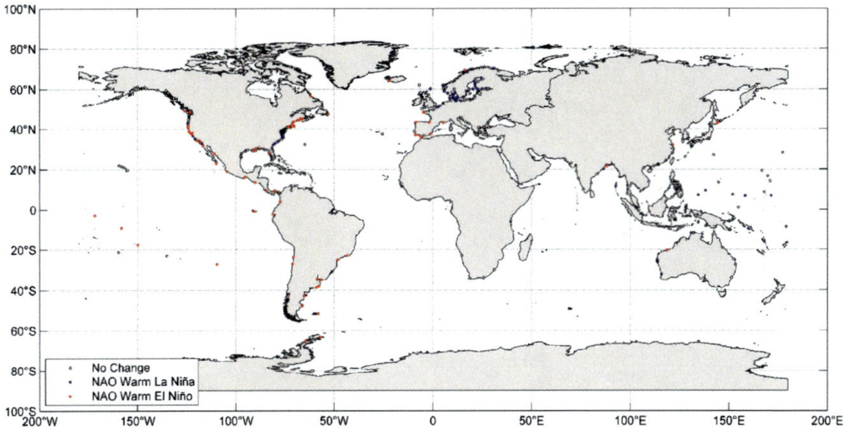

Figure 7.13 Changes in global sea levels attributed to NAO warm phase and two phases of ENSO based on Mann–Whitney U test.

NAO warm El Niño phase are depicted by the red dots, and sea levels not being influenced by either phase are depicted by the clear dots. The results noted a total of 133 stations throughout the world (~64.4%) to be influenced by NAO warm ENSO coupled oscillation. A total of 73 stations (~54.9%) are influenced by NAO warm El Niño phase, whereas a total of 60 stations are influenced by NAO warm La Niña phase (~45.1%). The remaining 241 stations throughout the world (~64.4%) are not influenced by NAO warm ENSO coupled oscillation.

In the United States, a total of 35 stations (~24.6%) are influenced by NAO warm ENSO coupled oscillation. A total of 25 stations (~71.4%) are influenced by NAO warm El Niño phase, and a total of 10 stations (~28.6%) are influenced by NAO warm La Niña phase. The rest of the stations in the United States, 107 stations (~75.4%), are not influenced by this coupled oscillation. The results show that less than 40% of stations throughout the world are influenced by this coupled oscillation. In the United Sates, NAO warm phase poses a low risk (~24.6%) of sea level rise when it is coupled with ENSO.

11. TREND ANALYSIS OF REGIONAL MONTHLY DATA: UNITED STATES

Evaluation of monthly sea level data for the United States using SK test is described in this section. Fig. 7.14 shows the locations for the months of January, February, March, and April where sea levels are increasing with

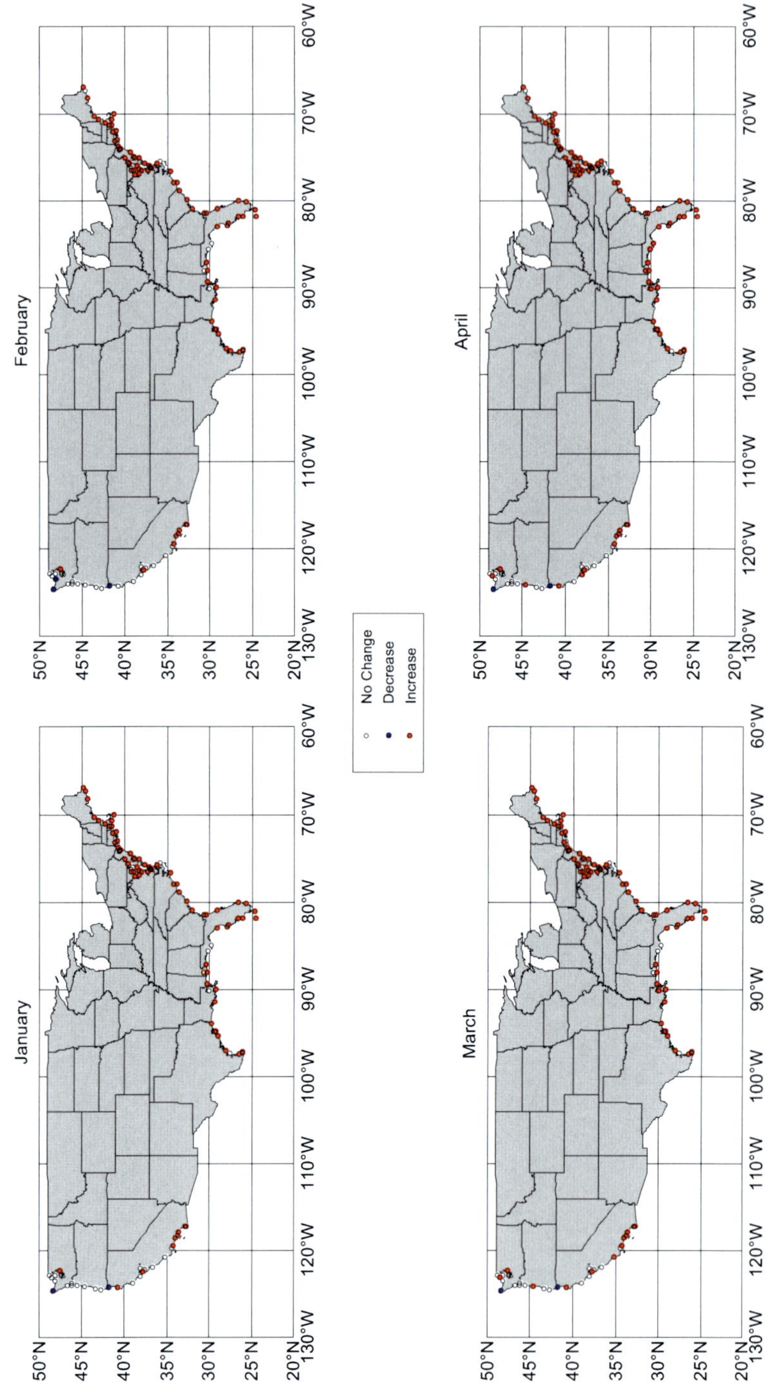

Figure 7.14 Sea level trends for the United States based on data from four different months using Spearman's Rho test.

time (red dots), decreasing with time (blue dots), and neither increasing nor decreasing with time (clear dots). The US results showed that for the month of January a total of 106 stations (~74.7%) have increasing or decreasing trends and these trends are statistically significant. From the 106 stations, 96 (~90.6%) of them showed increasing trends, and only 10 (~9.4%) showed a decreasing trend. The remaining 36 stations (~25.3%) show no change in sea level for the time series.

For the month of February, a total of 107 stations (~75.4%) have an increasing or decreasing trends, and these trends are statistically significant. From the 107 stations, 95 (~88.8%) of them showed increasing trends, and only 12 (~11.2%) showed a decreasing trend. The remaining 35 stations (~24.6%) show no change in sea level for the time series. For March, 111 stations (~78.2%) were found to have increasing or decreasing trends, and these trends are statistically significant. One hundred of these stations (~90.1%) showed increasing trends, and only 11 (~9.9%) showed a decreasing trend. The remaining 31 stations (~21.8%) show no change in sea level for the time series. For April, 116 stations (~81.7%) were found to have increasing or decreasing trends, and these trends are statistically significant. A hundred and five of these stations (~90.5%) showed increasing trends and only 11 (~9.5%) showed a decreasing trend. The remaining 26 stations (~18.3%) showed no change in sea level for the time series. Fig. 7.15 shows an overall view of the locations for the months of May, June, July, and August where sea levels are increasing with time (red dots), decreasing with time (blue dots), and neither increasing nor decreasing with time (clear dots). The US results showed that for the month of May a total of 129 stations (~90.9%) have increasing or decreasing trends and these trends are statistically significant. From the 129 stations, 115 (~89.2%) of them showed increasing trends, and only 14 (~10.8%) showed a decreasing trend. The remaining 13 stations (~9.1%) showed no change in sea level for the time series. For the month of June, a total of 129 stations (~90.9%) have increasing or decreasing trends, and these trends are statistically significant. From the 129 stations, 116 (~89.9%) of them showed increasing trends, and only 13 (~10.1%) showed a decreasing trend. The remaining 13 stations (~9.1%) show no change in sea level for the time series.

For the month of July, 132 stations (~93.0%) were found to have increasing or decreasing trends, and these trends are statistically significant. One hundred and nineteen of these stations (~90.2%) showed increasing trends and only 13 (~9.8%) showed a decreasing trend. The remaining 10 stations (~7.0%) showed no change in sea level for the time series.

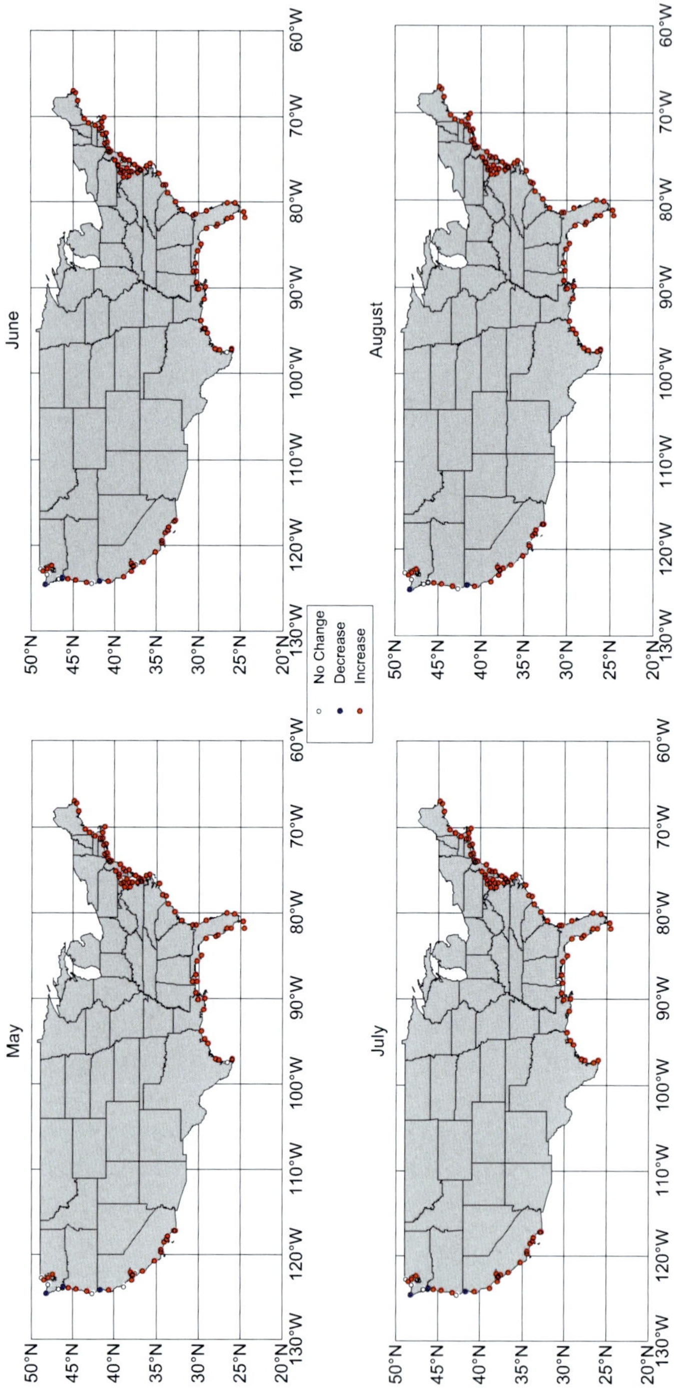

Figure 7.15 Sea level trends for the United States based on data from four different months using Spearman's Rho test.

For the month of August, a total of 128 stations (~90.1%) have increasing or decreasing trends. From the 128 stations, 117 (~91.4%) of them showed increasing trends, and only 11 (~8.6%) showed a decreasing trend. The remaining 14 stations (~9.9%) show no change in sea level for the time series. Fig. 7.16 shows an overall view of the locations for the months of January, February, March, and April where sea levels are increasing with time (red dots), decreasing with time (blue dots), and neither increasing nor decreasing with time (clear dots). The US results showed that for the month of September a total of 123 stations (~86.6%) have increasing or decreasing trends and these trends are statistically significant. From the 123 stations, 112 (~91.1%) of them showed increasing trends, and only 11 (~8.9%) showed a decreasing trend. The remaining 19 stations (~13.4%) showed no change in sea level for the time series. For the month of October, a total of 120 stations (~84.5%) have increasing or decreasing trends, and these trends are statistically significant. From the 120 stations, 109 (~90.8%) of them showed increasing trends, and only 11 (~9.2%) showed a decreasing trend. The remaining 22 stations (~15.5%) showed no change in sea level for the time series. For the month of November, 113 stations (~79.6%) were found to have increasing or decreasing trends, and these trends are statistically significant. One hundred and two of these stations (~90.3%) showed increasing trends and only 11 (~9.7%) showed a decreasing trend. The remaining 29 stations (~20.4%) showed no change in sea level for the time series.

For the month of December, 107 stations (~75.4%) were found to have increasing or decreasing trends, and these trends are statistically significant. Ninety-seven of these stations (~90.7%) showed increasing trends, and only 10 (~9.3%) showed a decreasing trend. The remaining 35 stations (~24.7%) showed no change in sea level for the time series.

11.1 Summary of Monthly Results for United States

The trend analysis results for the monthly data at US stations showed that many coastal areas, especially many regions in the east coast (from Figs. 7.14–7.16), are at a risk of being impacted by sea level rise. Table 7.9 provides a summary of the results. During the months of May through August, coastal area sites show increasing trends for sea levels. In the months of July and August, the sea levels had the highest percentages of increasing trends, whereas, the months of February and March had the lowest percentages of increasing trends and the highest percentages of decreasing trends. In general, all monthly trends suggest that more than 80% of them are increasing.

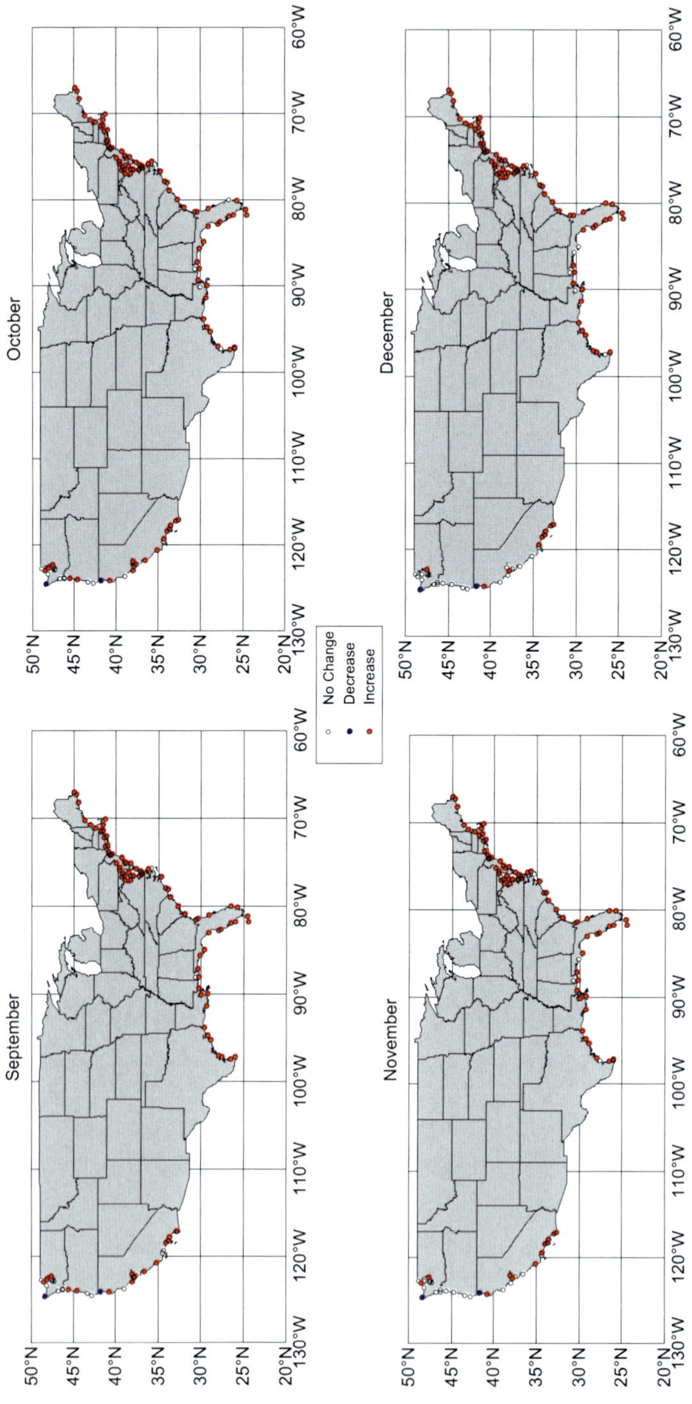

Figure 7.16 Sea level trends for the United States based on data from four different months using Spearman's Rho test.

Table 7.9 Monthly United States Trend Analysis Summary Overview

Month	Not Statistically Significant (%)	Statistically Significant (%)	Increasing (%)	Decreasing (%)
January	25.4	74.6	90.6	9.4
February	24.6	75.4	88.8	11.2
March	21.8	75.4	88.8	11.2
April	18.3	81.7	90.5	9.5
May	9.2	90.8	89.2	10.8
June	9.2	90.9	89.9	10.1
July	7.0	93.0	90.2	9.8
August	9.9	90.1	91.4	8.6
September	13.4	86.6	91.1	8.9
October	15.5	84.5	90.8	9.2
November	20.4	79.6	90.3	9.7
December	24.7	75.3	90.7	9.3

11.2 Summary of Results at Global Scale

Sea level evaluated using SK and MK tests show that approximately 60% of global tide gauges point to increasing trends. These observations also suggest that several coastal regions throughout the world are vulnerable to future sea level rise effects such as coastal flooding, beach erosion, etc. Similarly, trend analyses conducted for the sites around United Sates show that over 70% of tidal gauges have increasing trends. Interestingly enough, the majority of tide gauges on the east coast of the United States show that the sea level has been rising. The summary of the individual impacts on sea levels of ENSO, AMO, PDO, and NAO is represented in Table 7.10. ENSO is known to affect the sea levels throughout the world. The major effects of El Niño phase are shown in the southern portion of Europe, whereas, La Niña phase affects the northern portion of Europe. In the United States, ENSO has a great influence on sea levels during El Niño phase. The sea levels in South America are also highly affected by El Niño.

From all the individual effects of the four coupled oceanic and atmospheric oscillations, AMO is found to have the highest influence on sea

Table 7.10 Summary of All Oscillation Results

Teleconnection/Oscillation	Phase
ENSO	EL Niño (107) La Niña (73) None (194)
AMO	Warm (222) Cool (58) None (94)
PDO	Warm (104) Cool (157) None (113)
NAO	Warm (126) Cool (84) None (164)

levels throughout the world. AMO's warm phase affects the majority of coastal areas in the United States as well as the entire European continent. The pacific sea levels also seem to be affected by AMO's warm phase. Countries, such as Japan and China, show that sea levels most of the coastal sites are influenced by AMO's warm phase. PDO effects are also found to have a tremendous impact on global sea levels. Both phases affect the global sea levels almost evenly. For example, in the United States, the sea levels on the east coast and gulf coast are mainly affected by the cool phase. However, the sea levels on the west coast are primarily affected by PDO's warm phase. In Europe, the sea levels on the southern side are affected by PDO's cool phase, whereas, the sea levels in the northern portion are affected by the PDO's warm phase. In the Pacific Ocean, the main observation is that Japan's sea levels are affected by PDO's cool phase.

NAO effects are found to have the lowest global influences on sea levels. Instead, NAO influences most of the sea levels in the United States. NAO's cool phase affects the sea levels in most of the west coast region, while NAO's warm phase affects the sea levels in most of the east coast and gulf coast regions. The summary of the coupled impacts on sea levels of AMO, PDO, and NAO with ENSO are provided in Table 7.11. When AMO's warm phase is combined with ENSO, global sea levels are found to be mainly influenced by La Niña. In the United States, all regions that are affected by this coupled oscillation were entirely found to be affected by AMO warm La Niña phase. Similarly, the entire European continent is affected by AMO warm La Niña phase. AMO's cool phase combined with ENSO is not found to affect the global sea levels.

Table 7.11 Summary of Results From Coupled Influences of Oscillations

Coupled Oscillations	Phase
AMO & ENSO	AMO cool EL Niño (45) AMO cool La Niña (15) None (314)
AMO & ENSO	AMO warm EL Niño (29) AMO warm La Niña (166) None (179)
PDO & ENSO	PDO cool EL Niño (111) PDO cool La Niña (39) None (224)
PDO & ENSO	PDO warm EL Niño (37) PDO warm La Niña (168) None (169)
NAO & ENSO	NAO cool EL Niño (101) NAO cool La Niña (62) None (211)
NAO & ENSO	NAO warm EL Niño (73) NAO warm La Niña (60) None (241)

Only 40% of global sea levels are affected by the combination of PDO cool phase and ENSO. In the United States, all regions that are affected by this coupled oscillation were entirely found to be affected by PDO cool El Niño phase. The sea levels for the southern European region are mostly affected by PDO cool El Niño phase. While the sea levels for the northern European region are mostly affected by PDO cool La Niña phase. More than half of regions are affected by the combination of PDO warm phase and ENSO. At a global scale, PDO warm La Niña is prevalent, affecting phase on sea levels. Regions such as Europe, Pacific Ocean, the Middle East, and the United States sea levels are affected by PDO warm La Niña phase.

NAO's warm phase combined with ENSO is not found to affect the global sea levels. Around 40% of global sea levels are affected by NAO's cool phase combined with ENSO. The sea levels at different sites in southern European region are mostly affected by NAO cool El Niño phase, while the sea levels at different sites in the northern European region are mostly affected by NAO cool La Niña phase. The majority of the regions influenced by this coupled oscillation in the United States are affected by NAO cool El Niño phase.

Several limitations of the analysis reported in this chapter can be identified, and they are:

- Varying lengths of the observed sea level data: Gauges throughout the world have been installed at different times in history. Since the data used in the analysis is separated based on coupled oceanic and atmospheric oscillations linked with two temporal windows (i.e., warm and cool phase), lack of equal length of data at all stations limits a comprehensive evaluation of climate variability effects on sea levels. The unequal data length may influence the results of the nonparametric tests that are used to assess the differences in sea levels in two different phases of oscillations.
- Prevalence of missing data: Many stations had data gaps in different periods of record due to equipment malfunction or maintenance. In this study, if a year had missing data, data from that particular year was not used for analysis. Since the length of the data was already unevenly distributed, discarding data may impact the overall findings.
- Nonuniform spatial gauge distribution: The gauges across the world are not evenly distributed. Some regions, for example, Southern Europe and Africa have a total of 32 tidal gauges. The lack of gauges in some regions limits the understanding of the sea level changes. Thus, the analysis reported in this chapter cannot fully evaluate influences of climate change and climate variability in those regions.

12. TREND ANALYSIS OF REGIONAL MONTHLY DATA: JAPAN

Monthly sea level data from 14 tidal gauges in the coastal regions of Japan are analyzed for trends using parametric and nonparametric tests. The locations of these gauges are shown in Fig. 7.17. Parametric trend tests are carried out by fitting linear trend lines using regression. However, it was noted that at many sites trend analysis cannot be carried out due to issues with assumptions associated with linear regression analysis. Fig. 7.18 shows variation of sea level data and normal probability plot for data at site #1. Goodness-of-fit tests were conducted to confirm the normality of the data. Null hypothesis was accepted for two tests, Chi-Square and Lilliefor's tests, and it was rejected for Kolmogorov-Smirnov (KS) and Jarque-Bera tests. Issues with the residuals (i.e., the difference in observed-fitted values) were noted when Tukey-Anscombe and autocorrelation plots were generated. These plots are shown in Fig. 7.19.

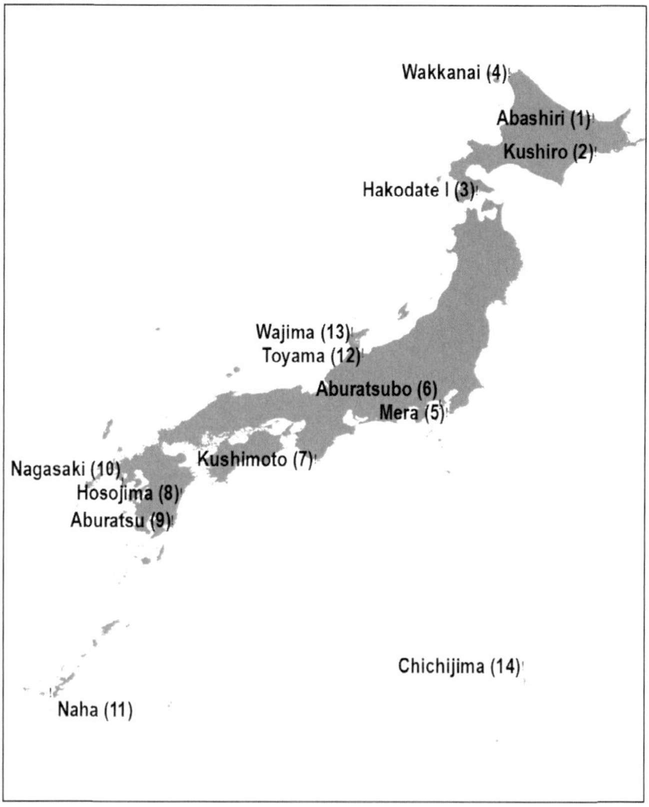

Figure 7.17 Location of sea level measurement gauges surrounding Japan.

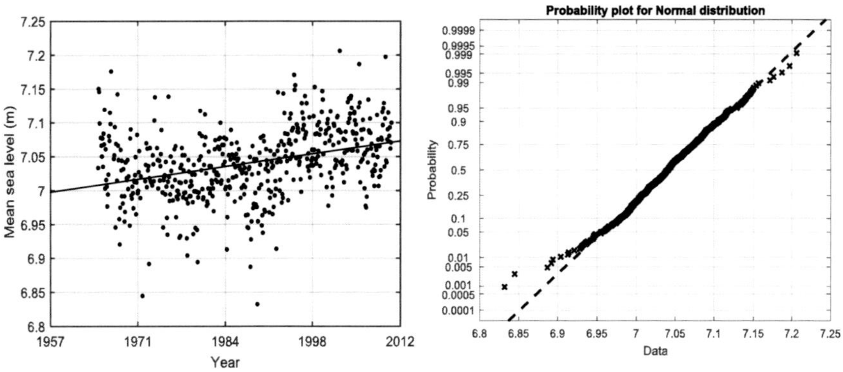

Figure 7.18 Variations of sea level at site #1 and normal probability plot of observations.

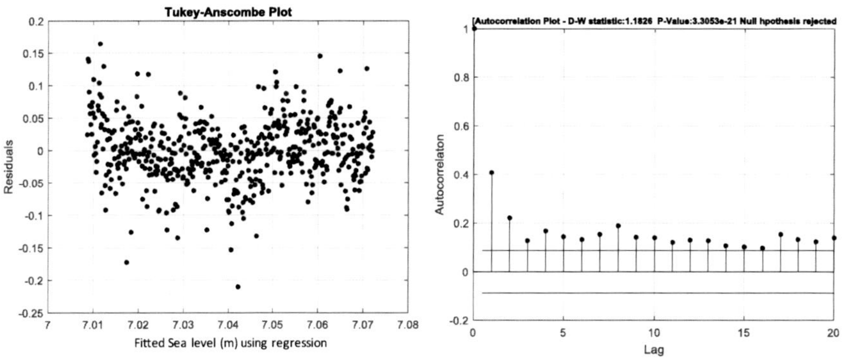

Figure 7.19 Visual evaluation of residuals obtained from observed and estimated values of sea level a using linear regression model.

Moderate heteroskedasticity was noted for residuals and null hypothesis was rejected when Durbin–Watson test confirmed the existence of autocorrelation in residuals at several lags. Considering the limitations of parametric trend analysis, trends at all sites are evaluated using a nonparametric SR test. Results from this test are shown in Fig. 7.20. Analyses at 11 out 14 sites show statistically increasing trends in sea levels. Strong serial autocorrelation is noted for observations at all sites. Interestingly, strong serial autocorrelation was also noted for monthly data at different sites.

12.1 Summary of Results for Japan

Increasing sea levels along with potential landfall of tropical and extratropical storms may raise the risk of coastal flooding in Japan. Trend analysis

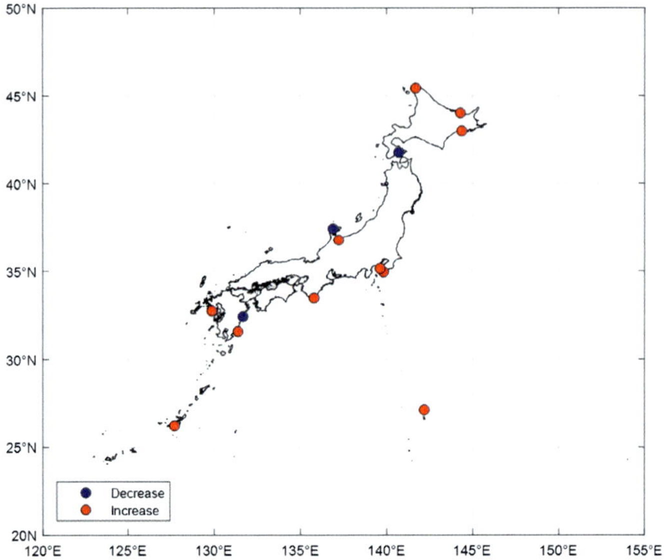

Figure 7.20 Spatial variation of sea level trends at different locations near coastal regions of Japan.

is carried out using SR trend test (or Spearman's rank correlation test). Evaluation of the influences of ENSO on sea levels indicates that cool phase (i.e., La Niña) contributes to higher levels. This is noted via unpaired two-sample t-tests and Mann–Whitney U tests. Evaluation of the influences of PDO on sea levels indicates that cool phase contributes to higher levels. This is observed based on results from unpaired t-tests and Mann–Whitney U tests.

13. DIRECTIONS FOR FUTURE WORK

An extensive analysis of long-term global sea level trends and variations due to possible climate change and variability is presented in this Chapter. A number of recommendations can be made for extensions of the analysis reported in this chapter and directions for future work can be provided. One recommendation is to use shorter timescales for both climate change and climate variability analyses to assess the risk of sea level rise around the coastal regions. By doing so, shorter-term analyses can provide contemporary results providing indications on the more recent changes in sea levels. Additionally, trends can also be evaluated using data from different starting points in time. A good starting point would be the year NOAA or any other agency began to track sea levels using altimeter measurements or the year from when all stations have sea level measurements.

Another suggestion for future studies is to analyse monthly sea levels for the entire world. In the analyses reported in this Chapter the data available at several sites in coastal regions of the United States were separated into each month. The monthly results were able to explain which months had a higher risk of sea level rise and had the lowest risk for sea level rise. If this analysis is applied to entire global sea level data, the results can help predict which months have a higher risk of sea level rise globally. Finally, it information about the other oceanic—atmospheric oscillations can be used to find their influences on sea levels. Since this study looks at the sea level globally, other oscillations also can also potentially affect the sea level results. Some suggestions of oceanic—atmospheric oscillations that can be used in future studies are North Pacific Oscillation (NPO), Interdecadal Pacific Oscillation (IPO), and Arctic Oscillation (AO). Additionally, the coupled effects of NPO, IPO, and AO with ENSO on global sea levels could be evaluated to find if ENSO is a limiting factor on climate variability with these three oscillations.

Future studies should also focus on quantifying the risk of tidal flooding combined with inundation issues caused by extreme storm events in the coastal regions. Temporal occurrences and frequency of nuisance flooding (i.e., high tide flooding) events should be investigated. The high tide flooding is increasingly becoming common in many coastal regions of the world and is exacerbated due to increasing sea levels. Frequent nuisance flooding (Sweet et al., 2014) or clear-sky flooding events can lead to public inconveniences and road closures and infrastructure damages. Hide tide flooding is also attributed to land subsidence and loss of natural coastal barriers. Statistical assessment of combined tidal and storm-related flooding events should be carried out.

14. CONCLUSIONS

Increasing sea levels in addition to potential landfall of tropical and extratropical storms may raise the risk of coastal flooding in many regions of the world. This chapter has focused on assessments of trends and variations in sea levels around the world. Trend analysis is carried out using SR and MK trend tests. Results from trend tests reveal that more than 50% of the sites globally are experiencing increasing trends. It is important to note that the regional tidal gauge density will bias the results suggesting the a few regions in the world are experiencing increasing sea levels than others. However, it can be concluded that many coastal regions in the world are under severe threat of increasing sea levels. Regional analyses at 142 sites in the United States and 14 sites in Japan show statistically increasing trends in sea levels at 109 sites and 11 sites respectively. Strong serial autocorrelation in monthly observations is noted at many sites located close to these two countries. Interestingly, strong serial autocorrelation was also noted for several months at different sites. Sea level variations for the United States indicate AMO, PDO, and ENSO phases have spatially varying statistically significant influences on sea levels. These observations are confirmed via nonparametric Mann—Whitney U tests. Similar variations were noted for sea levels at different sites across different coastal areas of Japan. Risk of tidal flooding (nuisance flooding) and storm-related flooding is higher in specific phases of oscillations, AMO warm phases in coastal areas of the southeastern United States and El Niño in the western United States.

REFERENCES

Ablain, M., Cazenave, A., Valladeau, G., Guinehut, S., 2009. A new assessment of the error budget of global mean sea level rate estimated by satellite altimetry over 1993—2008. Ocean Sci. 5 (2), 193—201.

Beenstock, M., Felsenstein, D., Frank, E., Reingewertz, Y., 2015. Tide gauge location and the measurement of global sea level rise. Environ. Ecol. Stat. 22 (1), 179—206.

Birk, K., 2009. The Interannual Variability of Midwestern Temperatures and Precipitation as Related to ENSO and PDO (M.S. thesis). Universsity of Missouri, Columbia, Missouri.

Birk, K., Lupo, A.R., Uinan, P., Barbieri, C.E., 2009. The interannual variability of midwestern temperatures and precipitation as related to the ENSO and PDO. Atmosfera 23 (2), 95—128.

Calafat, F.M., Chambers, D.P., Tsimplis, M.N., 2012. Mechanisms of decadal sea level variability in the eastern North Atlantic and the Mediterranean Sea. J. Geophys. Res. 117, C09022. https://doi.org/10.1029/2012JC008285.

Cazenave, A., Cozannet, G.L., 2014. Sea level rise and its costal impacts. Earth's Future 2, 15—34.

Cazenave, A., Llovel, W., 2010. Contemporary sea level rise. Annu. Rev. Mar. Sci. 2, 145.

Cazenave, A., Henry, O., Munier, S., Delcroix, T., Gordon, A., Meyssignac, B., Backer, M., 2012. Estimating ENSO influence on the global mean sea level, 1993—2010. Mar. Geod. 35 (Suppl. 1), 82—97.

Chao, B.F., Wu, Y.H., Li, Y.S., 2008. Impact of artificial reservoir water impoundment on global sea level. Science 5873, 212—214.

Chen, X., Dangedorf, S., Narayan, N., O'Driscoll, K., Tsimplis, M.N., Su, J., Mayer, B., Pohlmann, T., 2014. On sea level change in the North Sea influenced by the North Atlantic Oscillation: local and remote steric effects. Estuarine Coastal and Shelf Science 151, 186—195.

Cheng, Y., Hamlington, B.D., Plag, H., Xu, Q., 2016. Influence of ENSO on the variation of annual sea level cycle in the South China Sea. Ocean Eng. 126, 343—352.

Church, J., White, N., 2011. Sea level rise from the late 19th to early 21st century. Surv. Geophys. 32 (4), 585–602.
Church, J.A., Woodworth, P.L., Aarup, T., Wilson, W., 2010. Understanding Sea Level Rise and Vulnerability. Blackwell, New York.
Domingues, C.M., Church, J.A., White, N.J., Peter, G.J., Wijffels, S.E., Barker, P.M., Dunn, J.R., 2008. Improved estimates of upper-ocean warming and multi-decadal sea-level rise. Nature 453, 1090.
Emery, K.O., Aubrey, D., 1991. Sea Levels, and Levels and Tide Gauges. Springer: Verlag.
Ezer, T., Haigh, I.D., Woodworth, P.L., 2016. Nonlinear sea-level trends and long-term variability on Western European Coasts. J. Coast. Res. 32 (4), 744–755. https://doi.org/10.2112/JCOASTRES-D-15-00165.1.
Folland, C.K., Cloman, A.W., Rowell, D.P., Davey, M.K., 2001. Predictability of northeast Brazil rainfal and real-time forecast skill 1987–98. Climate 14, 1937–1958.
Goldenberg, S.B., Landsea, C.W., Mestas-Nunez, A.M., Gray, W.M., 2001. The recent increase in Atlantic hurricane activity: causes and implications. Science 293, 474–479.
Gray, S.T., Graumlich, L.J., Betancourt, J.L., Pederson, G.T., 2004. A tree-ring based reconstruction of the Atlantic Multidecadal Oscillation. Geophys. Res. Lett. 31, L12205.
Hamilton, B.D., Leben, R.R., Strassburg, M.W., Nerem, R.S., Kim, K., 2013. Contribution of the Pacific Decadal Oscillation to global mean sea level trends. Geophys. Res. Lett. 40 (19), 5171–5175. https://doi.org/10.1002/grl.50950.
Hamilton, B., Leben, R., Kim, K., Nerem, R., Atkinson, L., Thompson, P., 2015. The effect of El Niño-Southern Oscillations on U.S. regional and coastal sea level. J. Geophys. Res. Oceans 120, 3970–3986.
Han, G., Huang, W., 2008. Pacific decadal oscillation and sea level variability in the Bohai, Yellow, and East China Seas. J. Phys. Oceanogr. 38 (12), 2772–2783. https://doi.org/10.1175/2008JPO3885.1.
Hobbs, W.R., Willis, J.K., 2013. Detection of an observed 135-year ocean temperature change from limited data. Geophys. Res. Lett. 40, 2252–2258.
Hurrell, J.W., Loon, H.V., 1995. Decadal variations in climate associated with the north Atlantic oscillation. Clim. Change 31, 301–326.
Janin, H., Mandia, S.A., 2012. Introduction. In: Janin, H., Mandia, S.A. (Eds.), Rising Sea Levels an Introduction to Cause and Impact. McFarland and Company Inc., North Carolina, pp. 3–8.
Kerr, R.A., 2000. A north Atlantic climate pacemaker for the centuries. Science 288 (5473), 1984–1986.
Llovel, W., Becker, M., Cazenave, A., Jevrejeva, S., Alkama, R., Dechame, B., Beckley, B., 2011. Terrestrial waters and sea level variations on interannual time scale. Global Planet. Change 75 (1), 76–82.
Mantua, N.J., Hare, S.R., 2002. The pacific decadal oscillation. J. Oceanogr. 59 (1), 35–44.
Mantua, H.S., Zhang, Y., Wallace, J., Francis, R., 1997. A pacific interdecadal climate oscillation with impacts on salmon production. Bull. Am. Meteorol. Soc. 78 (6), 1069–1079.
Meier, M., Dyurgerov, M., Rick, U., O'Neel, S., Pfeffer, W., Anderson, R., Glazovsky, A., 2007. Glaciers dominate Eustatic sea-level rise in the 21st century. Science 317 (5841), 1064–1067.
Morice, C.P., Kennedy, J.J., Rayner, N.A., Jones, P.D., 2012. Quantifying uncertainties in global and regional temperature change using an ensemble of observational estimates: the HadCRUT4 data set. J. Geophys. Res. Atmos. 117, D08101.
Nicholls, R.J., Cazenave, A., 2010. Sea-level rise and its impacts on coastal zones. Science 328 (5985), 1517–1520.

NOAA, 2015. https://www.ncdc.noaa.gov/monitoring-content/sotc/national/2015/aug/sweet-marra-nuisance-flooding-2015.pdf.

NOAA, 2017. U.S. Department of Commerce. Retrieved January 31, 2018, from NOAA Web Site: http://oceanservice.noaa.gov/.

Pacific Climate Futures, 2017. Understanding Climate Variability and Change Australia.

Pirazolli, P.A., 1993. Global sea level changes and their measurement. Global Planet. Change 8, 135–148.

Scaife, A., Folland, C., Alexander, L., Moberg, A., Knight, J., 2008. European climate extremes and the North Atlantic oscillation. J. Clim. 72–83.

Si, Z., Xu, Y., 2014. Influence of the pacific decadal oscillation on Regional Sea level rise in the Pacific Ocean from 1993 to 2012. Chin. J. Oceanol. Limnol. 32 (6), 1414–1420. https://doi.org/10.1007/s00343-014-3363-4.

Solomon, S., Qin, D., Manning, M., Marquis, M., Averyt, K., Tignor, M., Chen, Z., 2007. Climate Change 2007; the Physical Science Basis; Contribution of Working Group I to the Fourth Assessment Report of the Intergovernamental Panel on Climate Change. Cambridge University Press, Cambridge.

Sutton, R.T., Hodson, D., 2005. Atlantic ocean forcing of North American and European summer climate. Science 309, 115–118.

Sweet, W.V., Park, J., Marra, J.J., Zervas, C., Gill, S., 2014. Sea level rise and nuisance flood frequency changes around the United States. In: NOAA, Technical Report NOS COOPS 73, 53 pp.

Tootle, G.A., Piechita, T.C., Singh, A., 2005. Coupled oceanic-atmospheric variability and U.S. streamflow. Water Resour. Res. 41 (12). W12408-n/a.

Trenberth, K.E., Hurrell, J.W., 1994. Decadal atmosphere-ocean variations in the Pacific. Clim. Dynam. 9 (6), 303–319.

Warrick, R.A., Oerlemans, J., 1993. Sea level rise. In: Warrick, R.A., Wigley, T.M., Barrow, E.M. (Eds.), Climate and Sea Level Change: Observations, Projections and Implications. Cambridge University Press, New York pp. (9) 3–4.

White, N., Haigh, I.D., Church, J.A., Koen, T., Watson, C.S., Pritchard, T.R., Watson, P.J., Burgette, R.J., McInnes, K.L., You, Z.-J., Zhang, X., Tregoning, P., 2014. Australian sea levels-trends, regional variability and influencing factors. Earth Sci. Rev. 136, 155–174.

Zhang, X., Church, J.A., 2012. Sea level trends, interannual, and decadal variability in the Pacific Ocean. Geophys. Res. Lett. 39 (21) https://doi.org/10.1029/2012GL053240 n/a.

FURTHER READING

Anderson, D., Sweeney, D., Williams, T., 2011. Hypothesis tests. In: Anderson, D., Sweeney, D., Williams, T. (Eds.), Essentials of Modern Business Statistics. Cengage Learning, pp. 365–413.

Ayyub, B.M., Kearney, M.S., 2012. Sea Level Rise and Coastal Infrastructure: Prediction, Risks and Solutions. ASCE, Reston, Virginia.

Azadeh, R., 2013. Response of Flood Events to Land Use and Climate Change: Analyzed by Hydrological and Statistical Modeling in Barcelonnette, France. Springer, Netherlands.

Cazenave, A., Dieng, H.-B., Meyssignac, B., Von Schuckmann, K., Decharme, B., Berthier, E., 2014. The rate of sea-level rise. Nat. Clim. Change 4 (5), 358–361.

Cohen, J., 1988. Statistical Power Analysis for the Behavioral Sciences, second ed. Academic Press, New York.

Corder, G.W., Foreman, D.I., 2014. Nonparametric statistics. In: Corder, G.W., Foreman, D.I. (Eds.), Nonparametric Statistics. John Wiley and Sons, pp. 139–171. Incorporated.

Dinse, K., 2011. Climate Variability and Climate Change What is the Difference? Michigan Sea Grant.

Dougals, B.C., 1991. Global sea level rise. J. Geophys. Res. Oceans 96 (C4), 6981–6992.

FitzGerald, D., Fenster, M., Argow, B., Buynevich, I., 2008. Coastal impacts due to sea level rise. Annu. Rev. Earth Planet. Sci. 36, 601–647.

Hallegete, S., Green, C.N., Corfee-Morlot, J., 2013. Future flood losses in major coastal cities. Nat. Clim. Change 3, 802–806.

Hamed, K.H., Rao, A.R., 1998. A modified Mann-Kendall trend test for autocorrelated data. J. Hydrol. 204, 182–196.

Holgate, S.J., Matthews, A., Woodworth, P.L., Rickards, L.J., Tamisiea, M.E., Fode, P.R., Pugh, J., 2013. New data systems and products at the permanent service for mean sea level. J. Coast. Res. 29 (3), 493–504.

Hughes, C.W., Milne, G.A., Tamisiea, M.E., Gehrels, W.R., 2009. Identifying the causes of sea-level change. Nat. Geosci. 2 (7), 471–478.

Leatherman, S.P., Kearney, M.S., Douglas, B.C., 2001. Sea Level Rise: History and Consequences, vol. 75. San Diego Academic Press.

Mann, H.B., 1945. Nonparametric tests against trend. Econometrica 13 (3), 245–259.

McGranahan, D.A., Balk, D., Anderson, B., 2007. The rising tide: assessing the risks of climate change and human settlements in low elevation coastal zones. Environ. Urban. 19, 17–39.

NCZMP, 2017. National Coastal Zone Management Program. Retrieved from: https://coast.noaa.gov/czm/about/.

Rahman, M.A., Yunsheng, L., Sultana, N., 2016. Analysis and prediction of rainfall trends over Bangladesh using Mann-Kendall, Spearman's rho test and ARIMA model. Meteorol. Atmos. Phys. 1–16.

Sagarika, S., 2014. Identification of Long-term Changes and Evaluation of the Relationships Among Streamflow Variability and Oceanic-Atmospheric Indices. Las Vegas: Report.

INDEX

Note: 'Page numbers followed by "f" indicate figures, "t" indicate tables.'

A

D

E

H

N

O

P

T

Made in the USA
Coppell, TX
25 November 2025

63973748R00259